Electronic Packaging of High Speed Circuitry

Electronic Packaging and Interconnection Series

Charles M. Harper, Series Advisor

ALVINO • *Plastics for Electronics*

CLASSON • *Surface Mount Technology for Concurrent Engineering and Manufacturing*

G. DI GIACOMO • *Reliability of Electronic Packages and Semiconductor Devices*

GINSBERG and SCHNOOR • *Multichip Modules and Related Technologies*

HARPER • *Electronic Packaging and Interconnection Handbook,* 2/e

HARPER and MILLER • *Electronic Packaging, Microelectronics, and Interconnection Dictionary*

HARPER and SAMPSON • *Electronic Materials and Processes Handbook,* 2/e

HWANG • *Modern Solder Technology for Competitive Electronics Manufacturing*

LAU • *Ball Grid Array Technology*

LAU • *Flip Chip Technologies*

LICARI • *Multichip Module Design, Fabrication, and Testing*

Related Books of Interest

BOSWELL • *Subcontracting Electronics*

BOSWELL and WICKAM • *Surface Mount Guidelines for Process Control, Quality, and Reliability*

BYERS • *Printed Circuit Board Design with Microcomputers*

CAPILLO • *Surface Mount Technology*

CHEN • *Computer Engineering Handbook*

CHRISTIANSEN • *Electronics Engineers' Handbook,* 4/e

COOMBS • *Electronic Instrument Handbook,* 2/e

COOMBS • *Printed Circuits Handbook,* 4/e

J. DI GIACOMO • *Digital Bus Handbook*

J. DI GIACOMO • *VLSI Handbook*

GINSBERG • *Printed Circuits Design*

JURAN and GRYNA • *Juran's Quality Control Handbook*

JURGEN • *Automotive Electronics Handbook*

MANKO • *Solders and Soldering,* 3/e

RAO • *Multilevel Interconnect Technology*

SZE • *VLSI Technology*

VAN ZANT • *Microchip Fabrication*

Electronic Packaging of High Speed Circuitry

Stephen G. Konsowski

Arden R. Helland

Drawings by Darnetta Anderson
and Robert L. Thing, Jr.

McGraw-Hill

New York San Francisco Washington, D.C. Auckland Bogotá
Caracas Lisbon London Madrid Mexico City Milan
Montreal New Delhi San Juan Singapore
Sydney Tokyo Toronto

Library of Congress Cataloging-in-Publication Data

Konsowski, Stephen G.
Electronic packaging of high speed circuitry / Stephen G.
Konsowski, Arden R. Helland.
p. cm.—(Electronic packaging and interconnection series)
Includes index.
ISBN 0-07-035970-9
1. Electronic packaging. 2. Microwave devices—Design and
construction. I. Helland, Arden R. II. Title. III. Series.
TK7870.K647 1997
621.381'046—dc21 97-1904
CIP

McGraw-Hill
A Division of The ***McGraw-Hill*** *Companies*

1 2 3 4 5 6 7 8 9 0 FGR/FGR 9 0 2 1 0 9 8 7

ISBN 0-07-035970-9

The sponsoring editor for this book was Stephen S. Chapman, the editing supervisor was Stephen M. Smith, and the production supervisor was Clare B. Stanley. It was set in Century Schoolbook by Dina E. John of McGraw-Hill's Professional Book Group composition unit.

Printed and bound by Quebecor/Fairfield.

Contents

Preface

Advances in electronic devices, circuit designs, materials, and circuit packaging technologies are together responsible for the constantly improving performance and affordability of electronics today. Examples of state-of-the-art products are cellular telephones, laptop computers, notebooks, and subnotebooks with fax capability. In addition, faster computing with less power consumption in smaller and lighter-weight devices continues to be the trend. High-definition television (HDTV) and hand-held communication equipment that can reach any location on earth via satellites are among the products under development and/or being introduced for sale. Certainly, future products will make these seem commonplace.

This book shows how advances in electronic devices, circuit design, materials, and circuit packaging have become intertwined in a manner that requires an integrated design approach to take full advantage of those technologies. This design approach depends on a fundamental understanding of signal transmission theory, digital and microwave circuit design, and the role of packaging in circuit performance. This book takes the reader through these disciplines in an easy manner that includes only the necessary mathematics required to understand and design the packaging, manufacturing, and test of high speed and microwave circuitry. Of particular significance in this scenario are certain packaging features of high speed signal processing and microwave technology that can be blended to achieve complementary performance. That is specifically what is required in telecommunications and mixed-signal (microwave and digital signals within the same circuit) products.

This book is about the analysis, design, and packaging of high speed digital and microwave systems. The fast rise times and high clock rates of newer generations of digital processing have increased the signal frequency bandwidth into the rf and microwave regions. The incredible increase in the level of integration of transistors and memory into silicon chips means that the performance of systems is often primarily determined by the packaging design and the interfaces between units. This book deals with both the packaging issues and the characteristics of high speed digital transmission lines. It is anticipated that integrating digital and microwave material in this book will facilitate communication and learning between the designers and users of high speed technologies, which are now found in high-performance systems in such areas as personal computers, communications, and entertainment. This book includes a generous number of illustrations that are intended to provide a better insight into the concepts than the words alone, and to emphasize that there are alternatives to consider in the design and development of high speed systems.

The first five chapters of this book are written from the perspective of the physical properties of packaging, with emphasis on special materials properties and assembly technologies required to achieve very high performance systems. The remainder of this book deals with high speed signal distribution from the perspective of digital transmission lines. The limited scope of this book omits significant detail in such related areas as microwave circuit design, transmitters and receivers, antennas, fiber optic communications, digital logic design, and the design of integrated circuits. Chapters 6 to 14 are related to other books dealing with digital systems, especially *Transmission Lines for Digital and Communication Networks* (McGraw-Hill, 1969) by Richard E. Matick, and *Signal and Power Integrity in Digital Systems* (McGraw-Hill, 1996) by James E. Buchanan, a coworker at Northrop Grumman (formerly the Westinghouse Electric Corporation Electronic Systems Group). Readers interested in an emphasis on the basics and developments in the rapidly expanding area of computers are referred to the *Computer Engineering Handbook* (McGraw-Hill, 1992) edited by C. H. Chen. Although these previous books (and others) deal with some of the same basic principles, there is little overlap of the more detailed information with this book.

The two parts of this book are written from somewhat different perspectives, but deal with common basic principles. Each chapter is intended to stand on its own in dealing with specific areas of interest. This means that there is some overlap of material in different chapters, but the reader may go to specific topics without reading other chapters. References are provided with most chapters to permit further reading of more detailed background informaiton on specific topics.

The first five chapters deal with electronic packaging of microwave and digital circuits. Electronic packaging can be thought of as a technology made up of various engineering disciplines: electrical, mechanical, materials, chemical, thermal, and others as well. The term packaging has evolved from the requirement that all parts must fit into some form such as a container (TV set or phone).

Chapter 1 introduces the high speed packaging issues of coupling, the influence of materials on circuit performance, and interconnections. Digital and analog circuits do not operate in the same manner in terms of inputs and outputs, but they do require very similar care in their electromechanical packaging. This is especially true as the clock frequencies increase for digital systems, because the interconnect media such as printed circuit boards require special materials and design techniques.

Chapter 2 discusses some fundamentals of electricity and magnetism, including definitions of the logarithmic scale named decibels (dB) for a compact representation of the ratio of numbers, Maxwell's equations and their meaning in electromagnetics (Gauss' laws, Faraday's law, and Ampère's law), wave behavior and waveguide principles, and the concepts of impedance, loss, permittivity, dielectric constant, and skin depth.

Chapter 3 describes various technologies upon which the efficient packaging of high speed digital and microwave circuitry depends, including materials and interconnection techniques. Among these technologies are various high-performance packages (multichip modules that are featured in high-end computer workstations), materials with special dielectric properties, composite materials, and integrated circuit or chip interconnection techniques that enhance the ultraminiaturization required in high speed systems.

Chapter 4 provides various microwave packaging design details, including discrete, hybrid, and monolithic microwave assemblies. Practical design features and the tradeoffs encountered in system packaging are discussed. These include component and materials selection, circuit partitioning, and integration of all the assemblies required to comprise a system. Also discussed are thermal control and environmental isolation (control of shock, vibration, and outside thermal influences).

Chapter 5 provides details on development and assembly of microwave circuit components and interconnections. This chapter deals with many considerations that determine the success or failure of a project, such as the degree of prototyping versus modeling required to assure circuit performance, accessibility for maintenance and/or repair, and the importance of proper grounding techniques at all transitions.

The second part of this book begins with Chap. 6, which describes transmission-line effects. The basic terms of rise time, frequency, and line length are defined. Signal overshoot and undershoot are defined and technology trends are discussed. The components of hollow waveguides and optical fibers are briefly described, and compared to wired interconnections such as coax, twinax, and stripline conductors for high speed interconnections. The basic transmission-line equations for line impedance, propagation, and attentuation are described.

Chapter 7 introduces the ideal line conditions as a practical approximation for analysis of transmission-line propagation. The equations for line impedance for basic conductor configurations are provided, with conversions and reasonable approximations for more complex configurations, such as differential pairs. An example is provided of the use of design of experiments (DOE) analysis for determining the sensitivity of line impedance to the dielectric constant and the dimensions of stripline conductors.

Chapter 8 describes the ends of the transmission line and the effects of termination resistors at the source and the load ends. Scaling factors for loading and line length are described. The importance of balanced source termination for control of crosstalk reflections is described. This chapter emphasizes the importance of including the ground return and cable shielding as part of the transmission-line interconnections. Examples of layer stackup for multilayer printed circuit boards are illustrated, with emphasis on assuring continuous signal return paths.

Chapter 9 describes the reflection factor and its relationship to the ratio of termination resistance to line impedance, which is also the voltage standing wave ratio (VSWR) for long lines. This chapter describes the relationship between the rise time of digital signals and frequency bandwidth. Examples of the effect of line losses on the signal waveshape on long lines are illustrated.

Chapter 10 deals with the issues of transmission-line drivers and loads that are not completely linear. The use of reflection diagrams to determine the signal levels on long lines are described and illustrated, including forward conduction and reverse recovery characteristics of junction diodes. A small resistance in series with both the driver outputs and clamp diode inputs will be shown to provide nearly ideal transient response without the higher transient currents or the static power dissipation of parallel load termination.

Chapter 11 provides the analysis of signal levels on lines with a linear termination at either the load or the source end, or both. The duals of different line terminations that result in the same waveform at the load are described. Transmission lines with the same ratio of the line impedance to the termination resistance at both ends are analyzed to illustrate the effects of increasing the mismatch ratio. This chapter presents the power series that estimates the overshoot and undershoot ringing deviations from the final levels, including scaling effects for lines that are shorter than the signal rise time.

Chapter 12 deals with the issues of the routing of critical signals, with an emphasis on loop routing where all drivers and loads are connected both ways in a loop without ends. This is usually applicable to lines with multiple loads and also may have multiple sources on a party line bus. It is shown that a loop routed line is equivalent to a single line with one end at the active source. The waveforms for party lines with multiple drivers not necessarily centered between two physical ends are described and illustrated.

Chapter 13 emphasizes the high speed characteristics of emitter coupled logic (ECL) interfaces that are capable of significantly higher rates than practical with TTL, CMOS, or BiCMOS technology. The limitations of the data-transmission rate as a function of line length are summarized. The use of parallel Schottky diodes for ECL line termination is described and illustrated. This chapter also discusses the characteristics of multiple ECL drivers on a party line bus, as well as multiple power-supply limitations.

Chapter 14 describes the characteristics of differential lines for high speed signal transmission on longer lines. Differential lines are usually terminated line-to-line to isolate signal currents, and differential receivers reduce the vulnerability to common mode interference, but balanced source termination is required to control conversion of common mode to differential noise. The use of parallel Schottky diodes for differential ECL termination is described and illustrated. Issues for ECL translators to TTL and CMOS logic, including multiple power-supply applications, are discussed.

Stephen G. Konsowski
Arden R. Helland

Acknowledgments

This book would not have been possible without the support and patience of my wife, Sheila. Her encouragement when the project often got sidetracked was always strong and positive. For her perseverance, I want to express my full appreciation.

At the same time, I want to thank Charles Harper, my long-time friend and associate, who conceived of this book and brought Arden, my co-author, and me together to make it a reality. His suggestions and experience in editing and his review of the manuscript were invaluable. I am most grateful to him for his help.

I also want to acknowledge John Gipprich of the Northrop Grumman Corporation, who advised me on various microwave-related topics, and Darnetta Anderson, who assisted with preparation of some of the drawings.

Stephen G. Konsowski

I offer many thanks to the associates I have had the privilege of working with over many years at the Electronic Sensors and Systems Division of the Northrop Grumman Corporation, which was previously part of the Westinghouse Electric Corporation. I have also had the opportunity to benefit from working relationships with several other companies. Anaren Microwave deserves special recognition for demonstrating the integration of high speed digital and microwave technology.

Special thanks are due to Jim Hudson and Jack Peters for the opportunity to develop much of the material that is the basis for this book, for the privilege of being assigned to challenging tasks, and for providing continuing education opportunities. The Northrop

Grumman Corporation is thanked for permission to publish this book. Jim Buchanan deserves special credit for emphasizing the analog characteristics of digital circuits and the importance of good design practices and adequate design margin for successful production and operation of high speed systems. Jim is also thanked for providing inspiration and guidance for this book.

The foundation for my professional career is composed of the encouragement provided by many other people, and of an education that began in country schools. I would like to give special thanks to Virgil Miller, a personal friend and inspiration, and formerly a college instructor at North Dakota State University, who introduced me to digital logic circuits many years ago. In addition, I want to thank my family for its encouragement and patience during the time I spent on this book.

Bob Thing deserves a great deal of credit for the excellent drawings he produced for this book. It is hoped that these illustrations will help to convey some of the concepts that are difficult to communicate in words only and often seem mysterious to many engineers.

Arden R. Helland

Conversion Factors

1 inch (in)	2.54 centimeters (exact definition)
1 foot (ft)	12 inches
1 foot	0.3048 meter (30.48 centimeters)
1 U.S. statute mile	5280 feet
1 U.S. statute mile	1609 meters (approximately)
1 nautical mile	1852 meters
1 mil	10^{-3} inch
1 mil	25.4 microns
1 meter (m)	39.37 inches (approximately)
1 kilometer (km)	10^{3} meters
1 centimeter (cm)	10^{-2} meter
1 millimeter (mm)	10^{-3} meter
1 micrometer (micron; μm)	10^{-6} meter
Typical human hair diameter	100 microns (roughly)
Wavelength of visible blue light	0.5 micron (approximately)
1 nanometer (nm)	10^{-9} meter
1 hour (h)	3.6×10^{3} seconds
1 week	168 hours
1 microsecond (μs)	10^{-6} second
1 nanosecond (ns)	10^{-9} second
1 picosecond (ps)	10^{-12} second

Electronic Packaging of High Speed Circuitry

Chapter

1

High Speed Digital and Microwave Packaging

The subject of packaging electronic circuits is very broad, covering such items as features and specifications of the components that comprise the circuits, the materials from which the interconnect media are made, design rules for efficient packaging, thermal considerations, processes to encapsulate or otherwise protect the circuit, testing, and special design characteristics germane to the nature of the circuit. These characteristics will be different if the circuit is involved in power conversion (for instance, ac to dc) or if the circuit is analog and operates at frequencies below several megahertz. When a circuit is digital and its clock speed is above 10 MHz, or when an analog circuit operates at 100 to 100,000 MHz (100 GHz), it can be considered as high speed or microwave, respectively. In these cases, very special packaging techniques are employed because interactions between circuits on the same or neighboring substrates or printed wiring boards can occur which could deteriorate circuit performance. Circuit performance could be altered by the undesired coupling of a signal from one part of the circuit to another, as one example. In these cases, an understanding of waves and their propagation is important because the conductors and even components are of similar size to the electromagnetic waves they control. The phase of a signal varies along a length of a line from 0 to 360°, and the magnetic and electrical field components of that signal vary in value similarly. This has a direct bearing on the location of a component along a conductor because the phase of a signal entering that component may be required to have a certain value at a precise moment in time.

High speed digital circuits and microwave circuits have a considerable commonality. In the final analysis, even digital circuits are analog in their fundamentals. Digital and analog circuits do not operate in the same manner in terms of inputs and outputs, but they do require similar care in their electromechanical packaging. This is especially true as the clock frequencies increase for digital systems, because the interconnect media such as printed circuit boards require special materials and design techniques. Both are susceptible to radiation of their signals and the unwanted reception or coupling of this radiation to neighboring circuits unless precautions are taken. Grounding and shielding are important; signals must be carried on transmission lines, and terminations, inductance, and capacitance must be carefully calculated and used to enhance rather than impede performance.

The range of applications of high speed digital circuitry has grown significantly since 1985 in commercial computers (mainframe and especially personal computers and laptops), and microwave circuitry has seen a like increase with cellular communications, global positioning satellite (GPS) equipment, real-time satellite video transmission and reception, and developments in automotive radar for traffic and vehicle control.

This book deals with the special requirements of both high speed digital and microwave packaging and is intended to assist designers as well as manufacturing and quality assurance personnel in their efforts to bring products to market rapidly through a better understanding of the special electronic packaging requirements of high speed digital and microwave circuits and by application of that understanding to accomplish error-free performance in their individual design and production responsibilities. Technologies and design practice for microwave circuit packaging are emphasized in Chaps. 1 through 5 (Chap. 3 covers both high speed digital and microwave packaging technologies). Considerations germane to high speed digital circuits are dealt with in Chaps. 6 through 14.

1.1 Packaging

Packaging of electronic components began to have a separate and very important role in the design and manufacture of electronics about 1960. At that time, transistorized electronics had become firmly established in commercial equipment such as portable hand-held radios and military hardware of all sorts. All the associated components such as capacitors, resistors, and inductors were shrunk in size to be compatible with transistors. This miniaturization was responsible for the lightening of weight and reduction of size of electronic equipment. Not long afterward engineers realized how to pack more features and performance into electronics without size, weight, or power penalties.

Both digital and analog circuitry were similarly affected by miniaturization. Digital computer circuitry, which had occupied thousands of cubic feet when composed of tubes, could fit into several hundred cubic feet when converted to transistors. Airborne fire control radar receivers, which were able to search and track only one target at a time, had their capability enhanced to track several targets simultaneously. Automotive electronics became more complex, with windshield wiper controls; seat belt, door, and other sensors; and ignition systems being added to a vehicle whose electronics previously consisted of only a radio.

As these remarkable changes, driven by the advent of the transistor, were taking place, another development was under way which would revolutionize the way people would live. That development was the integrated circuit, the single most dramatic factor in this trend to smaller physical size. The early devices contained several functioning circuits which had been comprised of many individual components such as resistors and transistors. Subsequent improvements in semiconductor technology permitted smaller component sites and interconnections on and within the integrated circuit such that the component count per integrated circuit rose rapidly. It continues to rise at this time, making possible entire processors on a single chip. Initially centered primarily around digital circuitry, the integrated circuit, or IC (or chip, as it is popularly called), was able to contain all the components associated with a particular function, such as a flip-flop or shift register, within itself. Resistors, transistors, and even small-value capacitors were grown upon or diffused into the top surface of the IC. This invention effectively caused the volume of transistor circuits to be shrunk by a factor of 300 to 1, taking into account the interconnect medium such as wiring.

All these changes in component character and function required new techniques for packaging in order to use space efficiently. Whereas tube circuits were connected by wires usually no smaller than 0.020 in in diameter, the transistorized circuits did not need such "heavy" wiring. Instead, 0.008-in-diameter wire or even ribbon (0.005 × 0.008 in) was used. When ICs emerged, they were placed in small packages with flat ribbon leads protruding from opposite sides or from all four sides. This planar or two-dimensional packaging approach was in contrast to the "can," or three-dimensional package, used for transistors, which resulted in a compact packaging form factor that is still in use, namely, the printed wiring board (PWB). The PWB has components arranged in a flat array on the board, which serves as both mechanical support and interconnect between the components it supports. Interconnections to other boards (if required) are made via connectors at one or more edges of the board. As the PWB is slid into position in a rack, these connectors plug into their mating halves, which are located in a backplane. Provisions are made for dissipating unwanted heat from components through metal plates under the components or by air blown over them (most frequently the former). This planar technique had been so successfully used and proved for digital circuitry that it was later made applicable also to analog and microwave packaging. Even power supplies contained or embedded within the larger circuit function (e.g., receiver) are often packaged in this way, which makes the entire function modular. In the case of microwave circuits, shielding and structure is provided around the entire board in the form of a box or case. This case can have radio frequency (rf) as well as power and sometimes digital connectors located on one edge. These connectors allow the case to be plugged into a backplane which contains mating connectors and wiring. As with digital boards, there is a slot/guide which facilitates mating of the connectors and assures accurate positioning of the case or module. Captive mechanical fasteners lock the case in place. Cases or modules can be placed side by side in a rack. In some assemblies, digital modules comprise a part of the rack's modules while the remainder are rf or microwave in nature.

1.2 Specifics of Microwave Packaging

Microwave systems have unique requirements for packaging technology as a result of the relatively short wavelength of the electromagnetic energy, the functions performed by the systems, and the type of devices used. The microwave band usually refers to the centimeter wavelength range of the spectrum (300 MHz to 30 GHz).

The major applications of microwaves are communication systems, including point-to-point line-of-sight microwave radio, satellite relay systems, point-to-multipoint radio for cellular mobile systems, avionics systems, navigation systems, landing systems, and weather radar; defense systems, including ground-based and airborne radar, space-tracking radar, and guidance systems; industrial systems, including drying, identification, and speed measurement; and consumer equipment, including intrusion alarms, cooking equipment, and home satellite receivers.

A major difference between microwave and lower-frequency systems is that circuit components must be considered as distributed elements rather than as lumped elements because of the relatively short wavelengths. There are exceptions when very small IC components are used. Design of packaging and interconnects for microwave and high speed digital systems requires extreme care to maintain impedance matching or low return loss to minimize detrimental effects on performance. Further, since typical circuit configurations approach a significant fraction of a wavelength or are several wavelengths in physical size, radiation of electromagnetic energy may occur under certain conditions, which leads to increased loss and crosstalk. The major packaging issues that must be addressed in any microwave and high speed digital system are coupling and interconnects.

1.2.1 Coupling

Coupling is the mechanism that enables microwave energy to be extracted from or added to a device or circuit. Coupling may be accomplished using a direct connection or as the result of coupled fields with no physical connection. A direct connection is used for coupling to a device such as a transistor or for joining an input or output coaxial connector to the package. Field coupling is used when a probe connected to an input or output waveguide or coaxial connector is used to couple to the fields within a microwave resonator or cavity. The coupling must provide a low-loss connection between a package and a source or load, or between a device and its circuit, with the appropriate bandwidth and impedance level with minimum parasitic effects. The input and output ports for microwave band packages at power levels up to 5 W utilize miniature coaxial connectors. The ports for packages in the 60-GHz and higher frequencies utilize rectangular waveguide.

1.3 Requirements for Microwave Materials

Microwave circuitry differs from dc and low-frequency circuits in that the signals carried in both the conductors and the dielectrics surrounding the conductors are influenced strongly by the electrical properties of the dielectric at high frequencies. For this reason the dielectric properties of packages which contain microwave circuits are quite important. A brief discussion of the development of materials for microwave circuits is presented here.

Early microwave printed wiring technology was derived from microwave power dividers for antennas and was then developed into flat coaxial configurations. In the 1940s a configuration was developed which had a single conductor and a single ground plane with a solid dielectric. This was called microstrip. Later a central-conductor, two-solid-dielectric, two-ground-plane system emerged which was called Triplate. This configuration is known today as stripline, although the original name stripline referred to an air dielectric arrangement.

The dielectric materials which were used at that time were fiberglass followed by unreinforced lower dielectric constant and lower loss plastics that were simply matrix materials. Reinforcement in the form of woven glass fabric provided mechanical strength but at the same time detracted somewhat from the quality of the dielectric. Less lossy reinforcement (quartz and randomly oriented short glass fibers) was introduced to restore the performance while still providing mechanical strength. This increased the design flexibility considerably for a wide variety of applications (microwave components such as couplers, filters, dividers, and combiners) and power distribution manifolding in phased-array antennas. The component manufacturing community continued to seek higher-performance materials and began to experiment with microstrip dielectrics such as quartz, aluminum oxide, sapphire, beryllium oxide, and some titanates. Since these materials were not organic printed circuit boards, a technique to deposit and adhere a conductor system to them was required. Thin film technology was the most convenient since it was developed and refined to produce integrated circuits and hybrid components and was well characterized. Sapphire, because of its original cost and the extensive machining required to render a usable substrate finish for thin film metallizations, did not survive as a major substrate material. Quartz had similar machining costs and found limited though continued usage. Aluminum oxide, easily produced in reasonable purity levels, emerged as a major material of choice. Beryllium oxide, with its high thermal conductivity and availability, was also used extensively.

As dielectric technology continued to be refined and a commercial market for microwave components developed in the communications field, considerable effort was spent on reducing the costs of both the materials and the processes to produce microwave circuit components. This required a different approach to metallizing the substrates. While an industry was under development for the production of analog and digital hybrid circuits, technologies emerged which began to be applied to microwave circuits. Thick film metallizations which could be applied by stenciling techniques to alumina or beryllia substrate surfaces and subsequently baked and fired into the ceramic had potential usage in the microwave arena, but the electrical conductivities of the metallizations were generally too low for most microwave applications. This issue was resolved with the introduction of high-electrical-conductivity thick film pastes by DuPont, EMCA, ESL, and others in the late 1960s. At this point microwave circuitry moved into the realm of affordable high performance.

Solid-state microwave receivers and transmitters of low power (less than 500 W) had their hardware made up of many microwave components such as low-noise receivers, circulators, amplifiers, mixers, and phase shifters, connected to each other with miniature coaxial conductors, referred to as coax or "hardlines." This arrangement was bulky and costly to fabricate. A better integration of the components was required. Systems producers began to combine various subfunctions into the same package rather than interconnect individually packaged subfunctions with coax. This resulted in more compact electronics equipment that became lighter in weight and easier to install and maintain.

Subfunctions were generally constructed on ceramic substrates and placed together into large packages where they were interconnected by ribbons of gold or by gold wire bonds. Solder and diffusion bonding were the methods of interconnecting the substrates. Although this approach was economical, some subfunction designs needed to be larger than the standard ceramic sizes in vogue. Larger ceramic plates were available, but handling and assembly issues dictated 2 in on a side to be an optimum size. Plates larger than that were too fragile to survive assembly and testing processes. Also certain shapes were required to achieve efficient size packaging. These shapes took the form of the container which held them in the final assembly, such as curved edges to fit within a missile or torpedo. This form was not easily achieved with brittle ceramics. A solution to the size limitation of ceramic substrates was offered by the 3M Company with its Epsilam 10, which was a double-sided copper cladding on a proprietary dielectric composed of a high dielectric constant (K) filler dispersed within a lower K matrix with a combined effective K of approximately 10. The value of 10 was chosen to simulate the K of 99+ percent alumina, so the Epsilam 10 could be substituted directly for it. This clad material was available in sheets 10 by 10 in, making it possible to construct large integrated microwave assemblies because the copper cladding could be patterned into the desired conductor configurations through photoetching. The copper cladding on the bottom side served as the ground plane. This material did not have the handling and breakage characteristics associated with alumina and therefore offered more versatility. Irregular shapes and sizes could be achieved by the user through the use of printed circuit technology, namely, machining and routing. Etching and plating was possible since the dielectric was compatible with standard chemical plating and etching processes. A short time after the introduction of Epsilam 10, several other manufacturers provided materials with similar characteristics.

This discussion relates primarily to microstrip circuitry.

1.4 Interconnects

The interconnect is the medium that transports microwave energy within the package. It also functions as a circuit element used for resonators, tuners, dc blockers, and chokes. The circuit elements define the range and bandwidth of operation of the active devices, including microwave transistors and diodes. Theoretically any form of microwave waveguide, including coaxial line, rectangular or circular waveguide, or planar waveguide, may be used for interconnects. However, modern microwave circuits with solid-state devices generally utilize planar waveguides, which have the appropriate size and shape factors to ensure good coupling with excellent reliability and performance for moderate power levels. Planar waveguides have higher loss than rectangular waveguides and are therefore best suited for broadband circuit applications. An additional advantage of the planar media is low manufacturing cost through the use of photolithographic fabrication.

Three levels or interconnects are discussed here. Interconnects exist between components (level 1), between structures which contain groups of components (level 2), and between those structures and the "outside world" (level 3). In digital circuits, level 1 interconnects for the most part have been accomplished by printed wiring boards with component leads soldered to pads on the PWBs. Level 2 connections were made for some time by direct wiring of pins on the mating portions of the backplane connectors. As digital computing speed (clock speed and signal rise time) increased, it was necessary to provide an interconnect technique that provided controlled impedance. Twisted pairs of signal and ground wires were introduced to serve that need.

Microwave circuitry has seen the use of controlled impedance for signal transmission for some time owing to the nature of the frequencies at which the circuits operate. Level 1 interconnects have been accomplished using microwave-compatible dielectric (insulator material) and PWB conductor technology. These interconnects, classed as microstrip and stripline, are described in detail in Chap. 2. Level 2 interconnects between boards were generally performed with coaxial cable and coaxial connectors. These were not only bulky but awkward and costly to use. Coaxial lines were made from metal jackets over a dielectric which encapsulated a central conductor. They were formed into specific shapes depending on the applications and required large radii where bends were needed. Consequently, considerable effort was expended on replacing them with a more efficient technology. Their replacement was a method of using floating coaxial connectors mounted in a backplane. The backplane contained stripline conductors connected to the center conductor of the connector. Level 3 microwave interconnects consist of rf connectors which are connected to the stripline conductors of the backplane and which mate to connectors on cables or other assemblies.

1.5 Materials for Microwave Packaging

Materials developments have played a large part in electronic packaging, particularly in the transmission of very high speed digital and microwave signals. Conductors have been chosen for their electrical conductivity and compatibility with the dielectric to which they are attached or bonded. Early ceramic dielectrics were made from aluminum oxide (alumina) over which a thin coating of glass was applied and reflowed. The glass was used to provide a smooth surface for the conductors, which were vacuum-deposited chrome and gold or aluminum. Chrome was used to provide an adherence layer over which the gold or aluminum was then deposited. As ceramic technology advanced, techniques were devised for preparing the alumina surfaces to be as smooth as their glazed predecessors so that the glass coating was no longer required. Complex large circuits have been made from as many as a dozen ceramic plates about 2 in on a side grouped together. Since the ceramic substrates were only double-sided, that is, printed with a backplane on one side and a conductor pattern on the other, they conformed to the definition of microstrip.

Continued microwave circuit materials development made it possible in the mid-1980s for designs to include several layers of conductors and dielectrics in the same structure, much as a printed wiring board is configured. This development concerns ceramic-like materials known as low-temperature cofired ceramics (LTCC). The cofiring of conductors and dielectrics at less than 1000°C allows the use of high electrical conductivity metals for top surface conductors and buried layer conductors as well. This means that microstrip and stripline-like configurations can be constructed and used in conjunction with lower-frequency circuitry and digital circuits, all within the same substrate. Such flexibility offers the designer opportunities to combine, within the same substrate, circuit functions that should be located in close proximity to each other. It also allows more compactness and has the potential to improve circuit performance because of fewer interconnections. With a wide variety of dielectric properties, LTCC materials are opening new vistas for their application. This includes components such as resistors and capacitors printed onto buried layer dielectric layers. Certain passive components can be included within the multilayer structure to increase overall component density because the materials that constitute resistors and low-value capacitors have demonstrated compatibility with cofired LTCC systems.

Chapter

2

Fundamentals of Electricity and Magnetism Applied to Electromagnetics

This chapter presents in easy to understand terms the primary concepts and examples of phenomena that form the foundation of electromagnetic theory. An understanding of electromagnetic theory is dependent on a grasp of the basics. Keeping this in mind, we have avoided cumbersome mathematics and instead substituted simpler mathematical relationships as much as possible. More detailed and rigorous treatments of electricity and magnetism fundamentals exist. The reader is advised to refer to the references sections at the end of most chapters for these more comprehensive descriptions. Among the more important topics are the four relationships (Maxwell's equations) that are based on electric and magnetic fields, and a general description of waves and how understanding their behavior helps to quantify and predict what happens in circuits at microwave frequencies. These phenomena are especially important as those frequencies increase and their corresponding wavelengths decrease. Another very important topic is that of the movement of signals through what are called transmission lines, which can have various shapes such as wires, flat conductors, and even devices called waveguides. In order to appreciate how signals pass along transmission lines of various kinds it is necessary to understand how electromagnetic waves move from one location to another. This is so because there is an interdependent relationship between electricity and magnetism. This relation can be depicted by using examples of direct and alternating current with which nearly everyone is familiar. Electromagnetic waves are exactly what the name implies: They have at once both electrical and magnetic characteristics. The way electromagnetic waves behave has been effectively represented by the concepts of fields, which are covered in this chapter. The chapter then treats transmission lines in general and introduces the fundamental concepts of impedance, inductance, loss, resistance, conductance, and capacitance.

2.1 Decibels

The decibel scale is used extensively in the mathematics describing microwave performance. The scale is mostly for comparisons of power and voltage. The decibel originated with the study of sound and hearing and is attributed to Alexander Graham Bell, from whom it takes its name, with the scale based on powers of 10. It is a logarithmic scale with which ratios are used. It has no dimensions and is used only when two like values are being compared. The decibel (dB) is defined as $10 \times \log_{10} (P_1/P_2)$, where P_1 and P_2 are two different power levels. Since the dimensions of P_1 and P_2 cancel out, this formula can apply to voltage as well as power. These quantities typically appear in expressions such as "the output power is up by 20 dB." This translates to comparing two power levels (for example, one amplifying circuit output with a second amplifying circuit) where the first is 100 times larger than the second, regardless of the absolute value of either. However, the use of the decibel for voltage comparisons requires squaring the voltage and is therefore not as popular.

To familiarize oneself with practical terms, it is useful to note that if the ratio and dB scales are compared, the following relationship can be seen: On the ratio scale, 1 is equivalent to 0 dB, 10 is equivalent to 10 dB, 100 to 20 dB, and so forth. It is also important to note that several ratios are very commonly used.

Figures 2.1 and 2.2 show the ratio and dB scales aligned with each other. These figures can be used as a reference to familiarize oneself with the relationship between a power ratio and dB.

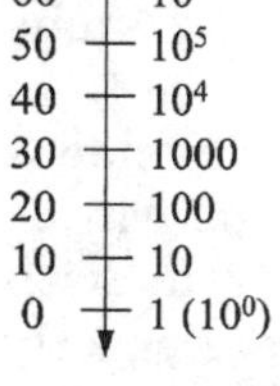

dB	ratio
70	10^7
60	10^6
50	10^5
40	10^4
30	1000
20	100
10	10
0	1 (10^0)

Figure 2.1 The decibel scale compared to power ratios on a logarithm scale.

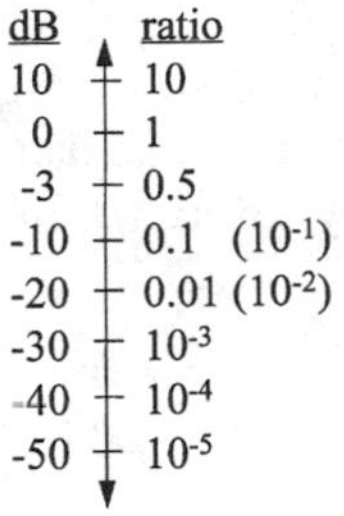

dB	ratio
10	10
0	1
-3	0.5
-10	0.1 (10^{-1})
-20	0.01 (10^{-2})
-30	10^{-3}
-40	10^{-4}
-50	10^{-5}

Figure 2.2 Comparison of decibel scale and power ratio for numbers less than 1.

A few examples of the ratios of the two power levels expressed in dB follow. We shall use log to represent $\log_{10}$.

Case 1: If the output of one circuit is 60 W and that of a second circuit is 20 W, then by the definition of dB,

$$\text{dB} = 10 \log \left(\frac{P_1}{P_2}\right) = 10 \log \left(\frac{60}{20}\right) = 10 \log (3)$$

$$= 10 \times 0.477 = 4.77, \text{ or approximately } 5$$

The ratios in dB are intended to be used for comparisons. Since 4.77 is very nearly equal to 5, we call it 5 because that number is easier to remember and is close enough for this purpose.

Case 2: The wattage of one circuit is 500 W. A second circuit's is 100 W. Since the first power is 5 times that of the second, the ratio is 5:1. Repeating the procedure, dB = 10 log (5) = 10 × 0.7 = 7. (See Table 2.1.) Therefore, the first power is up by 7 dB over the second power. The answer would be the same if the numbers were 50 and 10 W, or 5000 and 1000 W, since dB is a ratio and does not depend on the absolute values of the individual powers. It is nothing more than 10 times the logarithm of a ratio.

TABLE 2.1 The Relationship between dB and Power Ratios

Ratio	dB
2:1	3
3:1	5
5:1	7
8:1	9
10:1	10

In some cases the comparisons are intended to be diminishing rather than increasing. Before we proceed to those examples, a few rules will be stated. These derive from rules for logarithms and are presented here for review purposes and to show how they apply to the decibel.

Rule 1: The *product* of two ratios or numbers is expressed as their sum when their dB equivalents are used. For example, 10 dB + 20 dB = 30 dB. Referring to Fig. 2.1, $10 \times 100 = 1000 \rightarrow 30$ dB.

Rule 2: The *division* of two ratios or numbers is expressed as the difference of their dB equivalents. For example, 20 dB − 10 dB = 10 dB. Again referring to Fig. 2.1, $100 \times 10 = 10 \rightarrow 10$ dB.

Figure 2.1 is actually a segment of a scale that extends upward without limit and downward without limit. For practical purposes, the zone of this scale that is used is generally between +140 and −140 dB. Figure 2.2 shows a portion of the scale referring to logarithmic numbers less than one but greater than zero. (A logarithmic scale does not contain zero.) An example will assist the reader in using this scale: We wish to express a ratio of two numbers that is less than 1, in dB. We choose 0.2, and in ratio form, 0.2 = 2/10. Taking their equivalent dB values ($2 \rightarrow 3$ dB, $10 \rightarrow 10$ dB), 3 dB − 10 dB = −7 dB.

Sometimes it is convenient to express the dB number as a pure number. To do this one simply divides the dB number by 10 (see the definition of dB), which then becomes the power to which one raises the number 10. For example, to convert 10 dB to a pure number, divide by 10. This yields 1, which is then the exponent to which 10 is then raised: $10^1 = 10$. Referring to Fig. 2.1, 10 on the log scale is 10 dB. To convert 7 dB to a pure number, divide by 10. Then 0.7 becomes the power of 10 for the answer: $10^{0.7} = 5$.

2.2 The Symbol dBm

In solid-state microwave circuits the power is usually low (as opposed to high-power microwave circuits associated with amplifying devices such as traveling-wave tubes, magnetrons, and even klystrons). Instead of many watts, low-power solid-state circuits deal in milliwatts.

A milliwatt (mW) is one-thousandth of a watt. Because it is important to distinguish between a power ratio and an absolute power level in mW, dBm is used to represent power relative to 1 mW. Some may use dBW for power relative to 1 W, but dBm is more popular and is simply 30 dB greater than dBW. Conversion of power to dBm may be accomplished in the following manner:

State the power in milliwatts (1 W = 1000 mW, 4 W = 4000 mW, 0.05 W = 50 mW, etc.). Convert that number to dB and add an "m" to the dB so that dBm represents power relative to 1 mW. For example, to convert 5 W to dBm: 5 W = 5000 mW. Then 5000→37 dB, so 5 W→37 dBm, or 30 dB greater than 5 W→7 dBW.

2.3 The Neper

A unit that appears in some literature is called the neper. Recall the definition of the decibel: dB = $10 \log_{10} (P_1/P_2)$. The number 10 is the base of an exponent to which it is raised. In other words, $10^X = N$ represents a number which results from raising the number 10 to the power X. The number 10 is the base of the logarithm, since if the logarithm is taken of both sides of the equation, $\log_{10} 10^X = \log_{10} N$ which is the same as $X = \log_{10} N$. To find X we simply take the $\log_{10}$ of N. There is, however, a number e (which is 2.718...) that is the base of the natural logarithm. The natural logarithm is written $\log_e$ or ln. In this system, the number e instead of the number 10 is the base of the logarithm. The natural logarithm has great importance in the physical sciences because the number is helpful in nature in expressing events and relationships, such as bacterial growth rates or radioactive and electrical signal decay rates.

It is not the purpose of this book to describe this number, since most readers are familiar with it, but the number e is the base of power ratios in expressions involving the neper. The definition is $\ln (P_1/P_2)$ = neper. The natural logarithm of a number (ratio) can be determined with a scientific calculator, from tables, or from a slide rule (not a very popular device today). Some examples follow:

> To express the ratio of 12 W to 3 W in nepers, one takes the natural logarithm of the ratio 12/3 = 4. Since ln 4 = 1.39, the answer is 1.39 Np.
>
> Find the value of the ratio of 2 to 8 (as two power levels): Since 2/8 = 0.25, ln 0.25 = −1.39. In this case, the answer is −1.39 Np.

2.4 Fields

The term *field* is used in the study of static electricity to describe the influence of an electrical charge q_1 located at a distance r from another electrical charge q_2. That influence is called an electric field. The field extends outward in all directions and its strength at a distance from the charge has the same value everywhere on an imaginary sphere at the same distance from the charge. Stated in another way, the value of the influence of a charge fixed in space is not changed if one measures it at all places on an imaginary sphere surrounding the charge. The field is therefore said to be spherically symmetrical about the charge. This is true if the charge exists as a point in space, which is the ideal case that is easy to imagine and understand. If the charge is moving, another type of field surrounds that moving charge, a field of magnetism called a magnetic field. When a charge is introduced into the electric field set up by another charge, both feel the presence of each other as an attractive or repulsive force acting on each according to the polarity of the charges, their strengths, and the distance of their separation. If the polarity (positive or negative) is the same for both charges, the force is repulsive, and if it is different they are attracted to each other. The strength (magnitude) of one charge may be equal to or many times greater than that of the other. The field strength of a charge is said to be in direct proportion to the magnitude of the charge. The smallest unit of charge (negative charge in this instance) is that value ascribed to the charge of a single electron, which is about 10^{-19} coulomb. A negative charge can be many orders of magnitude larger than the smallest unit, which means that the number of electrons it contains is just as many orders of magnitude larger than a single charge. Conversely, if that charge is positive, it is missing that many electrons. A good example of the extremely large value of charges that can exist in relative proximity to each other is that which one experiences in an electrical storm where discharges of electricity can be seen to pass from one cloud to another with the ensuing report of thunder. The charges on one cloud in this simplified example "sense" the presence of (lack of charges of similar polarity) opposite charges in the other cloud, and vice versa. Being of opposite polarity, they attract each other, each unit of charge influenced by every charge in the opposite cloud. This reciprocal force of attraction under the right conditions causes the electrons and water droplets in the cloud containing the excess electrons to move toward the cloud deprived of its needed electrons. When this takes place, a great deal of stored energy is released in a very short time, and each cloud ends up more nearly neutral in charge afterward. The mathematical expression for the force each charge (independent of polarity) exerts on the other independent of charge polarity is

$$F = k \frac{q_1 q_2}{r^2}$$

This is Gauss' law for electric fields. As will be seen later, it is the first of the Maxwell equations.

2.5 Permittivity and Permeability

Two very important features of materials that affect electromagnetic waves are permittivity and permeability. Permittivity relates to electric fields and permeability relates to magnetic fields. Permittivity describes how a material under consideration allows itself to become polarized by an electric field imposed upon it from outside itself. Polarization refers to the realignment of charges within the dielectric along the lines of the imposed electric field. The charges in a dielectric are made to realign themselves or move from their resting place, with positive charges being moved in one direction and negative charges being moved in the opposite direction when an external electric field is imposed on them. The polarizability of a given material is expressed in relation to the polarizability of a vacuum or air. (For practical purposes the values of permittivity and permeability of air are the same as for vacuum, or what is commonly referred to as "free space.") This relationship is given as

$$\epsilon_r = \epsilon_0 k$$

or, stated another way,

$$k = \epsilon_r \div \epsilon_0$$

where k is known as the dielectric constant of a material. The value of $k = 1$ for air. The value of ϵ_0 is 8.85×10^{-12} faradays/meter. (The glossary at the end of the book gives definitions of these and all pertinent units.) The subscript r indicates that the value of the polarizability of a material is relative to the polarizability of air or a vacuum or free space.

In similar fashion, permeability is a measure of the magnetic property of a material that allows a magnetic field to be set up within itself by another magnetic field imposed upon it from outside. The relationship of the relative permeability of a material to the permeability of free space is

$$\mu = \mu_r \mu_0$$

where μ_r is the relative permeability of a given material. The relative permeability of air is very nearly 1.0, so the value of μ for air is μ_0, which is 4×10^{-7} henrys/meter. Values of the dielectric constant and relative permeability of commonly used circuit materials are given in Tables 2.2 and 2.3. It should be stated here that the dielectric constant of insulating materials can vary a great deal but their relative permeability is about the same, namely, 1. The relative permeability of most metals can be assumed to be approximately 1 except for iron,

TABLE 2.2 Typical Permittivities of Commonly Used Microwave Materials

Dielectric	Permittivity at 25°C, 1 kHz, expressed as *relative* dielectric constant
Polytetrafluoroethylene (PTFE)	2.04
Polyethylene	2.25
Glass-reinforced PTFE	2.55
Cyanate ester-glass	3.6
Polysulfone	3.8
Quartz	3.8
Polyimide-glass (woven)	4.2–4.6
Epoxy-glass (woven)	4.3–5.1
Low-temperature cofired ceramics	5.0–6.0
Ceramic-loaded PTFE	6.0–10.0
Beryllium oxide (beryllia)	6.9
99% aluminum oxide (alumina)	9.7

TABLE 2.3 Typical Permeabilities of Commonly Used Microwave Conductors

Conductor	Relative permeability μ_r
Gold	1
Aluminum	1
Tin	1
Copper	1
Silver	1
Nickel	600
Iron-nickel-cobalt	1100
Tantalum	1.001
Chromium	1.001
Titanium	1.01

nickel, and cobalt, which can be much greater than 1. These three metals are said to be ferromagnetic because they tend to behave magnetically very much like iron.

2.6 Impedance

In a simple dc circuit, the resistance to current flow is expressed as ohms. The ratio of the voltage (volts) across an element of a circuit to the current (amperes, or amps) in that element is defined as ohms. The more general use is that of impedance, which has the dimension of ohms. The ratio of the electric field strength to the magnetic field strength in a material or medium (air, ceramic, etc.) is defined as the characteristic impedance of that medium. The electric field strength is expressed as volts/length and magnetic field strength is amperes/length. Therefore, characteristic impedance is volts/amperes = ohms. The symbol for characteristic impedance is Z_0. Since the strengths of the magnetic and electric fields in a material are related directly to the material's permittivity and permeability, it turns out that

$$Z_0 = \sqrt{\frac{\mu}{\epsilon}} \tag{2.1}$$

This relation allows the characteristic impedance of any medium to be determined if μ and ϵ are known. The significance is that this concept allows us to realize that how electromagnetic waves travel in various media is determined according to the characteristic impedances of the media. The velocity of electromagnetic waves in a medium is

$$\nu = \frac{1}{\sqrt{\epsilon\mu}} \tag{2.2}$$

In air or a vacuum, the speed of light is

$$c = \frac{1}{\sqrt{\epsilon_0\mu_0}} \tag{2.3}$$

To determine the speed of electromagnetic waves in any other medium relative to their speed in air, one simply uses the dielectric constant and relative permeability in the relation

$$\nu = \frac{1}{\sqrt{k\epsilon_0\mu_r\mu_0}} \tag{2.4}$$

which yields

$$\nu = \frac{c}{\sqrt{k\mu_r}} \tag{2.5}$$

Since it is convenient and often important to know the speed of electromagnetic waves in various media, this relationship allows quick and simple computation of that speed in terms of the speed of light in air or a vacuum (3×10^8 m/s). The speed of electromagnetic waves in any medium other than a vacuum is always less than it is in a vacuum. Thus we often express the speed in a given medium as, for example, 80 percent of c, where c is known to be the speed in a vacuum.

2.7 Skin Depth

A phenomenon that can be extremely important at high frequencies is skin depth. It is the distance from the surface of a metal conductor to the depth at which the magnitude of a penetrating magnetic field falls to a value of $1/e$ (about 36 percent) of the magnitude at the surface. The letter e in this definition is the number e (~2.7), the base of the natural logarithm. The mathematical expression for skin depth is

$$\delta = \sqrt{\frac{\rho}{\pi \mu f}} \tag{2.6}$$

where f is the frequency of the changing magnetic field.

How waves propagate in a medium is dependent on the values of ϵ and μ for that medium. Another property of a medium that affects signal propagation is the conductivity σ, or its reciprocal, resistivity ρ. From the equation, it is seen that the skin depth becomes deeper as the resistivity increases and becomes shallower as the frequency and permeability increase. Although we have not yet shown how waves travel in devices structured to enhance this travel from one point to another (transmission lines), there is a point to be made about the conductivity of a metal when it is used as a conductor in a microwave circuit. Since the skin depth is larger with metals of poorer conductivity, the wave can penetrate more deeply into such metals, allowing some power from the wave to be absorbed into the metal. This power is then dissipated or lost as heat. Microwave power is always achieved at considerable expense because the conversion of raw dc power into microwave power involves losses due to efficiency considerations arising in the devices that do this conversion, so the portion of the original power is quite precious. The power lost in the conductor is lost because of the dc resistivity of the conductor. The dc resistivity of a conductor chosen for a microwave circuit should almost always be as low as possible (as we shall see later, this is often a challenging task). Certain designs may for various reasons perform better if a more lossy metal is used, but those are a rare exception. Referring to Eq. (2.6), it is easy to see that as resistivity increases, the alternating magnetic wave propagates more deeply into the conductor. Ideally, since power will be lost if any of the wave penetrates the conductor, the conductor should have no electrical resistance at all and should be a perfect conductor. In that case, no part of the conductor would offer any resistance to the wave because all of the energy in the wave would then be reflected away from the conductor.

However, this ideal case is not achievable, and we are faced with accepting the effects of Maxwell's eddy currents, which are generated in a metal in opposition to the imposed changing magnetic field.

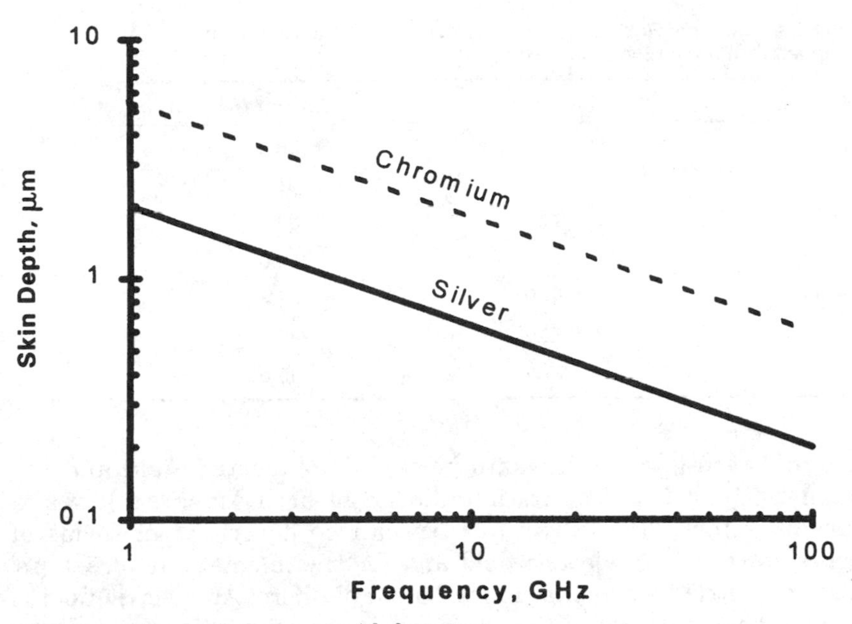

Figure 2.3 Skin depth decreases with frequency.

These currents dissipate a part of the wave's energy in flowing through the conductor and meeting resistance. If the currents flow deeply into a metal of relatively high resistivity, the portion of the conductor in which the flow of current takes place is greater if the frequency of the circuit is relatively low. As the frequency increases, the skin depth is shorter, and the eddy currents are then bunched into a decreasing thickness of a shell of the conductor, resulting in higher energy loss because of increased current density. This means that a poor conductor (high resistivity) can cause serious problems with power loss at very high frequencies. The skin depth δ for various conductors is shown in Fig. 2.3. A comparison of resistivity of pure gold with the inclusion of only 1 percent of another metal is shown in Table 2.4.

2.8 Gauss' Laws

We begin the consideration of electromagnetism with a brief summary of the pertinent electrical theory and mathematical foundation for microwave packaging. This discussion sets forth the observations of early experimenters regarding the behavior of electric fields, magnetic fields, and electric current and shows some relationships that have significance for the understanding of how signals behave in various

TABLE 2.4 Effect of Resistivity Increase in a Gold Conductor with 1 Percent Contamination of Any Metal Listed

Material	Resistivity	% increase in resistivity
Silver	2.82	28.2
Palladium	2.90	31.8
Cadmium	3.07	39.5
Platinum	3.30	50.0
Copper	3.59	63.3
Nickel	5.10	132.0
Tin	7.60	245.0

media. There is a relationship between the electric field and the magnetic field that has its foundation in both Ampère's law and Faraday's law. Those laws in turn can be described in terms of Gauss' laws for the electric field and for the magnetic field. Gauss' law for electric fields is paraphrased simply here: An electric field is created when a charge, either positive or negative, is introduced into a space. By convention, the field is said to emanate or diverge from a positive charge and to converge onto a negative charge. An electric field can diverge from an isolated point charge and never converge onto a negative charge. That electric field can exist as an isolated field. This is shown in Fig. 2.4. Gauss' law for magnetic fields is stated similarly: A magnetic field cannot be created by an isolated magnetic pole because isolated magnetic poles do not exist. Instead a magnetic field in space appears to begin at a source (pole of a magnet, for instance) and end at a "sink" (the other pole of that magnet). However, the magnetic field is in reality continuous, passing through the magnet and out the other end, and around in space to the other pole of the magnet. This is shown in Fig. 2.5. Ampère's law states that an electric current causes a magnetic field to exist. If the current is passing through a wire, the resultant magnetic field is tangential to a circle drawn around the wire where the plane of the circle is perpendicular to the wire (Fig. 2.6).

Faraday's law, on the other hand, states that an electric current in a circuit can be caused by a changing magnetic field. If a conductor such as a wire creates a closed path, and a magnetic field passes through that area bounded by the closed path, an electric current will flow in the wire if the magnetic field is changed in magnitude (strength) and/or direction. That electric current will be in direct proportion to the magnetic field. It is more accurate to say that the change in magnetic flux through the closed path with respect to

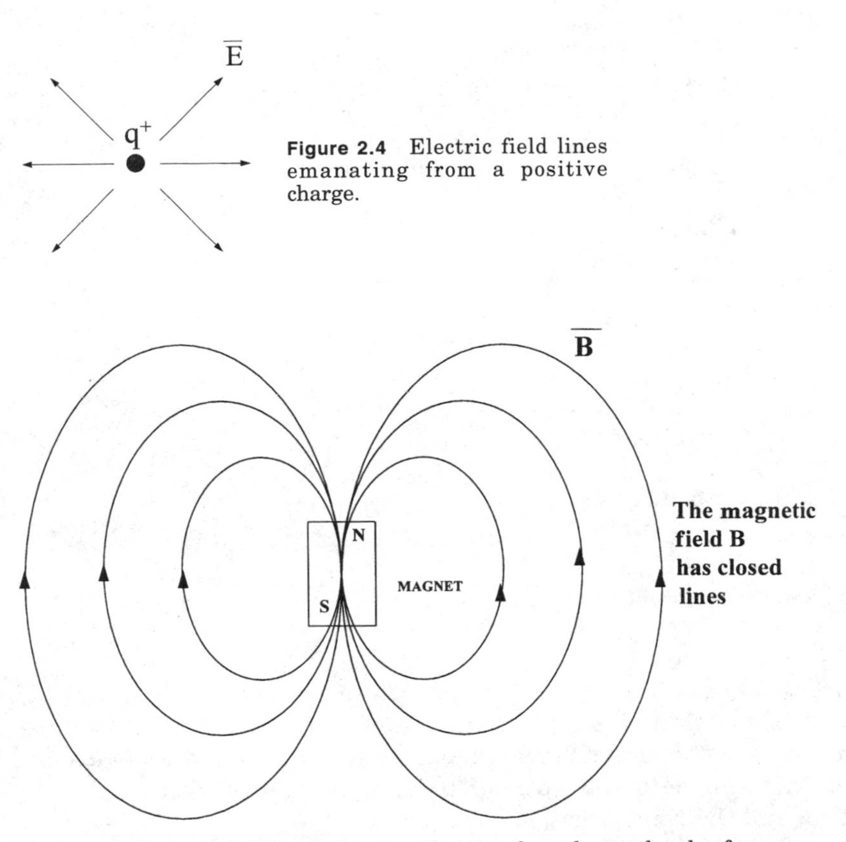

Figure 2.4 Electric field lines emanating from a positive charge.

Figure 2.5 Magnetic field lines shown emanating from the south pole of a magnet and returning to the north pole.

time induces an emf (electromotive force) in the circuit. This emf is the cause of current flow in the circuit. This is expressed mathematically as

$$E = -d\frac{\phi_m}{dt} \tag{2.7}$$

where E is the electromotive force, ϕ_m is the magnetic flux or field at the plane of the circuit, and the minus sign indicates that the induced emf opposes the changing magnetic forces.

The laws relating electric fields, magnetic fields, and changing electric and magnetic fields that were discovered by Ampère, Gauss, Henry, Oersted, Biot, Savart, and Faraday were stated concisely in the 1800s in four mathematical equations by James Clerk Maxwell, a British scientist. His equations relate the electric field vector **E** and the magnetic field vector **B** to their origins which are (1) electric

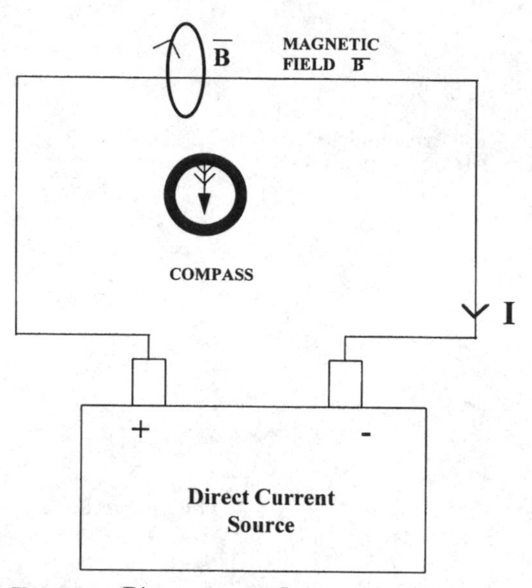

Figure 2.6 Direct current flowing through a wire causes a magnetic field to exist around the wire.

charges, (2) electric current, (3) changing electric fields, and (4) changing magnetic fields. Some simple examples will help to illustrate the electric and magnetic phenomena that led the early experimenters to make fundamental discoveries about electricity and magnetism and their interrelation.

A static electric field such as the electric potential of a battery when the battery terminals are connected to both ends of a wire will cause an electric current to flow in the wire. This current in turn causes a static magnetic field to exist around the wire along its length. This is readily demonstrated by placing a compass alongside the wire after the current has reached a steady value after it has been connected to the battery. The compass needle will align itself with the magnetic field which, as the compass is moved in a circle around the wire, can be seen to encircle the wire (Fig. 2.6). As in the example of the magnet with poles, the magnetic field is continuous and closed. If the compass is held steady in one position as the wire ends are removed from the battery and placed on the opposite terminals, the compass needle will rotate 180° and remain fixed in the new position because the magnetic field has been reversed (Fig. 2.7). The magnetic field strength reaches its original value but the polarity of the field is reversed. Now if an ac source is substituted for the battery, we no longer have a steady-state situation. The electric field will be continually changing from a value of zero to a positive maximum and back through zero to a maximum negative value, and then once again to zero. This change

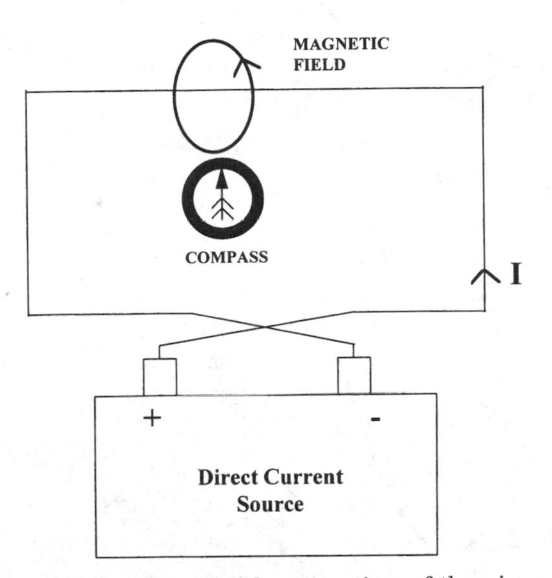

Figure 2.7 Switching the connections of the wire to the battery (Fig. 2.6) causes current to flow in the opposite direction and the magnetic field to reverse direction.

will be sinusoidal; that is, if the value of the electric field is plotted on a time scale, the resultant plot will be a sine wave. The significance of that fact will become evident when we discuss waves and their behavior in space and in dielectrics.

The magnetic field caused by the electric field will change just as the electric field changes, that is, in a sinusoidal manner. The compass needle will show this change in the magnetic field, first aligning itself with the earth's magnetic field when the current in the wire is zero. If the strength of the magnetic field caused by the current in the wire quickly and far exceeds that of the earth's magnetic field in the vicinity of the compass, then as the current builds up to a maximum value the needle will respond to the magnctic field of the current in the wire and become aligned with that field. The needle will follow the magnetic field of the current in the wire, and as the direction of the current in the wire changes back and forth, the needle will change direction, back and forth. This consequent or dependent behavior of the compass needle is a qualitative measure of the magnetic field as it changes, and also a qualitative measure of the changing electric field. The cases just described are shown in Figs. 2.8 through 2.11.

The alternating current in this example had a cause as well. If it were caused by a generator of an alternating electric field, there would be a flow of current in the wire if the resistance of the wire were anything less than infinity. The lower the resistance, the higher

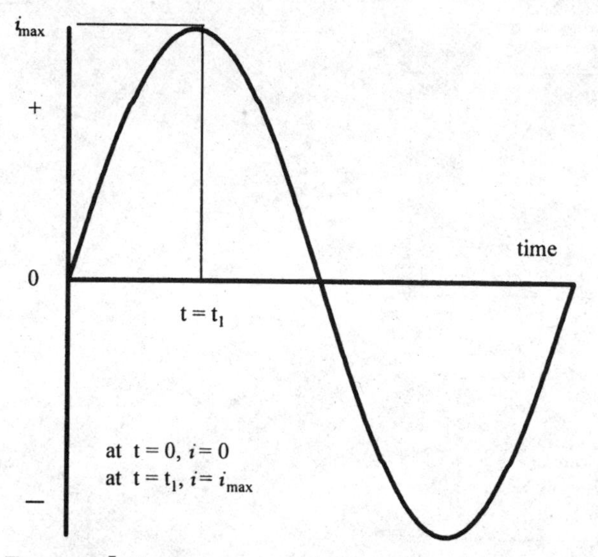

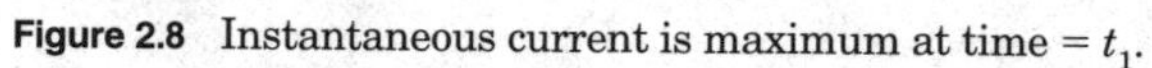
Figure 2.8 Instantaneous current is maximum at time = t_1.

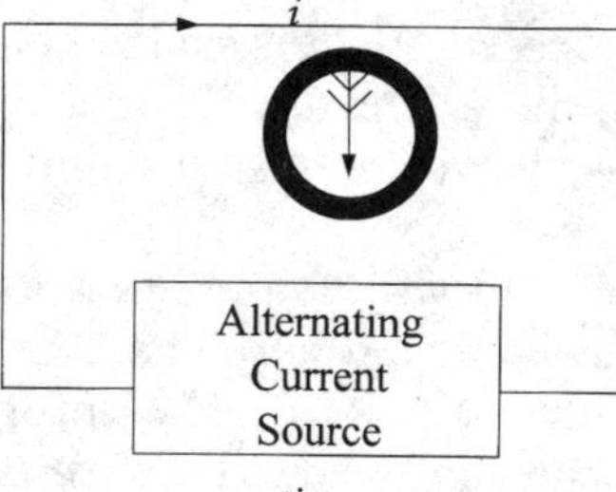

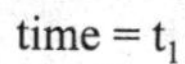
time = t_1

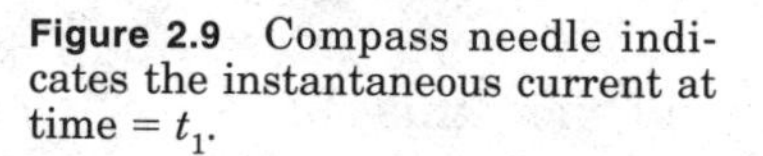
Figure 2.9 Compass needle indicates the instantaneous current at time = t_1.

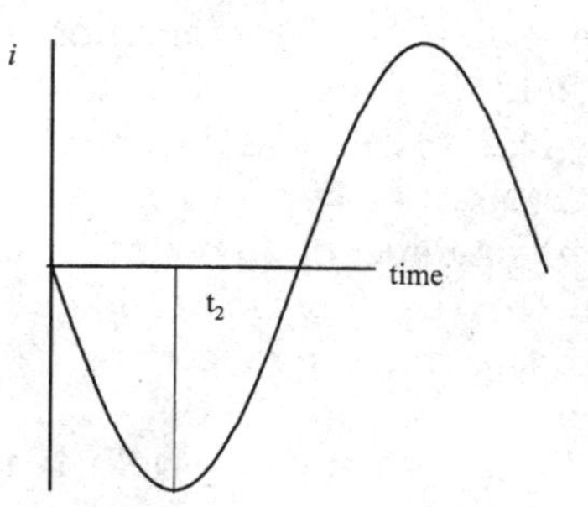

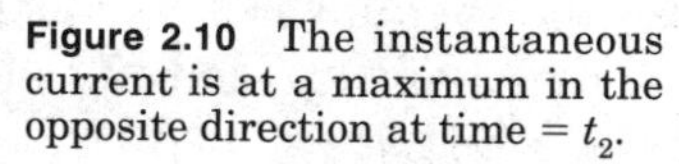
Figure 2.10 The instantaneous current is at a maximum in the opposite direction at time = t_2.

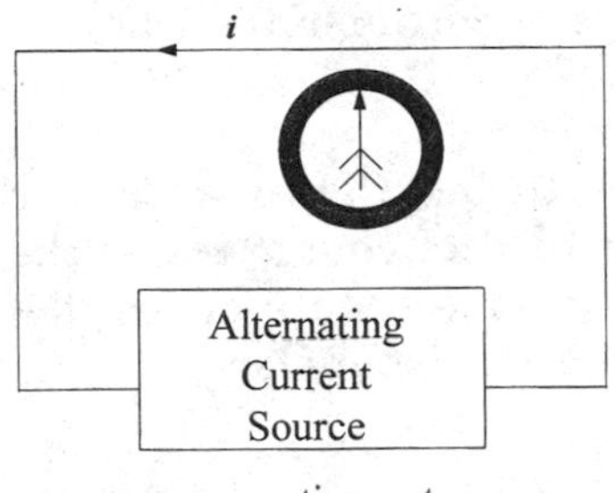

Figure 2.11 The direction of the current at time = t_2 as shown by the reversing of the direction of the compass needle.

is the current, according to Ohm's law. The higher the current, the greater is the magnetic field or flux, as will be discussed shortly. This interrelation between changing magnetic fields and changing electric fields is postulated in Faraday's law and Ampère's law. In its simplest form, Faraday's law relates the voltage (electric field) or electromotive force (emf) to the time rate of change of the magnetic field. In the same way, Ampère's law relates the magnetic field to the time rate of change of the electric field. Stated another way, a changing magnetic field will give rise to an emf or changing voltage, and a changing electric field will cause a changing magnetic flux.

The phenomena just described are best summarized in James Clerk Maxwell's famous equations. These equations are listed here in traditional order.

Gauss' law for electric fields:

$$\oint E \cdot ndA = \frac{1}{\epsilon_0} Q \tag{2.8}$$

where ϵ_0 is the permittivity of free space

Gauss' law for magnetism:

$$\oint B \cdot ndA = 0 \tag{2.9}$$

Faraday's law:

$$\oint E \cdot ndl = \frac{d}{dt} \int B \cdot ndA \tag{2.10}$$

Ampère's law with Maxwell's displacement current:

$$\oint B \cdot ndl = \mu_0 e_0 t + \mu_0 lt \frac{d}{dt} \int E \cdot ndA \tag{2.11}$$

The physical meaning of these equations is given here first for free space and then for bound charges in dielectric materials and the magnetization that exists in magnetic materials by virtue of their atomic currents.

Gauss' law for electric fields. This states that the net flux of an electric field through a closed surface is $1/\epsilon_0$ times the net charge inside the surface. The field intensity varies as the inverse square of the distance from the charge. The basis of this law is Coulomb's law

$$F = k \frac{q_1 q_2}{r^2} \tag{2.12}$$

because it describes how the field lines diverge from a point of positive charge and converge on a negative charge.

Gauss' law for magnetism. The flux of the magnetic induction vector through any closed surface is zero. This is another way of stating that the magnetic flux does not diverge from a point and therefore isolated magnetic sources or poles do not exist. If one traces the lines of a magnetic field, one always finds them describing closed curved paths. Even the magnetic fields around very large celestial bodies are not isolated but instead indicate "north" and "south" poles.

Ampère's law. A changing electric field (such as an electric current) causes a magnetic field.

Faraday's law. An electric field is generated by a change in magnetic flux. This describes how the electric field lines exist around an area through which the magnetic flux changes.

2.9 The Wave Equation

The amplitude of an electromagnetic wave traveling through space varies continuously as a function of both time and its distance from a source. A solution to the one-dimensional second-order partial differential wave equation describing the wave amplitude as a function of time and distance is

$$y(x,t) = y_0 \sin (kx - \omega t) \tag{2.13}$$

(ω is the angular frequency $2\pi f$). If we let $\omega = kv$, where v is the velocity and k is the wave number (wave 1, wave 2, etc., passing a fixed point in space), then the solution to the wave equation takes the form

$$y(x,t) = y_0 \sin (kx - kvt) \tag{2.14}$$

We can differentiate this function with respect to x:

$$\frac{\partial y}{\partial x}(x,t) = k\, y_0 \cos\,(kx - kvt) \tag{2.15}$$

Differentiating again:

$$\frac{\partial^2 y}{\partial x^2}(x,t) = -k^2 y_0 \sin\,(kx - kvt) \tag{2.16}$$

From Eq. (2.14), $y = y_0 \sin\,(kx - kvt)$, and

$$\frac{\partial^2 y}{\partial x^2}(x,t) = -k^2 y(x,t) \tag{2.17}$$

We can also differentiate the same expression with respect to t:

$$\frac{\partial y}{\partial t}(x,t) = -kvy_0 \cos\,(kx - kvt) \tag{2.18}$$

Taking the second derivative:

$$\frac{\partial^2 y}{\partial t^2}(x,t) = -k^2 v^2 y_0 \sin\,(kx - kvt) = -k^2 v^2 y(x,t) \tag{2.19}$$

Then combining Eqs. (2.17) and (2.19), we get

$$\frac{\partial^2 y}{\partial x^2}(x,t) = \frac{1}{v^2}\frac{\partial^2 y}{\partial t^2}(x,t) \tag{2.20}$$

which is the wave equation for harmonic waves.

This equation is very significant for microwave packaging because it can be used to relate changing electric fields to corresponding changing magnetic fields. In those instances y can be used to represent an electric field or a magnetic field.

2.10 Transmission Lines

To be effective, signals need to be controlled as they move about in circuits from one component to another. This control is made possible through the use of transmission lines. A transmission line can be defined as a device for guiding electrical energy from one point to another. Transmission lines are important devices that are used in all circuits, but they take special forms when used in both digital and microwave circuits. These devices facilitate the movement of signals by acting as conduits for those signals.

When properly configured, transmission lines move energy quite efficiently. *Efficiently* in this context means with a minimal loss of

Figure 2.12 Representation of a transmission line. V_s is the voltage source and R_s is the resistive load.

signal energy in the form of heat and/or radiation as the signals are guided along the transmission line. Heat can be generated through ohmic losses in circuit conductors and/or through dielectric losses because dielectrics are not perfect insulators, particularly at high frequencies. Radiation of signal energy can come about because of incomplete shielding. These phenomena will be discussed in more detail later in this chapter.

A simple transmission line can be represented by the circuit in Fig. 2.12. The signal source is V_s, R_s represents the resistance of the circuit, Z is the impedance, and the transmission lines are shown as two pipes or conduits. Signals travel along transmission lines as waves, so we use simple mathematics to describe what happens as waves move along lines. The amplitude of a wave passing a fixed point has continuously changing value as time passes. This change in amplitude can be described in terms of simple harmonic motion or with sine and cosine functions.

In the ideal case, there are no losses in the transmission line. If we let v represent the velocity of a wave, we can describe that velocity in terms of distance and time (velocity = distance × time).

$$V(\text{time, distance}) = f\left(t - \frac{x}{v}\right) \tag{2.21}$$

This can be related in standard terms for waves as

$$v(t, x) = A \cos (\omega t - \beta x) \tag{2.22}$$

where β is called the phase constant. The argument of the function is dimensionless, but it relates time and distance. Since ω is $2\pi f$,

$$v(t, x) = A \cos(2\pi ft - \beta x) \tag{2.23}$$

At $t = 0$ and $x = 0$ the cosine is unity, so the value of $v(x)$ is A. At time $t = t_1$ the wave has moved a distance $x_1 = vt_1$. At time $t = t_2$ (in this case one cycle) the value of $v(x)$ returns to the value of A since there are no losses in this example. In this time the wave has traveled a distance $x_2 = vt_2$. Since ωt_2 for this example equals 2π (one full cycle),

$$t_2 = \frac{2\pi}{\omega} \tag{2.24}$$

Then $vt_2 = \lambda$ or the length of one wave, and

$$v = \frac{\lambda}{t_2} = \frac{\lambda\omega}{2\pi} = f\lambda$$

Now $\omega = 2\pi f$ so $v_p = f\lambda$. Therefore, the velocity of propagation of a wave v_p is a product of its frequency and wavelength. From Eq. (2.23), if $x = \lambda$, $\beta x = \beta\lambda = 2\pi$, and $\beta = 2\pi/\lambda$. β is called the phase constant. Since $\lambda = v_p/f$,

$$\beta = 2\pi \frac{f}{v_p} = \omega/v_p \tag{2.25}$$

In the real world, no transmission line is without losses. These losses cause an attenuation of the amplitude of waves as they travel along a transmission line. An expression describing this condition is

$$v(t, x) = Ae^{-\alpha x} \cos(\omega t - \beta x) \tag{2.26}$$

The exponential term can be seen to diminish the amplitude of the wave as x increases and the cosine portion describes the cyclic behavior of the wave. The symbol α is defined as the attenuation constant.

Signals are carried by various transmission lines, depending on the type of circuit and the frequency of the signal. Simple circuits such as home wiring use a pair of conductors which provide a forward path that delivers energy to a load (vacuum cleaner motor, toaster heater elements, etc.) and a return path. Circuits operating at much higher frequencies use transmission lines that have different forms. Some digital circuits require that the conductor pair be twisted to reduce noise (the pair form a helix which allows the return ground to encircle the forward path and provide a shield). Coaxial cables are used to provide more effective shielding for the center conductor when that is required. This type of transmission line is used in cable TV. A small break in the shield (outer conductor or return path) allows stray signals to enter the dielectric medium of the cable and cause interference with the primary signal, as many TV viewers have found when the outer conductor has become separated from the connector at the rear of their TV. Waveguides are also used to carry signals at high frequencies. These can have several forms, such as rectangular, circular, and stripline, microstrip, and coplanar that are most commonly found in microwave circuitry. Examples of these are shown in Figs. 2.13 through 2.18. While all these are waveguides, the hollow guides are more commonly called waveguides while the others are called by their specific names: coax, stripline, microstrip, and coplanar.

The conduction of signals in waveguides is seen to be dependent on the way an electromagnetic wave will behave in the dielectric of the guide. In the case of the microstrip configuration, for example, the energy of the electromagnetic field travels in the dielectric according to how the top conductor is configured. That is, assuming that all the field is contained within the dielectric (not exactly the case) the field will move along a path defined by the top conductor, because the conductor and the ground plane beneath it form the guide. In reality, some of the field exists in the dielectric medium (air) above the top conductor. As the dielectric constant of the microstrip dielectric medium increases, more of the field will be contained within that dielectric. The condition just described is shown in Fig. 2.19.

2.11 Waveguide Principles

To understand how waveguides carry signals (waves), an understanding of the difference between a dc and an rf short circuit will help. Figure 2.20 shows a conductor which is clearly a dc short circuit. This conductor carries current from one terminal of the source (battery) to the other terminal. Unless the height h is very long and the width w is

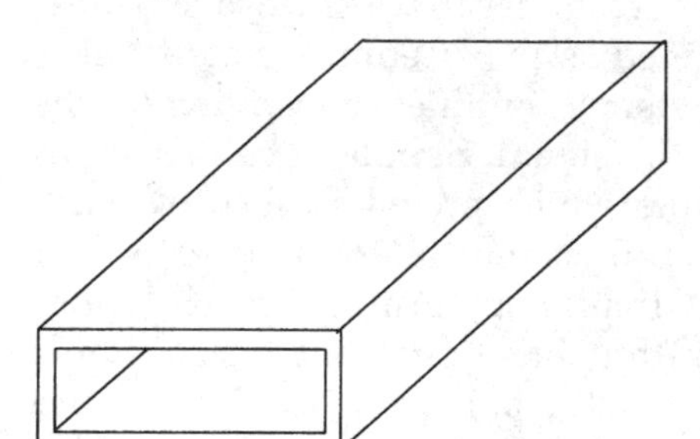

Figure 2.13 Rectangular waveguide.

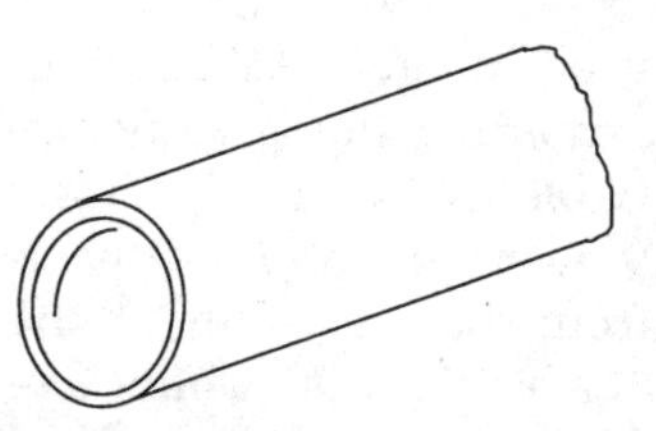

Figure 2.14 Circular waveguide.

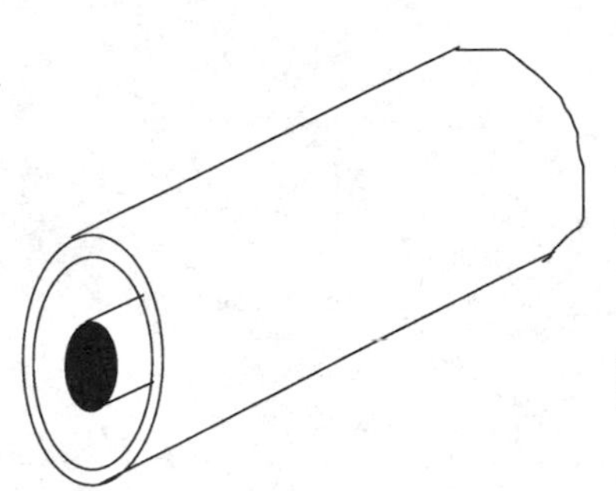

Figure 2.15 Coaxial cable showing the center conductor.

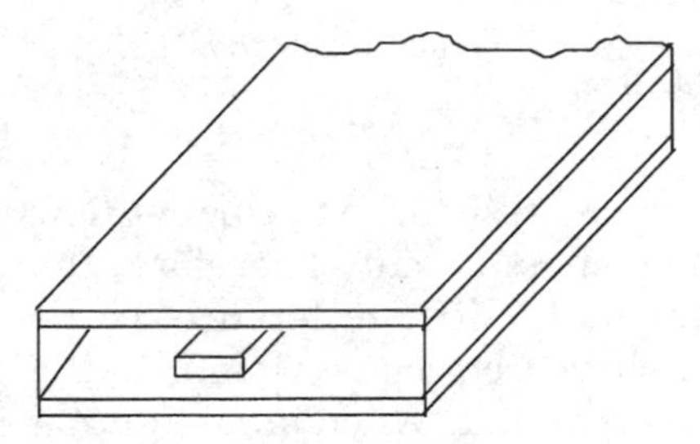

Figure 2.16 Cross section of stripline waveguide.

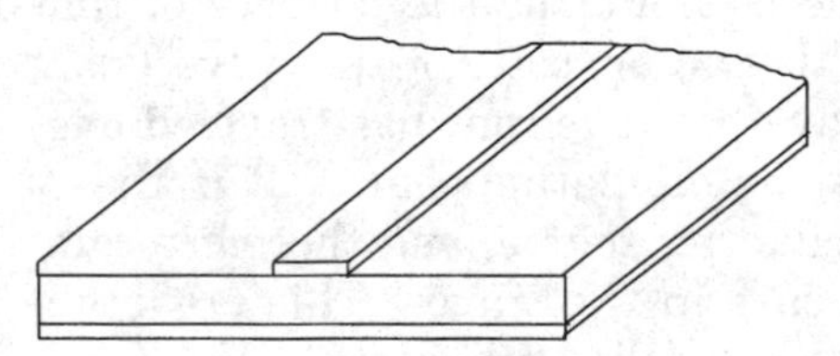

Figure 2.17 Microstrip waveguide showing top conductor above ground plane.

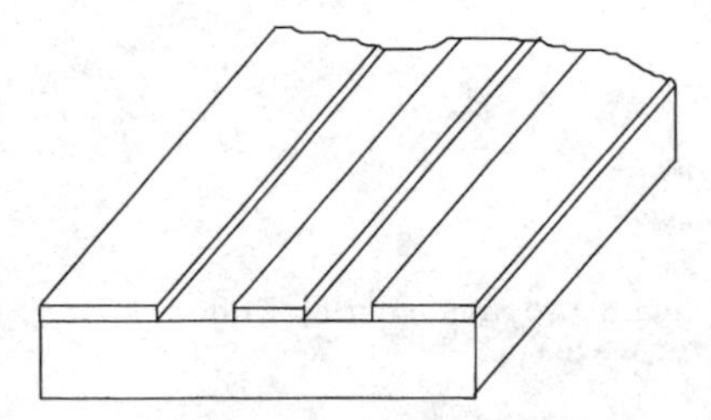

Figure 2.18 Coplanar waveguide with ground planes on both sides of top conductor.

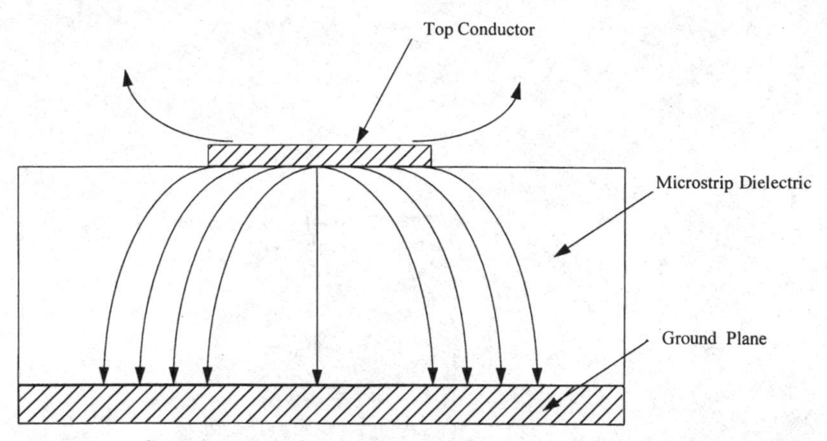

Figure 2.19 Cross section of microstrip conductors and ground plane showing fields concentrated in the dielectric under the strip conductor.

very small, the conductor will have very low resistance to current flow and will act as a dc and low-frequency short circuit. Therefore, the dimensions h and w of this conductor make little difference in the amount of current this conductor will short from one terminal to the other as long as the total path length is constant. When an oscillating voltage and current source is substituted for the battery[1] (Fig. 2.21), the situation is remarkably different. If the width w is one-quarter of a wavelength, the wavefront coming from the oscillator is shunted around to the other terminal of the oscillator as the polarity of that terminal becomes changed to exactly the opposite of the wave being shunted to it. Stated in another way, the wavefront has traveled one-half a wavelength when it reaches the oscillating source. It is therefore 180° out of phase with the oscillating source, and therefore completely destructive interference occurs in the waves. This creates a condition of infinite impedance, but only if the path is one-half a wavelength. Since the wavelength is a function of frequency, as the frequen-

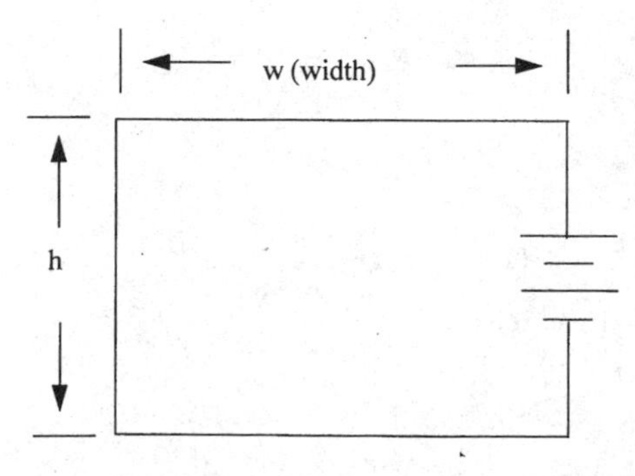

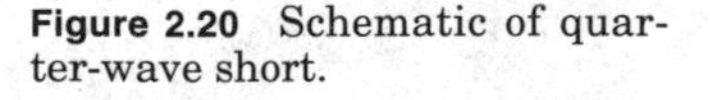

Figure 2.20 Schematic of quarter-wave short.

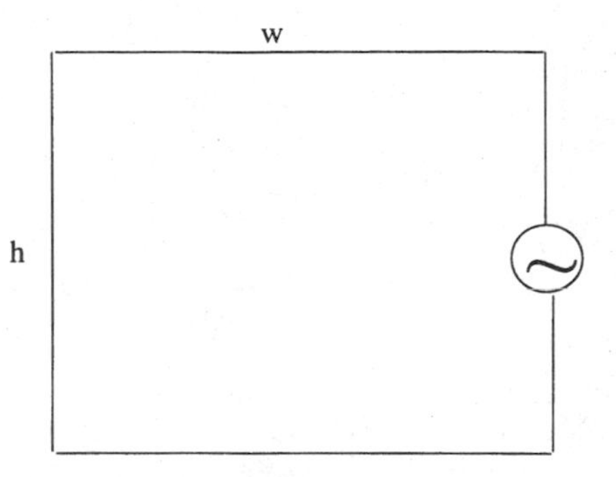

Figure 2.21 Schematic of quarter-wave short with an ac source. (Reprinted with permission from *Microwaves Made Simple*, edited by Stephen Cheung and Frederic Levien, Artech House, Inc., Norwood, MA, USA. http://www.artech-house.com.)

cy is increased, the path must become correspondingly shorter to maintain a condition of infinite impedance. No current will flow if the path $2w$ is exactly one-half wavelength ($\lambda \div 2$). In a practical sense it is not always possible to match the wave path exactly to the frequency so a minor current will be flowing. This brings us to the rectangular waveguide. Referring to Fig. 2.22, one can see that if a pair of parallel conductors of zero width are separated by a height h and connected by a quarter-wave short, adding such shorts on both sides of the conductor and up and down their lengths approaches the form of a rectangular waveguide. Figure 2.23 shows how this happens. As was stated, a quarter-wave short circuit does not affect the impedance which existed before it is connected to the conductors. If shorts 1, 2, and 3 are extended toward each other, so that neighboring shorts touch, a solid rectangle is formed. This can be seen from Fig. 2.24. It follows that since the conductors in the figure have zero width, a waveguide has width equal to $2w$ or one-half wavelength. A natural consequence of the waveguide geometry is that of cutoff frequency. No wavelength larger or frequency lower than that which meets the condition $\lambda = 2w$ will be carried by the waveguide. For this reason, rectangular waveguide is supplied in various sizes. The more popular ones are given in Table 2.5. Waveguides are used when the wavelength has small dimensions. A practical consideration is that of the wavelength of a 400-Hz signal. In aircraft applications, 400 Hz is the frequency at which ac power is usually supplied. This is a very low frequency in microwave terms, but it will help to illustrate a point. The wavelength of a 400-Hz signal is about 475 miles. This means that a waveguide which would carry that signal (one-quarter wavelength) would be 240 miles wide. Of course, this is totally impractical. Therefore, the transmission line needed to carry 400-Hz signals can be simply two wires separated by their insulation, which is many orders of magnitude smaller than the waveguide width. At 1 GHz, however, rectangular waveguide is a practical consideration. In this case the guide is

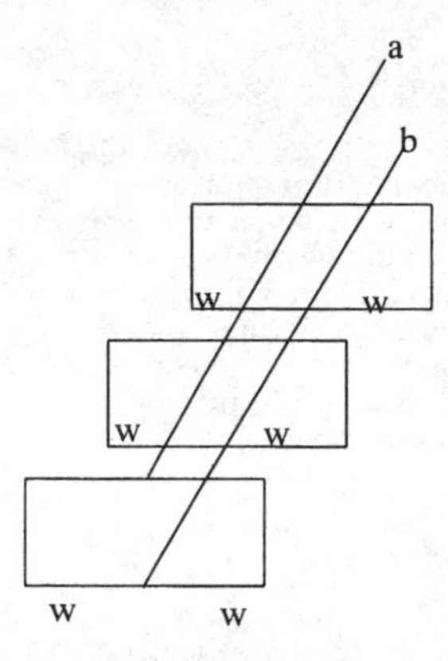

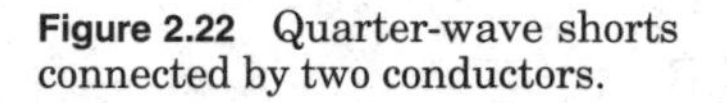
Figure 2.22 Quarter-wave shorts connected by two conductors.

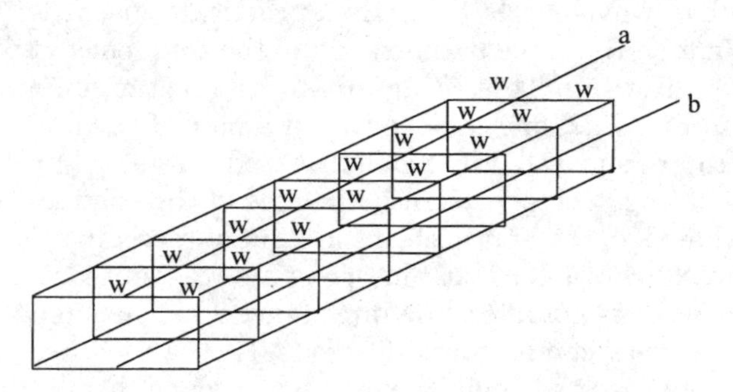

Figure 2.23 Quarter-wave shorts approaching a rectangular waveguide. (Reprinted with permission from *Microwaves Made Simple*, edited by Stephen Cheung and Frederic Levien, Artech House, Inc., Norwood, MA, USA. http://www.artech-house.com.)

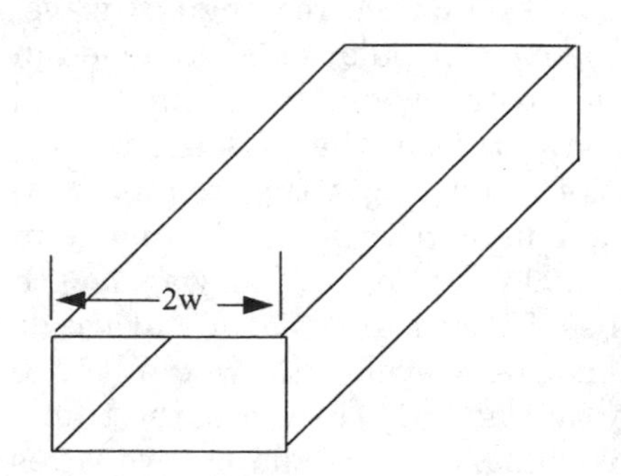

Figure 2.24 The rectangular waveguide results from the quarter-wave shorts being brought together.

TABLE 2.5 Waveguides and Their Respective Frequencies[2]

Band	Frequency range, GHz	EIA	IEC	UK	Width	Height	Cutoff frequency, GHz
S	2.60–3.95	WR-284	R-32	WG-10	7.214 mm (2.84 in)	3.404 mm (1.34 in)	2.080
G	3.95–5.85	WR-187	R-48	WG-12	4.755 mm (1.872 in)	2.215 mm (0.872 in)	3.16
C	4.90–7.05	WR-159	R-58	WG-13	4.039 mm (1.59 in)	2.019 mm (0.759 in)	3.71
J	5.85–8.2	WR-137	R-70	WG-14	3.484 mm (1.372 in)	1.580 mm (0.622 in)	4.29
H	7.05–10.00	WR-112	R-84	WG-15	2.850 mm (1.122 in)	1.262 mm (0.497 in)	5.26
X	8.20–12.40	WR-90	R-100	WG-16	2.286 mm (0.90 in)	1.016 mm (0.40 in)	6.56
P (Ku)	12.40–18.00	WR-62	R-140	WG-18	1.580 mm (0.622 in)	0.790 mm (0.311 in)	9.49
K	18.00–26.50	WR-42	R-220	WG-20	1.067 mm (0.420 in)	0.432 mm (0.170 in)	14.1
R (Ka)	26.50–40.00	WR-28	R-320	WG-22	0.71 mm (0.280 in)	0.356 mm (0.140 in)	21.1
Q	33.0–50.00	WR-22	R-400	WG-23	0.569 mm (0.224 in)	0.284 mm (0.112 in)	26.35
U	40.00–60.00	WR-19	R-500	WG-24	0.478 mm (0.188 in)	0.2388 mm (0.094 in)	31.4
V	50.00–75.00	WR-15	R-620	WG-25	0.3759 mm (0.148 in)	0.1879 mm (0.074 in)	39.9
W	75.00–110.00	WR-10	R-900	WG-27	0.2540 mm (0.100 in)	0.1270 mm (0.050 in)	59.0

SOURCE: Reproduced with permission of Hewlett-Packard Company.

TABLE 2.6 Comparison of Properties for Planar Waveguide Configurations[3]

Property	Microstrip	Stripline	Coplanar
Power capability	Medium	Medium	Low
Impedance, Ω	30–100	10–130	50–150
Loss by radiation	Low	Very low	Low
Unloaded Q (ratio of input energy to energy lost in the waveguide)	Medium	High	Low
Fabrication methods	Additive, subtractive	Additive with multilayer ceramic, subtractive with printed wiring board	Additive, subtractive
Fabrication difficulties	Access to ground	Access to center conductor	Straightforward

approximately 15 cm or 6 in wide. Planar waveguides are used extensively in microwave circuit packaging. A comparison of the more frequently used types is shown in Table 2.6. These are known by the names *microstrip, stripline,* and *coplanar.*

2.12 Modes and Signal Propagation

There are several ways in which electromagnetic radiation is carried in waveguides. The principal ones are transverse electromagnetic (TEM), transverse electric (TE), and transverse magnetic (TM) modes. TEM is not carried in rectangular waveguide and is generally ascribed to lower frequencies. In this mode the magnetic and electric vectors of the electromagnetic wave are both perpendicular to the direction of propagation. In free space, however, waves do travel in TEM. In the consideration of microwave circuitry the TEM mode is seldom encountered. The TE or transverse electric and TM or transverse magnetic modes, however, are commonplace. These two modes propagate readily within rectangular waveguide. The TE mode is one which has only the electric field perpendicular to the direction of propagation. The TM mode has only the magnetic field perpendicular to the direction of propagation. A few features of rectangular waveguides should be discussed here.

1. In the first place, an electromagnetic wave being carried by a waveguide by its nature contains a magnetic component that will, according to the fourth Maxwell equation, deal with eddy currents in the single-conductor internal walls of the waveguide. Therefore, to reduce the depth of penetration of the magnetic wave into the metal

walls, the electrical conductivity of the walls should be as high as possible (see Sec. 2.7). The skin depth is minimized when high-conductivity walls are used because as the conductivity of the metal decreases, the magnetic wave depth of penetration into the walls increases instead of being reflected off the walls. The eddy current within the metal meets resistance and the energy in the wave is converted to heat. This conversion of electromagnetic energy to heat detracts from the energy in the wave, which is always undesirable, so internal waveguide walls are sometimes made of copper plated with silver to a depth beyond the calculated skin depth. Both metals are nonmagnetic and have high electrical conductivity.

2. Since waveguide can carry waves whose frequency is above the cutoff frequency, the conductivity of the metal walls can be seen to be very important. The skin depth is reduced as frequency increases, which in effect forces the eddy currents into a decreasing conductor cross-sectional area. That simply causes more current to be carried by less conductor, and if the conductor has poor conductivity (high resistance), the current flow must generate heat according to the relationship power = I^2R. This general relationship between resistance R and current I holds for the rf case as well as dc. These losses are typically called I^2R losses.

3. The angle the incident wave makes with the walls of the waveguide can vary from zero up to nearly 90°. This means that the wave must travel a much greater distance within the waveguide when the incident angle is large because it reflects back and forth many more times than when the angle is small. Each reflection or encounter with the waveguide walls extracts energy from the wave in proportion to the resistivity of the metal wall. These features exemplify the need to keep the inner wall resistivity as low as possible. Put another way, the conductivity and thickness of the metallization should be as high as possible.

For generalized transmission lines, that is, transmission lines of all types, there is a common thread. The properties of these lines include inductance, capacitance, resistance, conductance, and impedance. The simplest case of a transmission line is that of two wires, one carrying the signal forward and the other a return. Since they are separated by a finite distance, the dielectric of the intervening medium is the insulator. All conductors have a resistance to current flow, because there is no perfect conductor with zero resistance. If the two wires are placed in parallel with each other and separated by a fixed distance from each other, we can represent their electrical properties by resistance, capacitance, and inductance. That is, a signal traveling along the wires will experience all of these features. There is resistance in the conductors per unit length r, resistance between conductors R

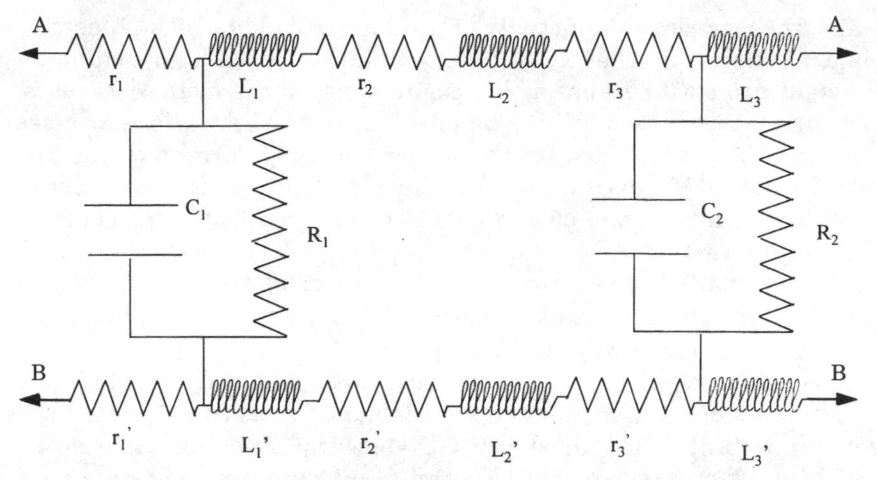

Figure 2.25 Schematic representing a transmission line.

because no dielectric is a perfect insulator, capacitance C between the conductors because they form "plates" of a capacitor, and an inductance per unit length of wire L. Even though there are no turns or coils in the wire, all wires have an inductance per unit length associated with them. These relationships between the wires are shown in Fig. 2.25. This represents a very small length of the line, but the properties repeat over and over along the lines. This is actually an equivalent circuit of two parallel lines.

When current flows in line A and returns in line B (the junction between the lines may be a dc short as described earlier or any of a variety of components or no junction at all), some of the energy from the current is dissipated in the line resistance ($r_1 + r_2 + r_3 + \cdots + r_n + r_n' + \cdots + r_3' + r_2' + r_1'$), some is used to become stored in the inductance ($L_1 + L_2 + L_3 + \cdots + L_n + L_n' + \cdots + L_3' + L_2' + L_1'$), some is used to charge up the capacitance ($C_1 + C_2 + \cdots + C_n$), and some is dissipated in the resistance of the dielectric between the two lines ($R_1, R_2, \ldots, R_n$). This resistance R is often distinguished from the line resistance r by using its reciprocal G, which is called conductance. This makes good sense since the dielectric between the lines is not a perfect insulator and does permit the conduction of some current from one line to the other. The amount of current conducted from wire A to wire B through the dielectric is dependent on the voltage across the dielectric and its resistance in a specific circuit. In practical terms for printed wiring boards, resistance between wires A and B is usually designed to be at least $10^{10}\ \Omega$ and can be as much as $10^{15}\ \Omega$.

If lines A and B are very long (such that there will be no effective return current flowing in the lines due to reflections at the end of the lines from impedance changes or mismatches) the characteristic impedance of the lines Z_0 is

$$Z_0 = \sqrt{\frac{r + j2\pi fL}{G + j2\pi fC}}$$

Since analysis of circuit performance is required prior to fabrication, simplifications and assumptions are usually made to reduce the cumbersome mathematical model to practical terms. The result of these approximations is generally sufficient for a first cut. It can be assumed that for most circuits the conductance is almost zero and line resistance is also very nearly zero.

The characteristic impedance then becomes

$$Z_0 \approx \sqrt{\frac{L}{C}}$$

the relationship that was encountered earlier in this chapter. Figure 2.26 depicts dependency of conductor line width on the dielectric constant for a fixed dielectric spacing.

2.13 Dielectric Constant and Loss Factor Measurement[2]

The relative dielectric constant of a material ϵ_r is defined as the ratio of the permittivity of the material to the permittivity of free space, or

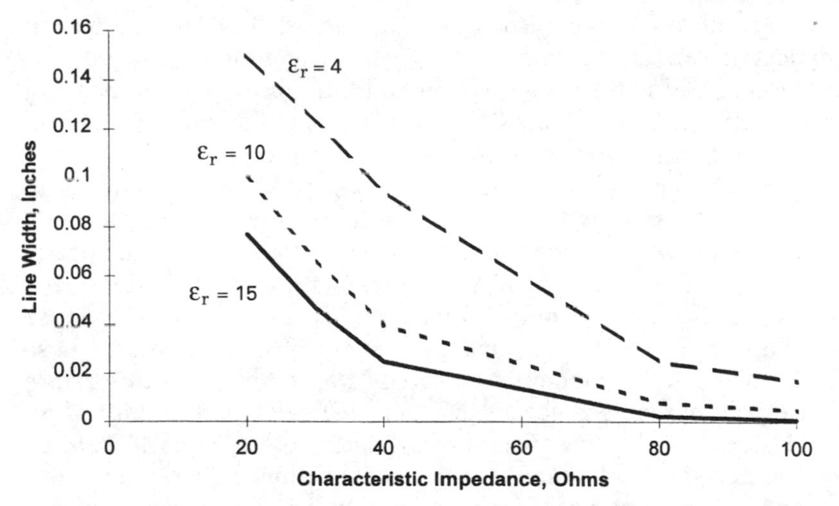

Figure 2.26 Plot of microstrip characteristic impedance showing variation of conductor line width on various materials.

$\epsilon_r = \epsilon/\epsilon_0$. For most discussions, when reference is made to the dielectric constant of a material, it is the *relative dielectric constant* that is being discussed. Permittivity is composed of a real component and an imaginary or complex one. The symbols are ϵ', the real component, and ϵ'', the imaginary component.

The real component is a measure of the amount of energy from an electric field imposed on the dielectric from an external source that can be stored in a dielectric. The imaginary part ϵ'' is called the loss factor. This loss factor is a measure of how much energy from the external field is dissipated or lost in the dielectric. The electric energy is converted to heat and lost to the surrounding environment. The dielectric constant ϵ_r of most solids can range in value from about 2 to several tens of thousands. The loss factor is ideally as small as possible. For microwave dielectrics, the loss factor ranges from 0.01 to 0.0001. There is a vector relationship between ϵ' and ϵ''. Since they are 90° out of phase, a vector diagram can be drawn with ϵ' and ϵ''. The angle between the vector sum and ϵ' is called δ. The tangent of that angle δ is then the ratio of the imaginary part to the real part, or the ratio of the energy lost to the energy stored. In the case of an alternating electric field, tan δ is the ratio of the energy lost per cycle to the energy stored per cycle. The loss tangent is also called the dissipation factor (often abbreviated D.F.), and that is the reciprocal of the quality factor *Q*. The dielectric constant of many materials varies with the frequency of the externally imposed alternating electric field.

There may be several polarization effects that together result in the overall permittivity. All polarizabilities have different resonant frequencies, and therefore as the frequency of the external field is increased, different resonances are experienced. For this reason, if the dielectric constant is measured as the frequency is raised, the dielectric constant can be seen to rise and fall and rise again and fall again, etc. Of equal importance, the loss factor will also vary with frequency. The dominant effects are usually atomic polarization, electronic polarization, and orientation polarization. Atomic polarization occurs when the positive and negative ions in a material become physically distorted under the external electric field. Electronic polarization occurs as the nuclei of atoms are displaced from their rest positions relative to their orbital electrons. Some molecules, because their atoms share electrons, form permanent electric dipoles. These molecules are usually randomly oriented in a material, and therefore the net dipole moment is essentially zero. Under the influence of an external electric field, however, these dipoles rotate so as to become aligned with that field. This is called orientation polarization. It is important to know the behavior of both ϵ' and ϵ'' in the range of frequencies the material is expected to experience in a circuit. Knowing these frequency-dependent values, the designer can compensate for undesirable variations with the design.

Measurement methods for determining the dielectric constant and loss tangent include the parallel plate, coaxial probe, transmission line, and resonant cavity techniques.[4] Each method has its own advantages and disadvantages. The methods are described briefly here:

Parallel plate method. This involves preparing a thin sample of dielectric such that it fits between the electrodes or plates of a test fixture to form a capacitor. The capacitance between the plates without the material under test and with the material are measured. The dissipation factor is also measured. The upper frequency limit of this method is about 30 MHz, but it is very useful for materials that are in film form and is not expensive to perform. If the material to be tested cannot be readily prepared in a film (0.005 to 0.020 in), this method should not be used.

Coaxial probe method. This method uses an open and coaxial transmission line (Fig. 2.27). The dielectric material under test is placed against the open end of the probe, and the fields penetrate the dielectric. The reflected signal is measured and provides a measure of the dielectric constant.

Transmission line method. The transmission line technique is one that requires preparing a sample of the dielectric such that it fits within a transmission line, usually a waveguide (either rectangular or coaxial airline). The reflected and transmitted signals provide the informa-

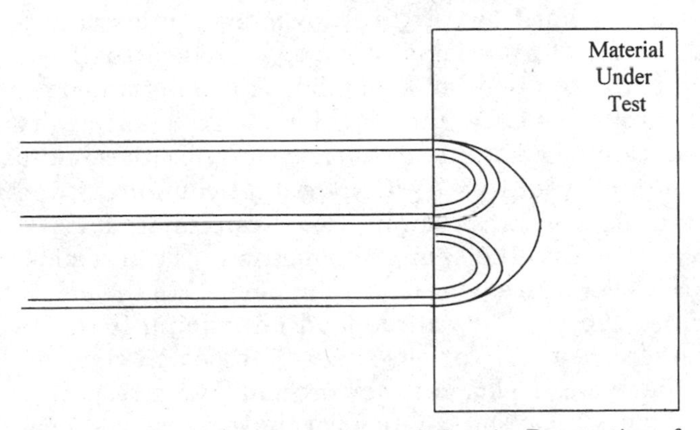

Figure 2.27 Coaxial probe testing a dielectric. Penetration of the electric fields is shown.

tion from which the dielectric constant and permeability as well can be computed. This method is useful up to 110 GHz.

Resonant cavity method. A resonant cavity is a structure that is used to measure the properties of a dielectric. The cavity is constructed so that it resonates at specific frequencies. To measure dielectrics, one calibrates the cavity when it contains only air. Then a dielectric is introduced into the cavity, and measurements are taken again. Cavities are high-Q in nature. They resonate at certain frequencies that are determined by their dimensions. The frequencies of resonance are narrow; that is, a clear peak resonance can be measured and the center frequency very accurately determined. Any material introduced into the cavity will shift the center frequency and the Q factor. Since the Q factor is the reciprocal of the dielectric loss factor, when the loss factor is very small, as it is with low loss dielectrics, the Q factor is a large number. It is the change in center frequency and Q factor that enables calculation of the dielectric constant and loss factor. The advantages of this method are that it is very accurate, and this high degree of accuracy enables detection of very small changes in the loss factor ϵ'' especially when the loss factor itself is very small. This feature is important when the loss factor of the dielectric is critical to circuit performance such as low-noise amplifiers. On the negative side, this method cannot determine dielectric constant and loss factor for a wide frequency band. That feature must be provided by other methods such as coaxial airline or rectangular waveguide. The coaxial airline is quite broadband and the rectangular waveguide is somewhat narrower in band, but samples to be measured are usually easier to machine and prepare when they are in rectangular form. It is difficult and expensive to make them in cylindrical form with a hollowed cylindrical portion to accept the center conductor, especially at higher frequencies. There must be a tight fit over the center conductor and against the walls of the coaxial airline, so that virtually no gaps exist. Therefore, the rectangular waveguide is usually preferred in spite of its narrower frequency band. While the resonant cavity method is more accurate at a specific frequency, deriving those data can be cumbersome and requires a PC (personal computer) and appropriate software. To avoid this complication, samples need not be machined to fit as tightly into the sample holder as in the transmission line method. However, placement of the sample within the cavity is important because the results will depend on placing it in the region of maximum electric field. Various shapes can be accommodated (strips, spheres, bars, etc.) so long as they are small with respect to the overall cavity dimensions. The cavities can have various shapes as well.

One advantage of the resonant cavity method over the transmission line method is that the magnetic permeability of a material does not have to be 1.000+ (very nearly unity). This is so because of cavity features which cause the magnetic field to be zero in the space where the electric field is a maximum. For this reason ferromagnetic materials can be evaluated for their dielectric constant and loss factor by this method even though their magnetic permeability is far from 1.000. However, the requirement still holds that the sample must be very small compared to the cavity size.

2.14 References

1. Stephen Cheung and Frederic Levien (eds.), *Microwaves Made Simple,* Artech House, Inc., Norwood, Mass., 1985.
2. *Basics of Measuring the Dielectric Properties of Materials,* Application Note 1217-1, Hewlett-Packard Company.
3. Based in part on Richard Brown, *Microwave Hybrids,* Technical Monogram, ISHM, 1989.
4. *Dielectric Materials Measurements—Solutions Catalog of Fixtures and Software,* Hewlett-Packard Company.

Chapter

3

Packaging Technologies for High Speed Digital and Microwave Circuits

3.1 Multichip Modules

Multichip modules (MCMs) have been around since 1959 with the issuance of a patent to Jack Kilby of Texas Instruments for the integrated circuit concept. As soon as a circuit function was incorporated on a chip of silicon, engineers began to place two or more of them in a hermetic box (package) with a few other components, creating a hybrid circuit. These early multichip packages or modules were precursors of the more complex hybrid assemblies which followed. As integrated circuits became increasingly more integrated, they were called large scale integration (LSI) and later very large scale integration (VLSI). More and more functions were incorporated into chips, and engineers continued to place many of them in hybrid circuits. Hybrids have grown in complexity by virtue of the devices they contain as well as the internal connecting circuitry and number of output leads. For over 20 years one could count on the number of transistors on a chip to double about every year. This combining of chips of increased complexity into a higher-order circuit package or module has continued because higher integration of devices enhances multifunctional packaging; that is, high levels of integration are achieved without the need for expensive ASICs.

Today the complex hybrid circuit is called a "multichip module." Not only are the devices contained within it complex, but the internal interconnect systems are sophisticated and varied. The dielectrics may be ceramic, silicon, organic, or even diamond, and the conductors may be aluminum, copper, gold, or other relatively high electrical conductivity metals. Multichip modules are used for both analog and digital circuits.

There is a generally accepted classification of MCMs primarily by their dielectric materials. It is as follows:

MCM-D. The substrate is silicon or ceramic plus a deposited dielectric for additional layers. The dielectric is usually applied as a polymer and spun by centrifugal force across the silicon surface.

MCM-C. The substrate is ceramic, which could mean aluminum oxide or low-temperature cofired ceramic.

MCM-L. The substrate is a printed circuit board. The most dramatic advances have been made in this category since 1992, with very thin multilayer circuitry on laminate materials of decreasing thickness for packaging high-density memory chips used in PCMCIA cards (now designated printed circuit cards).

There are variations and combinations of the listed main groups as well in packages containing high-lead-count multifunctional chips.

Most recent developments of the MCM technology include rf and power-conditioning circuits. In these cases, the attractiveness of packaging densities offered by the use of MCM techniques has resulted in transmit and receive module miniaturization, and very densely packaged power-conditioning circuits. The latter consist of switching power supplies for personal computers and printers, and distributed and embedded power sources for aircraft equipment. Power-conditioning circuits involve the conversion of alternating to direct current or the inversion of direct to alternating current. Such circuits involve high current, occasionally high voltage, and almost always lost electrical power which must be dissipated. Weight and size reductions have occurred in this area because of modularity trends as well as the systematic elimination of bulky inductive components. This facilitates hybridization of circuit components. However, with this shrinking of size and increase of component density, operating temperatures of components may exceed design specifications. Consequently, special materials and techniques are required to remove heat efficiently from these circuits.

One reason for the existence of the new generation of digital hybrids and MCMs is electrical performance. A major use of the digital MCM is in personal computers, both desktop and laptop, and workstations, where clock rates are nearing 200 MHz and signal rise times are in picoseconds. At these performance levels, devices must be placed together as closely as possible, and the dielectric beneath signal conductors must have a dielectric constant as low as possible so as to minimize the times the signals require to get to their next location.[1]

The packaging of analog circuitry which deals with microwave and millimeter wave technology progressed in a similar fashion to the digital world. In this arena, techniques which preserve signal characteristics at high speeds have always been required. Thus concerns about the dielectric constant and loss tangent of a dielectric have resulted in very high quality materials development.

3.1.1 MCM-C

The MCM-C, where the C stands for ceramic, is a type of multichip module material. The term "ceramic" is applied to a variety of materials which include aluminum oxide, beryllium oxide, aluminum nitride, cordierite, silicon carbide, mullite, diamond, and glass ceramic. Glass ceramic includes the class of materials called low-temperature cofired ceramics. Thick film ceramics, where the conductors are made from glassy materials which contain conductive and dielectric particles, and the dielectrics are mixtures of glasses also considered ceramic in this scenario.

In the MCM-C type of multichip module, dielectrics consist of ceramic materials from the group just listed. MCM-C dielectrics can be divided into two categories: high-temperature firing (greater than 1500°C) and low-temperature firing (between 800 and 1000°C). While both processing temperatures are considerably above room temperature, the "low-temperature" material systems, that is, dielectric and metallization, can be processed in firing ovens and furnaces which process thick film circuits. This is a distinct advantage to those manufacturers that possess equipment capable of processing thick film materials—dielectrics and conductors.

3.2 The Cofired Ceramic Process

A cofired ceramic is one in which the dielectric and conductor are matched so that they "fuse" together at the same temperature to form the desired multilayer construction. In this approach, a ceramic powder is mixed with organic binders which hold the particles together, and solvents which allow fabrication of the ceramic into a flat tape. This is called a slurry. The slurry is placed on a ribbon of plastic (usually polyester) tape and the plastic tape is drawn under a knifelike gate called a "doctor blade." As the plastic tape is pulled under the blade, the slurry on top of the plastic tape is made to become a thin film about 0.006 to 0.008 in thick. As the plastic tape is drawn past the doctor blade, the solvents in the slurry evaporate rapidly and leave the dried ceramic-binder combination attached to the plastic tape. This system is then rolled up into convenient lengths for further processing.

When it is desired to process the ceramic, the tape is unreeled and the ceramic layer is removed from the plastic support. The ceramic material is then called a "green" or unfired tape. It is flexible yet able to support itself and can be handled like a sheet of paper. This form factor is ideal for the fabrication of a multilayer conductor and dielectric plate which can be an MCM-C module interconnect system. The sheet of ceramic and binder is made into a multilayer conductor system as follows: Dielectric sheets are drilled or punched to open holes which will contain conductor vias. They are then metallized by printing a metal slurry pattern on both sides of the sheet. This pattern will form conductors when it gets fired. Another printing places a via metallization into the open holes for further processing.

If a multilayer construction is required, the layers of tape are placed one atop the other such that layer 1 is the topmost one and layer *N* is the bottommost. Each tape layer has two conductor layers on it and each layer is aligned with the others during pressing through keying holes which fit over alignment pins in the press. The group is then pressed together so that the layers will become a monolithic slab after firing.

A major difference between the high-temperature and low-temperature systems is the metallization. In the high-temperature system, the metallization is usually tungsten. Tungsten is used because metals other than those which are refractory would either melt and/or oxidize so completely they would not retain the desired shape and dimensions nor would they any longer be conductors, if indeed they were not burned off. The reason the conductors would oxidize or melt is that high-temperature ceramic must be fired at temperatures well beyond the metal's melting point.

The electrical resistivity of tungsten is quite high compared to metals used in thick films and with low-temperature firing ceramics. In applications requiring narrow or fine conductor widths measuring many millimeters in length, the total dc line resistance can measure tens of ohms. For circuits depending on low-resistance conductors, this limits high-temperature ceramics somewhat in their range of application. Thick film and low-temperature ceramics allow the use of comparatively high conductivity metals which widen the range of applications. Consequently MCM-C modules are being fabricated from low-temperature ceramic and metal systems. It should be noted that not all ceramics have the same strengths, and this can be a factor in the selection of a material. High-temperature materials have greater flexural strength than low-temperature ceramics. Table 3.1 shows some mechanical and electrical properties of materials being

TABLE 3.1 Properties of MCM Ceramics

Material	Loss Dielectric constant	Thermal tangent $\times 10^{-4}$	Coefficient of linear conductivity, W/m-K	Surface expansion $\times 10^{-6}$/C	finish, μin
96% alumina (Al_2O_3)	9	7–8	28	6–7	15–25
99.5% Al_2O_3	9.8	2–3	30	6–7	3–5
Aluminum nitride (AlN)	8–10	5–20	150	4–6	10–20
Silicon carbide (SiC)	40	500	250	3.5	5–10

SOURCE: Based in part on Richard Brown, *Microwave Hybrids,* Technical Monogram, ISHM, 1989.

used for MCM-C multichip modules as well as other ceramics of interest in electronic packaging. Strength is not the only property of importance in MCM-C material selection. Dielectric constant and loss tangent are properties which strongly influence electrical performance of high speed digital and microwave circuits as well. In general, digital MCMs require a very low dielectric constant because the speed of a signal traveling along a conductor depends on the dielectric surrounding the conductor. The higher the dielectric constant, the slower the speed of the signal, and consequently, fewer computations may be performed in a unit of time if a signal must travel slowly. Microwave circuits, on the other hand, depend more on the dissipation factor of the dielectric than on the dielectric constant. Microwave signals become attenuated more per unit length of signal path when the dielectric has a high dissipation factor (usually a value above 0.005 is considered high). It is important to know the value of the dissipation factor at the frequency (or frequencies) of interest because as frequency increases, so does the dissipation factor to some extent. No dielectric material is a perfect insulator, but the more it behaves as a perfect insulator, not only in preventing electric current from passing through it but also in being not easily polarizable, the less electrical energy it causes to be lost from signals and converted to heat. A material with a high dissipation factor (0.01 or higher) is said to be "lossy." The lower the loss, the higher is the quality of the dielectric. In power transmission, much energy can be lost in a poor dielectric. However, this loss is also important in low signal level microwave circuits such as receivers. If a received signal is weak, it will become even weaker as it travels through a lossy dielectric. Although amplification of the weak signal is possible, this consumes more power and adds noise and distortion to the signal. In high-performance circuitry it is usually advisable to select the lowest-loss dielectrics consistent with other properties and cost.

It can be seen from the previous discussion on electrical conductivity of metallizations used in multilayer ceramics that a high-conductivity metal provides the greatest design flexibility. By the same token, a dielectric which has very low loss will afford the designer similar freedom. For these reasons many MCM-C designs have concentrated on dielectric materials which can be cofired with thick film metallization systems. Among these are the LTCC modules which IBM has chosen for its 390/ES9000 computers.[2] These modules consist of as many as 64 buried conductor layers. Materials are available today for MCM-C designs from a variety of sources with a large range of properties. LTCC metallization systems offer high conductivity and dielectrics are becoming low in loss characteristics. Compatible resistor systems can now be included in buried layers, leading to lower-cost multilayer structures.

3.2.1 MCM-L

The MCM-L style of multichip module was used principally in packaging digital circuitry until the early 1990s. In response to high demand for denser circuitry for cellular phones, pagers, notebooks, etc., rf and microwave circuits are often fabricated on printed wiring substrates in MCM-L form. The L stands for "laminate." Laminate refers to the layering of strips of materials atop one another. In this case it is the layering of multilayer printed circuit board (MLPCB) materials. This multilayer technology is highly developed for MLPCBs but less so for MCM technology. The reason for the existence of MCM-L is that surface mounting of components with high lead count and leads spaced more closely has driven printed wiring board technology to finer conductors, more conductors per layer, and finally to a point of component density comparable to MCM-C technology. The number of PCB manufacturers currently producing surface mount boards with line widths of 0.003 in either from fully additive methods or from copper cladding through subtractive methods provides sufficient evidence that MCM-L technology is a viable competitor with MCM-C for many applications. In order to achieve 0.003-in etched line widths it is necessary to begin with a cladding which is thinner than conventional PCB claddings, generally about ¼ oz (0.0035 in thick). Thicker claddings produce a pyramidal shape in cross section which can penetrate the dielectric above during lamination and cause an electrical short circuit. The dielectric material most used for standard MLPCB technology is epoxy-glass. Since the dielectric constant ε_r strongly affects signal transmission speeds, it is very important that the dielectric constant exhibit negligible variation from layer to layer. Some typical MCM-L candidate dielectrics are

TABLE 3.2 Properties of Typical Microwave Laminate Materials

Type	Dielectric constant	Loss tangent	CTE, ppm/°C in-plane
PTFE	2.3	0.0002	100–110
PTFE with woven glass	2.5	0.0007	30–40
Epoxy-glass	4.7	0.008	35–45

shown in Table 3.2.[3] Most of the signal layers of an MCM-L are buried within the multilayer structure and should be considered as stripline rather than microstrip for timing analyses. The same table shows that signals traveling in the stripline configuration require more time to traverse a given distance than those traveling in microstrip. Therefore, the importance of the dielectric constant cannot be emphasized too much. Care should be exercised, however, in the selection of the dielectric material because characteristics such as thermal cycling–induced microcracking of the resin and thermal coefficient of expansion can be major factors. The reason microcracking of the resin surface can be problematic is that fissures could open in the dielectric which would cause the fine conductors to become severed. This would in turn cause intermittent conduction of signals and power as well. The thermal coefficient of expansion is a major factor in MCM-L technology because of interconnections to the chips. Depending on how interconnections are made between chips and MCM-L substrates, differences in expansion coefficients between chips and the supporting dielectrics could cause failure of the chip mounting material and/or electrical connection. A chip mounting material which is tolerant of expansion differences will be more reliable if many temperature cycles can be expected in the application. Chip interconnects such as “flip chip” using solder bumps will experience considerable stress during thermal excursions and consequently could fail prematurely. The choice of not only the MCM-L substrate material but also the mounting material is very important from the standpoint of chemical compatibility with the devices to be packaged. Most of these devices are sensitive to potentially corrosive constituents in organic resins, particularly hardeners, as well as mobile ions which can influence device performance by virtue of their electrical charge.

A further consideration is the circuit line density requirement of MCM technology. Using through-hole approaches that require excessive routing space will force interconnects to require, in turn, an excessive number of layers. Staggered vias, on the other hand, can greatly reduce the number of layers, typically by 25 percent or more. In order to interconnect to devices, a line width and spacing consistent with pad-to-pad pitch on the devices is required. This could be 0.006 in or lower and would be an absolute requirement for flip chip mounting technology since the mounting pads on the substrate must match those on the device.

A technology is just emerging which could, however, make the MCM-L approach viable for the densest known applications. This is known as conductive polymer technology. Originally developed for devices, this technology is readily adaptable for MCM-L applications. These polymers are made conductive by chemical reactions with reducing or oxidizing agents to create charge carriers.[4] The stability and processing characteristics of these materials as well as relatively low cost to produce them are causing them to be considered for a variety of usage, of which MCM-L is a logical candidate.

3.3 Package Materials

The materials which are used to package MCMs generally come from the classes of ceramic or metal enclosures used to house hybrid circuits for some time. They are usually hermetic; that is, they meet or at least are intended to meet the provisions of MIL-STD 883, Method 1014.9. This means that they should not allow atmospheres to leak into or out of the cavity containing the circuit according to Table 3.3, which is excerpted from MIL-STD 883.

Most MCMs would fall into the second category because of their size. The most favored metal for package enclosures has been Kovar, which is a combination of iron, nickel, and cobalt. Aluminum oxide (alumina) has served as the main ceramic package material with beryllium oxide (beryllia) being used for applications requiring thermal conductivity greater than that of alumina. Recent materials (aluminum nitride, silicon carbide, and diamond) have emerged as substitutes for beryllia because that material has been regarded as

TABLE 3.3 Package Leak Rates as Specified in MIL-STD 883D, Method 1014.9

	Bomb condition			
Volume of package V, cm^3	lb/in^2 abs ± 2	Minimum exposure time t_1, h	Maximum dwell t_2, h	R_1 reject limit (atm cm^3/s He)
<0.05	75	2	1	5×10^{-8}
≥0.05–<0.5	75	4	1	5×10^{-8}
≥0.5–<1.0	45	2	1	1×10^{-7}
≥1.0–<10.0	45	5	1	5×10^{-8}
≥10.0–<20.0	45	10	1	5×10^{-8}

potentially toxic. As reported by Timothy Hudson,[5] the Defense Electronic Supply Center (DESC) has notified the electronics industry that it considers beryllium and its compounds sufficiently toxic that it elevated it to a higher position on its hazardous materials list (HAZMAT). This action has had two results. The first is that designers are somewhat reluctant to specify beryllium compounds in any new design. The second is that research into substitutes for beryllium compounds, and beryllium oxide in particular, has been intensified. This research has resulted in remarkable strides being made in aluminum nitride packages and substrates and in diamond dielectrics as coatings and substrates. Silicon carbide substrates have also undergone development and refinement.

3.3.1 Metal packages

Kovar and other alloys were first developed and used to fulfill the need for metal leads of vacuum tubes to match the coefficient of expansion of the glass envelope. These leads penetrated the glass as feedthroughs and expanded and contracted with the glass as the operation of the tube caused them to experience both a rise and a fall in temperature. The electrical conductivity of these metals was considered sufficiently high to serve most needs since they were generally used for short lengths, only to penetrate the hermetic envelope. Later it was found that miniature relays and similar electromechanical devices which encountered sparking as two conductors or electrodes at different electrical potentials were brought together eroded these electrodes prematurely because of the presence of oxygen. The placement of the entire relay assembly into a hermetic envelope greatly enhanced relay life. The envelope at first was glass, but Kovar was later used to provide a rugged assembly which would withstand handling more readily. To provide electrical paths, Kovar leads were used to penetrate the envelope with matching expansion glass beads surrounding them, which isolated them from the Kovar envelope or can. The leads were fused to the glass beads and then electroplated to provide protection from processing chemicals and the environments encountered in service, as well as to enhance solderability for wire attachment in subsequent assembly. This technology was adopted by the hybrid circuit industry and used for over 30 years for hybrid circuit packages. Improvements in glass feedthroughs and lead design have continually taken place such that this packaging method has proved to be dependable when applied correctly. Certain features of glass-to-metal seals are prone to problems, however. The principal one is cracking of the glass meniscus on the lead. This is especially troublesome with rectangular leads because of the extremely small radius at the corners of the lead. Minor flexing of the lead can cause tiny flakes of the glass to be cracked away, leaving exposed a small amount of bare Kovar which was previously protected by the glass. This could lead to corrosion of the lead and eventual failure of the device. Another potential problem is the cracking of the glass bead, which can be either radial or circumferential. Cracks can appear to be so deep as to penetrate all the way to the inside of the package. Radial cracks are specifically forbidden by MIL-STD 883. The ultimate tests of whether a crack is detrimental are the leak tests of this MIL-STD. Correlation of visual criteria and leak testing is troublesome and often inconclusive. For these reasons the trend toward ceramic packages has become very strong, although Kovar enclosures continue to be used in both commercial as well as military microelectronics.

3.3.2 Aluminum oxide (alumina) packages

Alumina packages have been used to house individual die as well as hybrid circuits successfully for some time. They have been constructed in several different configurations. First of all, they have been used very much like the Kovar envelopes, with a base which has a cavity to contain a chip or multiple chips, and a metal cover which fits over the walls of the cavity and is soldered to metallization atop the walls. This metallization is fired or sintered into the alumina surface to provide a hermetic seal. The metallization is a refractory metal or combinations of refractory metals. Typically, these are either tungsten or molybdenum-manganese. The walls may be attached to the base through a sealing glass or directly as a cofired layer. Electrical paths are provided by metallization patterns which pass beneath the walls. In the case of the cofired wall, the metallization pattern is cofired as well, that is, sintered along with the alumina. Most packages available today are constructed by cofiring layers of ceramic tape laminated together rather than with sealing glass. Cofired technology is well refined and practiced broadly both domestically and abroad. Alumina ceramic package sizes range from single-chip carriers measuring $0.180 \times 0.180 \times 0.036$ in thick, with 16 metallized and plated electrical contact pads (four per edge) to multichip packages measuring over 2 in on a side with many hundreds of pads on all four edges brazed to tin-plated or gold-plated Kovar leads.

3.4 Microwave Package Materials

This section considers the special requirements for packaging microwave circuits. Microwave circuitry differs from dc and low-frequency circuits in that the signals are carried in both the conductors and the dielectrics surrounding the conductors. For this reason the dielectric properties of packages which contain microwave circuits are quite important. The development of materials for microwave circuits as well as the techniques known as microstrip and stripline and their foundation in coaxial (coax) signal transmission are presented here.

Early microwave printed wiring technology was derived from microwave power dividers for antennas and was then developed into flat coaxial configurations.[6] ITT introduced a configuration which had a single conductor and a single ground plane with a solid dielectric that was a variation of the air dielectric power divider scheme which used two ground planes, and called this configuration microstrip.[7] Later Sanders Associates developed a central-conductor, two-solid-dielectric, two-ground-plane system called Triplate.[8] This configuration has emerged into what is known today as stripline, although the original name stripline referred to an air dielectric arrangement introduced by Airborne Instrument Laboratory.[9] The dielectric materials which were used at that time were fiberglass, followed by tetrafluoroethylene and Rexolite (trademark of C-Lec Plastics Inc.). Tetrafluoroethylene and Rexolite were unreinforced and as such were simply matrix materials. Reinforcement in the form of woven glass fabric provided mechanical strength but at the same time detracted somewhat from the quality of the dielectric. Less lossy reinforcement (quartz and randomly oriented short glass fibers) was introduced to restore the performance while still providing mechanical strength. This increased the design flexibility considerably for a wide variety of applications (microwave components such as couplers, filters, dividers, and combiners) and power distribution manifolding in phased array antennas. The component manufacturing community continued to seek higher-performance materials and began to experiment with microstrip dielectrics such as quartz, aluminum oxide, sapphire, beryllium oxide, and some titanates as well. Since these materials were not organic printed circuit boards, a technique was required to deposit and adhere a conductor system to them. Thin film technology was the most convenient since it was being developed and refined to produce integrated circuits and hybrid components and was well characterized. Sapphire, because of its original cost and the extensive machining required to render a usable substrate finish for thin film metallizations, has not survived as a major substrate material. Quartz has similar machining costs but has found limited though continued usage. Aluminum oxide, easily produced in reasonable purity levels, emerged as a major material of choice. Beryllium oxide, because of its high thermal conductivity and availability, was also used extensively.

3.5 Single-Chip Packaging Materials

The packaging of microwave circuits can vary from single-chip packages to highly integrated packages with dozens of devices contained and interconnected within one package. The most common single-chip microwave packages contain power transistors. These power transistors are mounted either onto the package base metallization or a metal which serves as the base of the package and is often used to provide mechanical mounting for the package. This metal is generally Kovar. The remainder of the package is alumina. A two-part frame of alumina is brazed to the Kovar base and a cover of either Kovar or alumina is soldered to the upper part of the frame. Kovar leads emanate from the two-part frame (which is really two frames glassed together with the two transistor leads sandwiched between them). The third contact is through the base.

The variety of microwave components used today require packages which are sometimes larger than the single-chip package and contain a hybrid microwave circuit rather than a single chip. This variety is extensive and represents different configurations with axial and radial leads, leads on one side only, leads on two sides, leads on three sides, and leads on all four sides.

Thermal considerations influence designs a great deal, and the engineer needs an insulating material for substrates and packages as well that is affordable yet high in thermal conductivity. One ceramic material that meets those requirements is aluminum nitride. Aluminum nitride (AlN) is a recent material innovation intended as a competitor for beryllium oxide for high thermal conductivity packaging. Packages can be made which measure over 4 in on a side and have as many as 10 layers of circuitry. The metallization is cofired tungsten and molybdenum which yields high-resistivity conductors. It should be noted that aluminum nitride has been successfully bonded to LTCC materials to form packages that have high electrical conductivity conductors for applications that require multilayer circuitry as well as high package thermal conductivity. This allows the use of a substrate with a low k (LTCC has a k of approximately 2 to 5 W/m-K) in conjunction with a package that has a k of 150. That combination greatly extends the applicability of AlN to a variety of high-power circuits.

3.5.1 High-power package feedthroughs

In place of the familiar fine leads usually seen on a hybrid package, power hybrid packages have 0.040- to 0.090-in-diameter pins emanating from the package walls. Electrical connections are made to the circuit contained within the package by means of these pins, which consist of a central conductor insulated from an outer metal shell by a bead of glass. The arrangement of conductor, glass, and shell is a delicate one by virtue of the glass, which must maintain a hermetic seal between itself and the shell as well as between itself and the central pin. The glass, whose coefficient of expansion is very close to that of the pin, is attached to the pin by a chemical bond between the oxides on the pin and the glass. In certain pin designs where the shell is not matched in thermal expansion to the glass, the shell and pin are assembled to the glass by heating all of them together until the glass flows. The shell contracts more rapidly than the glass, forming a compression seal between itself and the glass. The shell is sometimes called a ferrule. The ferrule material is of course different from the central pin material since it is desired to have the shell constrict more rapidly than the glass as they both cool. If the pin were the same material as the ferrule, it would reduce its diameter more rapidly than the glass, causing the bond between them to rupture. This kind of seal is reputed to be more achievable than a matched seal in which all elements of the feedthrough have approximately the same coefficient of expansion. For the copper package, it is probably more advantageous to use a compression seal with the ferrule having a coefficient of expansion less than copper but greater than the glass. This design distributes the expansion differences to each element (material) of the seal so as to minimize the tensile and compressive loads on any one interface.

Attempting to avoid the glass cracking issues associated with the design just described, some packages have recently been made with alumina used in place of the glass. This type of feedthrough generally consists of only the ceramic and the pin. The ceramic (alumina) is metallized and plated on its outer circumference and the walls of the central hole to allow it to be brazed to the pin and soldered or brazed to the copper package.

The central pin is by design required to carry high current, and it should therefore be made from a good electrical conductor. Copper would make a good choice from the conductivity standpoint, but its coefficient of expansion would preclude its selection for use with glass seals. This apparent dilemma is solved by the use of a metal cladding over a copper pin. This cladding is a match to the glass (usually Kovar or similar alloy such as Alloy 51 is used) yet bonded metallurgically to the copper pin. The cladding thickness is chosen to have sufficient strength to maintain a close match to the expansion of the glass while not reducing the electrical conductivity of the pin below design limits.

On the inside of the package electrical connections are usually made by wire bonds between the devices and the pins and/or substrate tracks and the pins. To accommodate wire bonding, the pins are frequently coined to have a flat spot on the top side of the pin inside the package. (Outside the package the pin cross section is round.) Wires from the substrates or devices are bonded to the pins on the flat spots. Both gold and aluminum wires are used, but the costs of using gold in the diameters required for the high currents carried in these packages dictate the use of aluminum wire.

3.6 Metal Matrix Composites

High strength, high thermal conductivity, low density, and good toughness are all available at low risk and low cost. This sounds hard to believe, but for the most part, these features which designers seek in the materials they specify for electronic packaging are possible to realize through metal matrix composites. While no one material has all the characteristics listed, it is possible to choose how much of each feature one can obtain in a given material. This happens when one or more materials and reinforcements are combined. For example, a metal with high thermal conductivity can be combined with a reinforcement which has a high elastic modulus and low density to achieve a lightweight, high strength, high thermal conductivity composite. In practical terms, combinations like these exist. They are combinations of copper and graphite or aluminum and graphite or aluminum and silicon carbide. These composites are being manufactured and marketed as electronic packaging materials.

To see why MMCs are so attractive as a design consideration, consider the properties of materials which match the CTEs of silicon devices. The CTE of aluminum is 22×10^{-6}/K and silicon is 3.8×10^{-6}/K. If one must mount the silicon on aluminum and expect the bond between the two to survive thermal cycling, it is necessary to use a number of different materials between the silicon and aluminum which have CTEs somewhere between those of aluminum and silicon. This is not a desirable solution. Another approach would be to modify the CTE of the aluminum with a material which would constrain the aluminum so that the combined CTE would more closely match that of silicon. One could use carbon fibers or particles of silicon carbide (SiC) for this purpose. The choice of which to use could be governed somewhat by factors other than the resultant CTE, such as cost. While commercially available carbon fibers have a Young's modulus of 130×10^6 lb/in^2 (this is about 12 times that of aluminum) and a thermal conductivity of 600 W/m-K ($1\frac{1}{2}$ times that of copper and 3 times that of aluminum), the cost of carbon fibers is many times higher than the cost of high-purity silicon carbide particles (carbon fibers are sold for over 100 U.S. dollars per pound while SiC particles are sold for between 10 and 25 U.S. dollars per pound. This in itself accounts for the popularity of SiC as an MMC reinforcement over carbon fibers. Other features to consider are (1) carbon fibers are generally used in long lengths and laid in the matrix where they are placed, and (2) the thermal conductivity of the fibers is much higher along the major axis than it is perpendicular to that axis. This causes the composite to have orthotropic physical properties. Particles of SiC are used to reinforce aluminum and to endow it with isotropic properties. By varying the volume percentage of the SiC in the matrix it is possible to tailor the composite's CTE from roughly 10×10^{-6} to over 20×10^{-6} per °C.

Metal matrix composites have been used in a variety of applications for electronics. The principal examples are for substrate supports or carriers and hermetic packages. An emerging application is the standard electronic module heat sink. Comparison testing between MMCs and PMCs for this purpose is under way. The main reasons for using MMCs as substrate carriers are (1) tailored matching to the CTEs of substrates which carry devices such as gallium arsenide and silicon and (2) improved thermal conductivity over the composite matrix material. Although these features are addressed to some extent by other technologies (PMCs), metal matrix composites are also required to become hermetic housings, and this brings several additional requirements into the picture. These are low cost, isotropy of physical properties such as expansion characteristics and thermal conductivity, and isolation from the surrounding environment through hermetic sealability. Two material combinations have extensive usage, and the experience gained with their use has been valuable in influencing further development. These are aluminum/silicon carbide and aluminum/graphite. These composites have helped to spur further research, resulting in carbon- or graphite-reinforced copper, which is a newer MMC with promise of outstanding thermal conductivity and very low in-plane CTE. This CTE is a close match to silicon and makes this composite attractive as a candidate for packages containing high-power dissipating silicon devices if the in-plane CTE is the same in all directions. Uniform in-plane CTE can be achieved by constructing the composite so that the reinforcing continuous fiber tows are placed at 0° and the next layer at 90°, the next at 0°, and so on. If the layers are all placed at 0°, the in-plane CTE will be quite low in the fiber length direction but much higher (nearly that of copper) in the transverse direction. In order to take full advantage of the benefits and properties the reinforcements can bring, it is important to use them in an orientation that results in essentially uniform behavior unless anisotropy is desired for the application.

3.6.1 Reinforcements for metal matrix composites

These are in common use, and the properties of the resultant composites are quantified to a large extent. It is important to realize that the values in the tables represent measurements of specific loading or volume percentages of reinforcements which can vary from 10 to as much as 60 percent. Composite properties vary as the amount of reinforcement in them is increased or decreased, whether that reinforcement is in the form of short filaments or fibers, continuous fibers, or particles.

More materials are used as reinforcements than were listed earlier. Some of these additional reinforcements are boron, aluminum oxide (alumina), and refractory metals. Bonding between the matrix material and the reinforcement is the key to performance of the MMC. This performance depends on the bond integrity, which in some materials combinations is enhanced by interface coatings to decrease reactivity between the matrix and reinforcement. Treatments of the reinforcement are sometimes used to decrease lubricity and prevent premature pullout of the fiber from the matrix.

An example of a reactive combination is the combination of graphite fibers and aluminum. This composite has enjoyed success in structural applications, but its use in electronics as a hybrid package material is nonexistent. Its low density (weight) and high thermal properties render it a candidate for heat sinks in electronic assemblies. To avoid the reaction between aluminum and graphite at liquid aluminum temperatures, vacuum deposition of the matrix onto the graphite fibers is used.

Alumina fibers are another material used to reinforce aluminum. To strengthen the bond between matrix and reinforcement in this case, small amounts of lithium are added to the melt.[10]

3.7 Interconnects

The term "interconnect" is used here to connote the electrical connection of devices such as semiconductors and passive components like capacitors, inductors, and resistors to an electrically conductive medium that in turn is connected to the next higher level of assembly. Semiconductor devices (chips) are varied in configuration and size and in number of inputs and outputs, and as might be expected, a variety of techniques is used to interconnect them. Semiconductor devices generally either are packaged in housings that contain one chip or are part of a multichip assembly which serves as the package. Another technique avoids the use of a package for the chip and places it directly on the interconnecting medium, for example, a printed wiring board. The other passive devices can exist in "chip" form or as encapsulated components. The reason these passive components can be procured as chips is that they can be combined with semiconductor devices to produce circuits which have components that are compatible in size and form factor.

The principal semiconductor devices that comprise most circuits are integrated circuits. These devices can be either analog or digital types. Analog types are used in applications that are not specifically computing functions. These include receivers, control circuits, amplifiers, and many microwave and millimeter wave circuits. Digital devices, on the other hand, are principally memory devices, gates, counters, buffers, and arithmetic units. The main feature that all these devices have in common is that many individual transistors, diodes, and passive components such as resistors and capacitors make up the devices; that is, a large number of components (sometimes many thousands of transistors, for example) are contained within a single integrated circuit. In the case of very complex digital devices, hundreds of electrical contact or attachment points are required for each integrated circuit. If a computer function circuit contains many of these integrated circuits, several thousand interconnects will be needed to tie all the points to their respective component's pads via the connecting substrate or printed wiring board. This can be an expensive task not only because of the large number of interconnects but also because with so many individual contacts it is possible to make mistakes in wiring. In addition to mistakes it is possible to make the bond incorrectly, that is, with the bonding parameters not controlled tightly. Repair of such device interconnects is very seldom accomplished successfully and the entire device must be removed and replaced. For these reasons it is necessary to use highly repeatable electrical connection techniques. It is also desirable that these techniques be capable of interconnecting many of the integrated circuit's pads simultaneously.

3.7.1 Device interconnection techniques

Devices can be connected by a variety of techniques. The main ones used today for interconnecting unpackaged devices are wire bonding, flip chip, tape automated bonding (TAB), and Z direction adhesives. A new technique recently introduced to interconnect individually packaged devices is solder-free interconnect.

Wire bonding. Wire bonding refers to the very early method of connecting the three portions of transistors—emitter, base, and collector—to the package containing the devices. There are three common types of wire bonding: thermocompression, thermosonic, and ultrasonic. The first to be described is thermocompression bonding. In this technique a gold wire previously formed to a very fine diameter (0.001 in and less) and wound on a spool is fed through a tungsten carbide tip which is part of a wire bonding machine. The wire protrudes below the tip (called a capillary) and is formed into a ball by moving the capillary over a tiny hydrogen flame or causing a spark to jump to the wire and melt it into a ball. The ball is several wire diameters in size and is too large to be pulled up into the capillary hole, which is only about twice the wire diameter. This feature is important because as the capillary moves down vertically it pushes down on the gold ball and causes the gold wire to unreel from the spool. The spool tension is adjusted to be low enough to keep the wire from paying out too swiftly and becoming tangled while not exceeding the elastic limit of the very soft gold wire. Early capillaries were energized by heating alone, and this type of bond is called a thermocompression bond.

A design improvement which added ultrasonic energy to the thermal energy of the tip has been in place for some time. This type of bonding is called thermosonic bonding. The improvement allows tip temperatures to be lowered and thereby improves process yield. The process of attaching the wire to the bonding pads of the devices does not differ for either thermocompression or thermosonic bonding except in the application of ultrasonic energy for the latter.

Connections are made to the devices by causing the ball of the gold wire to be pressed onto the pads of the devices. These pads are almost always aluminum. The heat and, in the case of thermosonic bonding, ultrasonic energy cause the gold and aluminum to alloy, forming a metallurgical bond. Purity of the metallic constituents is a stringent requirement for a successful long-term bond. Typically, the gold wire is 99.999 percent pure and the aluminum pad is 99.99 percent pure. Impurities can cause the formation of intermetallics which will, with elevated temperature, cause the bond joints to become embrittled and ultimately to fail.

As just described, the ball of the gold wire is pressed onto the aluminum pad of the device to form the electrical connection. However, this bond is only one-half of the connection. The other half is formed when the capillary is lifted from the bond on the device. At this point the capillary can be moved in any direction over the plane of the work area. This feature of ball bonding is especially important for producing wire bonds on substrates that contain large numbers of devices. The capillary is then moved to the desired bond pad on the supporting substrate or package and made to press the wire onto the pad. This time there is no ball under the capillary but rather the wire emanating from the hole in the capillary and bent under a portion of the capillary bottom surface (parallel to the substrate). After the capillary makes the bond, which is called a stitch bond, it is raised up from the substrate. As this transpires, the wire is clamped above the capillary and held tightly. This causes the wire to separate from the flattened wire portion of the bond. A portion of the wire separated from the bond protrudes below the capillary and is made to form a ball as indicated at the beginning of the cycle.

Ultrasonic bonding is the third form of wire bonding. This type uses aluminum wire instead of gold wire. Aluminum wire was introduced as an alternate to gold in the 1960s because many gold wire bonds were failing to remain attached to the aluminum pads on devices. The symptom of the failure was a condition known as "purple plague" wherein a small portion of the ball bond at the interface to the chip pad would become purple-colored. Not long after the purple material appeared, the bond would fail. The actual cause was quite elusive (it was actually a combination of causes which included contaminants in the wire introduced during wire production combined with high-temperature storage of the bonded devices), and while investigators struggled to find the cause, the alternate, aluminum ultrasonic bonding, emerged. In ultrasonic bonding, the tip is a grooved foot with the wire, which in this case is aluminum, passing under the foot and guided along the groove. The wire again is 0.001 in in diameter or larger. Whereas the aluminum bonding pads of devices are 99.99 percent pure, the wire contains about 1 percent silicon to improve its lubricity so that it does not drag on and become hung up on the extrusion die when it is being drawn into wire. This does not affect the quality of the wire bond. The reason thermocompression bonding cannot be used for aluminum wire bonding is that aluminum is an extremely active metal that forms oxides on its surface which do not get broken up sufficiently during thermocompression bonding and interfere with the formation of a diffusion bond between the wire and the pad. Ultrasonic energy, however, breaks some of these oxide bonds, but more importantly, the action of the tip scrubbing the wire back and forth opens up fresh unoxidized aluminum in both the wire and pad and permits an aluminum-aluminum diffusion bond to form readily between the wire and pad. No ball is formed at the free end of the aluminum wire. A disadvantage to this type of wedge bond lies in the requirement that the wire be dressed almost in a direct line away from the chip bond. Bonding hybrid circuits in which wires must go from the chip to the substrate in any direction is severely restricted when this type of bonding is used. Ultrasonic bonds can be made closer together than ball bonds (0.003 as opposed to 0.004 in). This allows chip pads to be placed 33 percent closer, which on a device with hundreds of outputs saves a great deal of chip area that can be used instead for the active portion of the device.

Aluminum wire brings with it certain economies over gold wire. Although the cost is only a little less than that of gold wire, the small difference can be significant when large amounts are used as in devices with hundreds of outputs or with power devices that carry large amounts of current requiring significantly increased diameters and multiple bonds to the same pad (it is not uncommon to use 0.020-in-diameter aluminum wire). Multiple bonds may serve the purpose for carrying high current in low-frequency and power circuits, but the excessive inductance associated with fine wires causes performance to become degraded in high-frequency circuits. In these instances relatively large gold-plated copper ribbons are frequently used instead.

Both gold and aluminum wire are used extensively today in packaging, with aluminum use being higher in individually packaged chips and gold use higher in hybrid style and MCM packages.

Flip chip bonding. In flip chip bonding, the devices are not mounted to the package or substrate right side up, as they are in wire bonding. Instead, they are mounted face down with their pads serving as the electrical and mechanical connections. The principal pioneer, developer, and user of this packaging technology is the IBM Corporation. Flip chip packaging is the densest packing technique that exists because it uses no real estate on the substrate beyond the boundaries of the chip to make the electrical connections. This fact is extremely significant for very high input-output devices used in hybrid multichip modules (MCMs). Extremely high speed circuits can benefit from the proximity of chip pads to substrate lands. The bonding pads on the chip are placed directly over the corresponding pads on the substrate.

Solder ball flip chip. The mounting of the chip is accomplished by an ingenious use of solder "bumps" or balls, which are plated onto the chip pads and reflowed when the chip is placed onto the substrate. The flip chip technology described here is the main interconnect method between chips and substrates in use at IBM. Flip chip was introduced in 1964 by IBM as a solder ball flip chip, or SBFC, interconnect, a main feature of its solid logic technology hybrid modules. This approach was very revolutionary and was watched by all packaging engineers but practiced essentially by only IBM until recently. Initially a copper ball was contained within the solder ball to act as a standoff and establish a minimum height or separation between the chip and thick film land and pad on the substrate because the edges of the chips were unpassivated. This was improved upon later by printing glass dams onto the substrates and eliminating the copper balls.[11]

At the time of its introduction SBFC held promise of delivering large improvements in reliability and cost reductions for chip interconnects by virtue of the simultaneity of joint creation for all the chips on a substrate. The promise was essentially fulfilled, and several other features have been established as well that make the technique attractive, especially for MCMs. Robustness of the joints in many environments and very efficient packaging densities have caused engineers to reconsider earlier rejections of the technique for their applications. Much work has been accomplished to mitigate concern over coefficient of expansion (CTE) differences between the chips and some of the substrates on which they are mounted. Stacked bumps which distribute the stresses involved in thermal cycling are reported to improve fatigue life substantially.[12,13]

A serious deterrent to the use of SBFC is the requirement that the chips be metallized with chrome and then chrome and copper (in a phased manner) followed by copper and tin. This metallization must be applied at the wafer stage of semiconductor fabrication. Since most hybrid circuits and MCMs are designed to use various manufacturer's chips, it is necessary that those manufacturers agree to process the devices with preparatory metallizations or with solder bumps.[14] This requirement makes it very difficult to have a SBFC technology under control because industry demand has not established a strong market for device manufacturers to address. To circumvent this problem, a Japanese company has developed a process that involves a solder wire which can be ball-bonded to aluminum chip pads to avoid the requirement for wafer metallizations applied atop the aluminum.[15] At this time, however, the wire is not available for sale, but a wafer bumping service is. The company involved is Tanaka Denshi Kogyo and its U.S. agent is Carmel Chemicals of San Jose, Calif. The success of this approach depends first of all on the ability of the wafer manufacturer and chip purchaser to negotiate an agreement that is mutually acceptable. Having achieved such an agreement, the chip user must still compare the additional costs of the services required for bumping with conventional bump production techniques to conventional wire bonding to determine the viability of this approach. On a production scale, this technique may be feasible in certain instances where the wafer producer is not able to furnish bumped devices. On a prototype scale, it appears more plausible if there can be an economically feasible approach to the issue when production ensues.

Z-axis adhesive flip chip. Flip chip innovations began to appear in the mid-1990s because of the difficulties involved with producing solder bumps on devices. Although the SBFC process has been proved to be a high-performance technology, the requirements for special metallizations to be applied at the wafer step in device production and special agreements with device producers have driven some to explore alternatives. One of these alternatives uses electrically conductive adhesives in place of the solder bumps.

Tape automated bonding (TAB). Tape automated bonding, or TAB, is a technique for interconnecting semiconductor chips of all kinds to substrates and packages by the attachment of beamlike conductors to both chips and substrates. It is an especially attractive interconnect method for microwave circuits because of its low inductance as compared with wire bonding. A comparison of the inductance of wire and ribbon is shown in Figs. 3.1 and 3.2. The beams are pho-

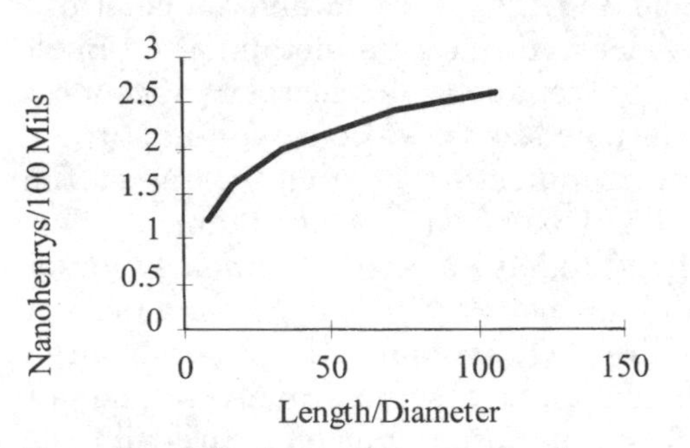

Figure 3.1 Inductance of round wire for any ratio of total length to wire diameter.[3]

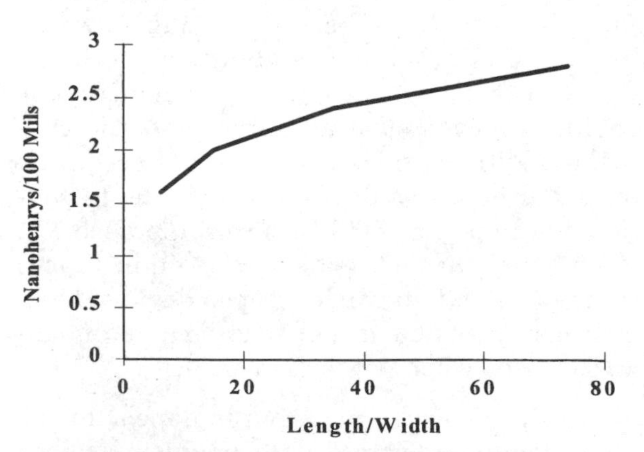

Figure 3.2 Inductance of rectangular ribbon for any ratio of length to width.[3]

tolithographically formed from clad or plated metal (usually copper) atop a polyimide film. Parts of the beams are cantilevered over openings or windows in the polyimide film. The beams are bonded to the pads of chips in a simultaneous fashion (inner lead bonding, or ILB), and later the other ends of the beams are bonded to substrate lands or package leads (outer lead bonding, or OLB). This technique emerged as a means to produce packaged chips economically on a large scale, because as the cost to produce the chips themselves became very low, producing the connections between the chip pads and the package pads became a major part of the total cost. It was necessary to reduce the cost of that part of the assembly procedure to keep pace with the automation which took place with the introduction of larger wafers (from 3 to 5 and 6 in diameter). Even as chips became more complex and larger, the wafer sizes grew to accommodate more devices per wafer and consequently more devices per lot. Device yields increased as better diffusion, epitaxy, photolithography, and clean room techniques were introduced. Therefore, lower costs were incurred to produce the devices as chips. As long as labor-intensive manual wire bonding remained as the primary chip interconnection technique (automatic wire bonders of that time were slow and cumbersome to use, not to mention very expensive even for manufacturers of millions of chips per year), packaged chips would remain expensive and the market for them would not grow quickly. This problem was partially solved by the introduction of TAB. There were two types of TAB. The first was used with chips that had plated bumps applied to the pads after the final metallization step. The second type used chips with no special metals applied after the final metallization and instead had bumps applied to the interconnect beams. This second type is called bumped tape automated bonding, or BTAB.

As devices became more complex and required many inputs and outputs, some manufacturers sought to reduce wafer real estate by using TAB because TAB generally allows chip pads to be smaller and placed closer together. This results in less chip area devoted strictly to interconnects and frees up some precious area for more active components to be placed on the wafer. Electrical considerations contribute to improved performance, such as better impedance control and lower inductance over wire bonds. Other features are more consistent physical placement of the beams with respect to each other and greater bond strength, resulting in higher reliability. TAB also allows more testing of the chip to take place because test pads can be printed onto the polyimide film that are consistent with larger dimensions required by probe pins. The conductors on the film can be made to fan out to pads on any desired centers.

The TAB process of bonding to chips (inner lead bonding). TAB bonding to chips involves first of all the preparation of a wafer containing chips. Whether a wafer is bumped or not, if the chips are to be bonded using TAB, the wafer is waxed to a silicone rubber or similar elastomer on its backside. It is then scribed in the avenues between the chips and the chips are separated from each other although still held onto the elastomer by the wax. The wafer is then placed in a TAB ILB machine where the foil strip containing the beams is brought over it. After the ends of the beams are aligned to their appropriate chip bonding pads, the heated bonding tool presses the beams onto the chip and a thermocompression bond (or a soldered joint in the case of solder bumps) is made. As the bonding tool is retracted, the chip is released from the wax bond beneath it because the wax melts as the ILB is made, and the chip is suspended from the beams on the foil. The foil is indexed to the next site and brought over the chip to which it is to be bonded, and the process is repeated. As the chips are bonded, the foil containing them is reeled up. The reel is later taken to an outer lead bonder where the other ends of the beams are bonded to packages or substrates.

The TAB process of bonding to packages or substrates (outer lead bonding). The process of outer lead bonding (OLB) begins with the placement of the reel of ILB bonded chips onto the outer lead bonding machine. The reeled up foil is fed to an excising tool which cuts the beams away from the foil near the edge of the window in the foil. At this time, a vacuum pickup placed directly above the chip picks up the inner lead bonded chip with its beams which have been freed from the foil and moves to a location directly above the site on the substrate or package where the outer leads are to be bonded. After alignment is complete, the OLB tool comes down and bonds the beams to lands or pads on the substrate or package.

Solder-free interconnects. Solder has served as the principal interconnection technique for electronic assemblies for over 70 years. At the time of electron tubes and other large components it was an ideal method to effect a reliable electrical contact between two or more wires and socket pins or other component mechanical supports and connection terminals. The diameters of the wires and other terminals were large with respect to the amount of solder required to form the electrical connection. Even as components became miniaturized and had some increases in their pin count (from one dozen to about four dozen) their size and spacing remained large enough so that they could be located 0.100 in from the center of one lead to the center of the other. However, the trend toward increased device speed and computational power caused the lead counts for individual chips to increase significantly. To retain the same lead sizes and spacing would cause the packages containing the chips to grow excessively large and thereby require correspondingly large footprint areas on the printed wiring boards on which they would be placed. Therefore, it became necessary to reduce both the lead width and the spacing. Leads were brought out of the package on all four edges or sides to minimize overall package size, and their size was reduced from 0.020 to 0.013 in width. In addition, lead center-to-center spacing went to 0.050 in. This change in lead spacing was not sufficient to keep up with the increase in lead count, so lead center spacings were driven downward to 0.025 in and then to 0.020 in. At this point it was discovered that soldering leads so small (and even smaller in special cases with corresponding spacing of only 0.006 in) presented assembly problems. Handling the devices required special carriers because the leads were very fragile and easily became entangled with each other. New techniques for soldering such fine leads had to be developed. These techniques included inert gas assisted lead tinning and solder reflow procedures such as vapor phase soldering, hot bar soldering, and infrared reflow.

As these technologies were developing, some effort was devoted to pin grid array and pad grid array packaging. Both approaches depended on bringing the electrical contacts out of the package through the bottom rather than the edges. In both cases, the contacts were placed on 0.050-in centers and formed an array of contacts on the bottom surface of the package. In the case of the pin grid array configuration, plated pins were brazed to the pads on the package bottom surface. The pad grid array configuration has only pads on the bottom of the package. Pin grid array packages were more acceptable to designers than pad arrays because the pins represented a known interconnect technology, that is, pin and socket. Pad arrays would be interconnected through soldering to their respective lands on printed wiring boards. This approach, although used in limited fashion, did not satisfy the requirement of military specifications that the joints be accessible for inspection. A further drawback was that the solder joints would be too short to allow flux and cleaners captured between the printed wiring board and the package to be flushed away. This condition would make it very difficult if not impossible for such populated assemblies to pass the low ionic contaminant residue requirement of MIL-P-28809 (currently known as MIL-C-28809). Consequently this type of package did not receive much support.

Several solder-free interconnect technologies that do not necessarily deal with high pin count packages should be mentioned. Aside from wire bonding, these interconnects are elastomers, epoxies, and memory metals.

Elastomeric interconnects. The most often used and the oldest elastomeric interconnect consists of a strip of silicone rubber slices that are alternatively insulative and conductive. The conduction is achieved by filling silicone with conductive metallic particles. Sheets of both conductive and nonconductive silicone are cast into uniform thicknesses (roughly 0.010 in) and cured. They are then sliced into squares typically measuring 0.100 in on a side and assembled and glued alternately together to form a stack of squares of insulator/conductor/insulator, etc. They function as a connection medium because they are placed between a conductor such as a land on a printed wiring board and a contact for a key pad in a calculator. Their application is not limited to the example, however. A variation of this technique has the conductive portions made smaller to accommodate fine-pitch applications.[16] Alignment between the elastomer and pads to be connected is not critical because the elastomeric connections are usually at a much finer pitch than the contacts being made. A disadvantage to the use of elastomers is the requirement to use relatively high pressure to maintain sufficiently low resistance to constitute a reliable joint. Another problem that ensues is that all elastomers relax with time, that is, they lose their elasticity to some degree and therefore the pressure initially applied to them is reduced. This in turn causes the resistance in the connection to increase as the pressure becomes lowered. Care must be taken to determine both the amount of relaxation expected in the elastomer with its corresponding increase in contact resistance and the effect of that increased contact resistance on the performance of the circuit under consideration.

Loss of contact force at cold temperatures was reported in recent studies done by Rajendra Pendse at Hewlett-Packard. This work cautions against using elastomers alone if significantly low temperatures will be encountered while the equipment is operating.[17] A drop or loss of contact force of over 37 percent on the elastomer was measured from room temperature to −55°C. Thermal hysteresis is another phenomenon to consider with elastomeric connections. When a clamped elastomer is heated from room temperature and then cooled back to the starting temperature, the force on the interconnect increases with temperature increase. When cooling takes place, however, the loss of force is greater for each degree of temperature change than the increase was for the same delta temperature. Under moderate temperature changes, typical losses of force were in the neighborhood of 50 percent.

To obviate these design difficulties, Pendse suggests using a metal spring (the softer the better) in series with the elastomer. In this way, when the elastomer relaxes under the fixed load, the spring pushes on it to minimize the loss of contact force. Comparative experiments demonstrated that force retention was as high as 92 percent with a soft spring (low spring constant), and 74 percent with a harder (1.6 higher spring constant) spring. With sufficiently low ratios of metal to elastomer spring constants, open circuits can be avoided even at temperatures as low as −55°C.

Epoxies as interconnects. Metal-filled epoxies are seeing increased usage as a means of effecting interconnects between surface mounted components and printed wiring boards. At first used to provide an electrical as well as mechanical function for use with chip capacitors, their success has led to use with leadless chip carriers and other leadless components as well. When only several components are being mounted onto printed wiring boards, the conductive epoxy is generally dispensed from a tube into which it has been placed after mixing of the constituents. Hypodermic syringes are used to perform this step because they can be depended upon to control the dispensing of small amounts of resin at one time. Although time-consuming and labor-intensive, the approach is quite effective if only a few sites are required to have the resin applied. On the other hand, if many components are to be connected to the printed wiring board by this technique, a method is called for which can dispense resin to many or all sites simultaneously. Stenciling or silkscreening are such methods. They are used to apply the resin to multiple sites on printed wiring boards. Masks are created with openings in them to allow the epoxy to be squeezed onto the boards. These masks are made from thin brass or stainless-steel shim stock, usually about 0.002 to 0.003 in thick. Holes are photolithographically etched into the metal to match the pads on the printed wiring boards. Masks are then placed in silkscreen frames and secured onto a silkscreen printer. Conductive epoxy is spread across the mask at one end and a mechanically driven squeegee then prints the resin onto the board. Pot life of the resin is important if more than several boards are to have the resin applied at the same time. While it may be convenient to have the epoxy become cured rapidly after components have been assembled to the boards, the speed of the cure should be selected so that enough time is allowed to place all the components onto their printed wiring board pads before a significant amount of curing occurs. The most advantageous type of epoxy to use is one which will not cure rapidly at room temperature but cures rather at elevated temperatures.

Memory metal interconnects. Memory metals fall into two classes which, when either heated or cooled to a specific transition temperature, change their crystalline structure and hence their volume. This feature is put to use in certain connectors which, when taken through the temperature range of −55 to +125°C, do not change crystalline state but at a transition temperature safely below the range in the one case or safely above in the other case will exhibit a change in structure. In the first case, within the temperature range noted, a connector socket staked into a printed wiring board will retain an inner diameter slightly less than the component lead or pin intended to fit within it. In order to assemble the component, one cools the sockets to the temperature of liquid nitrogen and places the component pins within the sockets, which upon cooling have opened their inner diameters sufficiently to accept the component pins. Upon heating back to room temperature, the component pins are tightly grasped by the inner diameters of the connector sockets. A memory metal collar which surrounds a beryllium-copper socket causes the socket to open and then close tightly around the pin. The behavior is similar for connectors that depend upon heating to be taken to a transition temperature except that the procedure is reversed. In this case, heating relaxes the socket's grasp on the pin.

One strong feature of memory metals is their ability to form gastight connections without the use of solder or welding. This is attributed to the high forces of the connection which are in the neighborhood of twice those of friction fits. Another feature is the lack of a requirement for special tools since the assembly is accomplished strictly with application of heat or the withdrawal of it. On the negative side, the cost of complicated sockets or other grasping devices makes the approach attractive for very special applications but not for common designs, which can be accomplished more readily by standard pin and socket technology.

Button board interconnects. Button board technology is a new approach to solderless interconnects. A thin plastic sheet is populated with buttons of fine wire wound around to form springlike cylinders which are reliable interconnects. The wire is randomly woven to form cylinders measuring 0.020 in diameter by 0.040 in long. This random weave produces a spring which does not relax after repeated loading or under thousands of temperature cycles. The wire consists of a molybdenum core with nickel and gold plating and measures about 40 mm in length before it is formed into the cylinder shape. The buttons are placed into holes in the plastic sheet and protrude above and below the sheet even when compressed. The buttons are placed at locations in the sheet matching the lands on the printed wiring board which are intended to be connected to the pads on leadless ceramic chip carriers. The sheet with the buttons in it is dropped into a hole in a structure bonded to the board, and the leadless chip carrier is then placed onto the sheet. A clamp is then applied to a plate above the leadless chip carrier, and the electrical connection is made between the leadless chip carrier and the printed wiring board. Some of the inherent features of this approach are listed here.

Compliance to out of plane conditions of the printed wiring pads is an inherent characteristic of the button board. Camber in the printed wiring board can be compensated by the button board because of the range of compression designed into the buttons. The connection between the button and the leadless chip carrier as well as the printed wiring pad is reliable because the compression force is 2 oz distributed over a very small area for a pressure of 500 to 800 lb/in^2.

Continuous connections have been maintained during high vibration levels using this technique. Since the button is completely or nearly completely compressed as assembled, it is very unlikely to have its contact with either connecting member interrupted during vibration.

Depending on the design, the cost to implement the button board approach can be considerably lower than a comparable soldered assembly. On a one-to-one basis, the cost has been found to be one-third lower using button boards over solder. In addition, since solder fluxes are not used, there is no need to use ozone-depleting or environmentally friendly cleaning solvents either. The cost of such cleaners including disposal and cleaning labor steps is not necessary. Solder inspection, a costly and no-value-added operation, is also eliminated.

3.8 Chip on Board (COB)

Chip on board refers to the practice of placing the bare die either directly on the printed wiring board or on a substrate which is then placed directly on the printed wiring board. The main purpose of doing this is to avoid the baggage which comes with using packaged devices. These have a higher cost and a much larger device footprint. With COB, assemblies are much smaller than they would be if all devices were placed in packages before assembly to printed wiring boards. The actual board size reduction can be as much as 50 percent. Packages in typical computer card applications account for slightly over 50 percent of total device costs, so the savings in both cost and area can be substantial. Package height also can be greater than the space available, and COB offers lower-profile assembly possibilities as well. In 1992, COB on both sides of the board emerged as a means of further increasing component density.

Substrate materials consist of FR-4 epoxy-glass, ceramic (alumina), silicon, and even films such as polyimide. Choices depend on performance and application. The die are attached most often by epoxy, and sometimes the conductive type is used if the circuit requires it. Die are often attached directly to the main printed wiring board as just mentioned, for simplicity. At times, as the design requires, cavities are prepared in the board into which devices are placed, thus allowing the board height to be reduced by the depth of the cavities. Electrical connections are made by wire bonding, with either gold or aluminum wire being the medium. Higher reliability is achieved, however, with gold.[18] Gold has been found to resist thermal cycling better than aluminum. More cost is involved with the use of gold because of the need to gold plate the bonding pads on the substrate or board. Environmental protection is usually provided by an overcoat of epoxy on the die and the wire bonds, which also gives some ruggedization to the board for handling during test. Since COB assemblies are highly unrepairable, ruggedization and process control are quite important.

In 1993, a further development called chip-scale packaging took place. The term has since been changed to chip-size packaging. The most prevalent technique for achieving packaging on the scale of the chip is ball-grid array technology, in which metallization is added to the top (sometimes the bottom) surface of a chip as pads in the form of an array. The pads are connected to the bonding pads on the chip. Usually solder is added to these pads for interconnection to a substrate. This technique allows large input-output chips to be interconnected in an extremely dense fashion, thereby eliminating the need for a package many times the size of the chip.

3.9 References

1. Howard Markstein, *Electronic Packaging and Production,* vol. 31, no. 10, October 1991, p. 40.
2. Rao R. Tummala and J. Knickerbocker, "Advanced Co-Fire Multichip Technology at IBM," *Proceedings of the International Electronics Packaging Conference,* San Diego, Calif., September 1991, p. 39.
3. Richard Brown, *Microwave Hybrids,* Technical Monogram, ISHM, 1989.
4. H. E. Saunders and K. F. Schock, Jr., *Proceedings of the 8th International SAMPE Conference,* Baltimore, Md., June 1992.
5. Timothy L. Hudson, "Selecting a Ceramic Substrate for Multichip Modules," *Electronic Packaging and Production,* June 1992, p. 68.
6. R. M. Barrett, "Microwave Printed Circuits—An Historical Survey," *IRE Transactions on Microwave Theory and Techniques,* MTT-3, March 1955, p. 1.
7. D. D. Grieg and H. F. Englemann, "Microstrip—A New Transmission Technique for the Kilomegacycle Range," *Proceedings of the IRE,* December 1952, p. 1644.

8. N. R. Wild et al., *Handbook of Tri-Plate Microwave Components,* Sanders Associates, Inc., Nashua, N.H., 1956, ASTIA AD110157.
9. H. S. Keen, "Scientific Report on the Study of Strip Transmission Line," *AIL Report* 2830-2, December 1955.
10. John V. Folz and Charles M. Blacmon, "Metal Matrix Composites," *ASME Engineered Materials Handbook,* vol. 1.
11. Karl J. Puttlitz, Sr., *International Journal of Microcircuits and Electronic Packaging,* vol. 15, no. 3, 3d Quarter 1992, p. 113.
12. R. Satoh, M. Oshima, H. Komura, I. Ishi, and K. Serizawa, "Development of a New Microsolder Bonding Method for Vlai," *Proceedings International Electronics Packaging Society Conference,* September 1987.
13. N. Matsui, S. Sasaki, and T. Ohsaki, "VLSI Interconnection Technology Using Stacked Solder Bumps," *Proceedings 37th Electronic Components Conference.*
14. P. A. Totta and R. P. Sopher, "SLT Device Metallurgy and Its Monolithic Extension," *IBM Journal of Research and Development,* vol. 13, 1969, p. 226.
15. John Tuck, "Concurrent Product Development," *Circuits Assembly,* August 1992, p. 20.
16. Robert A. Bourdelaise, "Solderless Alternatives to Surface Mount Component Attachment," 4th International SAMPE Electronics Conference, June 1990.
17. Rajendra D. Pendse, "Low Relaxation Elastomeric Pressure Contact System for High Density Interconnect," *International Journal of Microcircuits and Electronic Packaging,* vol. 15, no. 3, 3d Quarter 1992.
18. Elliott H. Newcombe, "COB Increases SMT Density," *Electronic Packaging and Production,* December 1992.

Chapter

4

Practical Design Details for Microwave Circuit Packaging

Microwave devices and their interconnecting and mounting together to form larger modules or assemblies (sometimes called integrated assemblies) are the primary considerations of this chapter. One of the characteristics of integrated assemblies is the close physical proximity of components. The tight packing of devices comes about primarily through a need for improved performance, although lower costs and size reduction go hand in hand with other improvements in devices. Improved performance results from fewer transitions (changes in the line or conductor carrying a current or signal) encountered in signals passing from one device to another. For example, in the case of discretely packaged devices connected by a cable or a printed wiring medium, a signal passes through the output leads a of device to an interconnecting medium such as a copper strip on a printed wiring board, through the leads of another device and into the other device itself. If the devices were part of an integrated assembly, the signal leaving one device would travel to the common interconnecting medium and then into the other device. Since at each transition from one medium to another there are reflections of signal waves and power, however small, it is necessary to conserve power by applying the most efficient packaging techniques available. In the integrated assembly example, losses associated with the device leads do not exist because those leads do not exist. In their place are direct connections to the printed wiring strip or track. Such measures of integration are commonplace in the packaging of microwave devices and integrated assemblies to effect a more power-conserving design. As will be shown later, placing devices in close proximity improves microwave performance in qualities other than preventing unnecessary power losses.

In the microwave frequency range of several hundred megahertz to several hundred gigahertz, the duration of an rf cycle is very short. If the frequency is high enough, the signal will travel a distance close to the dimensions of the physical circuit. That could cause the phase angle of the signal to be considerably different at one end of the physical circuit than at the other end. To understand what this means, consider a signal whose wavelength is about 1 in in the medium in which it is traveling (printed wiring board or a ceramic substrate). A full wave can be broken into 360°. A part ½ in away from another part is then separated by 180°, or if the signals were observed at both parts simultaneously, they would be 180° out of phase. Most microwave circuits have longer wavelengths, such as 2 to 10 ft. A wavelength of 2 ft means that every inch represents a difference of 15° of phase. This is quite different from dc circuits or circuits operating at low frequencies. How that phenomenon affects circuit performance depends on the circuit function and its relationship to other elements of the subsystem as well as the frequency. For example, in a state-of-the-art radar that transmits its signals through many individual solid-state transmitters simultaneously (a phased array radar), the transmission must be precisely timed to take place at the same instant from each individual transmit element. In addition, the phase of the signal (except for the special beam pointing modes of operation) must be at the same angle or portion of the cycle at each transmit element at the same moment in time. In other words, if the transmission of the signal is to be made so that the amplitude of a pulse rises from a minimum to a maximum value and falls back to the minimum value in a given period of time, the amplitude of the pulse must be at the same level in each transmit element at the same instant in time. Since the elements may be located several feet or even tens of feet from each other, maintaining the appropriate phase relationships for signals traveling varying distances in an array of transmit elements is a matter of great concern to the designer.

Another feature of high-frequency circuits, called the skin effect, is discussed in Chap. 2. Electrons (currents) tend to flow more in the outward portions of conductors than in the center regions as the frequency of the signal increases. This has several consequences that are not encountered in dc or low-frequency circuits. The rf resistance of a conductor is usually much higher than the dc resistance of the same conductor because of this bunching of electrons near the surface. As a result, high-frequency conductors are usually made from the highest-conductivity metals practical to use in a given application. In the case of copper conductors on printed wiring boards, it is necessary to coat the copper with a passivating metal such as tin. Since the conductivity of tin is poorer than that of copper, the rf resistance of such conductors located upon the outer surface of a printed wiring board is worsened by the skin effect.

Further electrical features or effects of microwave frequencies constitute the design domain of packaging for circuit operation in these frequencies. The most notable are signal attenuation, loss of signal energy through radiation and control of that loss through various forms of shielding, loss of energy due to imperfect dielectric properties (high values of loss tangent), signal reflection at transitions due to mismatches in impedance, dielectric constant effect on physical circuit features (conductor width and height above a ground plane), and undesired or parasitic coupling of energy from one conductor to another within the same circuit or to conductors of another circuit.

Mechanical, materials, and assembly considerations are additional features which require attention. These in themselves are not usually peculiar to microwave circuit packaging, but in some cases special care must be taken in materials selection and process control to enhance the electrical performance.

An example of the sensitivity of performance to materials properties will demonstrate the special attention that needs to be paid to detail when dealing with microwave circuit packaging. Loss tangent is a property of dielectric materials that is extremely important in microwave circuits because as the loss tangent increases, the dielectric is less and less an ideal insulator. This means that rf energy is lost throughout the dielectric because of either its poor dc electrical resistance or the relatively loose chemical bonds of molecules within the dielectric which cause absorption of rf energy, or both. When a circuit that is being designed requires that practically no power can be lost through inefficient materials, the selection of both conductor and dielectric materials is important. Particular attention should be paid to the loss tangent of the dielectric. Even if care has been taken to use a conductor whose material and thickness limit the conductor losses, the dielectric can account for the loss of sufficient power to be a major concern.

4.1 Passive Parts

Passive parts are electronic components that do not contain semiconductor junctions (a characteristic of active components such as transistors and diodes). Passive parts contain elements of the following: resistance, capacitance, and inductance that are assembled to become circuit functions. The elements may be discrete pieces such as resistors, capacitors, and inductors in chip form or printed onto substrates by either thick film or thin film techniques, or the elements may be formed from transmission-line topology, principally microstrip. Passive parts are interconnected with active parts to form circuit functions; the electrical characteristics of the parts, such as their input and output impedances, are unique. Microwave circuit designers are also concerned with the physical construction of parts they intend to use because the way a part is configured determines how much inductance or capacitance it has in addition to its resistance. No resistor is purely resistive, no capacitor is purely capacitive, and no inductor is purely inductive. A part may be selected for its resistance, but it also brings along some capacitance and some inductance as well. The same situation applies to capacitors and inductors. They all have some elements of the other two components of impedance.

4.2 Discrete Assemblies

A discrete assembly is a part that consists of passive parts and semiconductors such as transistors, diodes, and microcircuits. The semiconductors are usually enclosed in a hermetic package, although they may be encapsulated in a plastic material (silicone or more often epoxy). A hermetic package is one that is sealed with a gastight seal. The only gastight seals are metal-to-metal or metal-to-glass. Other seals range from dust cover quality, which lets moisture into the assembly, to fairly good (such as encapsulation that prevents moisture from collecting rapidly). For a full explanation of hermeticity in microelectronics, see MIL-STD 883. In most cases, microwave semiconductors are hermetically sealed. This is because an encapsulant changes the electrical environment of the semiconductor and requires special consideration for matching impedances to the signal line (transmission line). A discrete assembly is often used because it is commercially available and performs a standard function. An example would be an amplifier which would have the characteristics required for the particular design, such as linear response and fixed gain for the desired frequencies. Commercial catalog discrete assemblies are fixed in their geometry, and in many cases that is quite acceptable to the designer. In some cases, the designer requires a different form factor (the catalog part may be too thick or has input-output connections inconsistent with the other components of the larger assembly). The latter situation is then addressed by a custom design which meets the designer's needs. For custom designs, it is especially important to be sure that the semiconductors will perform as expected at all temperatures anticipated in operation. Operation of passive parts over the anticipated temperature range should be verified as well if the circuit is particularly sensitive. In military applications, the semiconductors and/or microcircuits must be verifiable of specified performance through screening testing at levels required to meet system requirements. This can include performance under certain environmental conditions. Such testing is usually done by the semiconductor manufacturer prior to shipment to the assembler. It is sometimes necessary that the entire discrete assembly be tested at the levels imposed on the semiconductors because the semiconductors are packaged within the same envelope as their associated passive parts. Although this configuration is referred to by many as a hybrid circuit, the definition is somewhat open. Our definition of hybrid circuits follows.

4.3 Hybrid Assemblies

Hybrid assemblies are parts that contain semiconductors in bare chip form, that is, not individually encased in a hermetic or plastic enclosure with their leads emanating from the enclosure. Hybrid assemblies are used in many types of consumer electronics like camcorders, personal computers, laptops, and cellular telephones where very compact packing of the electronic parts is necessary. The hybrid assemblies that exist in microwave and rf circuits are similar to the discrete assemblies described earlier, but the bare chip semiconductors do not occupy as much area as their packaged versions. The same circuit board technology may be used in both cases, but frequently a more innovative approach is found involving a combination of conventional printed wiring board and an overcoat of an unreinforced dielectric material that carries low-level signals. Wire bonding is usually used to make interconnections to the input-output pads of the semiconductors, but other forms of interconnect may be found as well in hybrids. Among these other forms are ribbon bonds, TAB (tape "automated" bonding), and various versions of inverted semiconductor electrical attachment generally referred to as "flip chip." These special interconnect technologies are described in Chap. 3.

Hybrid circuits are generally fabricated by depositing interconnects and circuits on passive substrates and attaching active devices and certain components such as capacitors to the conductor pattern defined on the surface of the substrate. Die bonding, soldering, epoxy bonding, and wire or ribbon bonding are used in the attachment process.

A major advantage of hybrid circuits is that a wide variety of device types made from different semiconductor materials can be used. The devices can be tested in advance of mounting to ensure good yield. The device may be used in packaged or unpackaged form. The hybrid circuit utilizes the semiconductor area only for active devices. In-circuit tuning is possible in hybrid circuits during production.

Disadvantages of hybrid circuits include high labor content in production assembly, variation in parasitics due to the manual operations in die mounting and wire bonding, increased parasitics due to the wire bonding, and somewhat decreased reliability due to bonding and mounting. Additive or subtractive approaches may be used in defining circuit patterns on substrates. The subtractive approach, or thin film process, incorporates photolithographic definition of patterns in metal layers that were deposited by electron beam deposition, sputtering, or resistive heating evaporation from crucibles. The additive, or thick film, process applies metal patterns by the printing of conductors through a screen and the subsequent firing of the inks.

The technological issues that must be considered are the selection of appropriate substrate and conductor materials, the surface finish of the substrates, the metal-deposition process to ensure good adherence to the substrates, microwave performance, and long-term stability of the metal system.

Hybrid assemblies may be sealed in an enclosure that is hermetic or sometimes coated with a plastic (epoxy or silicone), although microwave hybrids most often fall into the first category. If the hybrid is hermetically sealed, the electrical connections pass through the hermetic envelope in the form of feedthroughs consisting of metal conductors surrounded by a dielectric that forms a seal around those conductors. The thermal expansion of both materials must be very closely matched so that the seal is maintained. Typical materials combinations are iron and cobalt alloys and glass or aluminum oxide ceramic if the leads are round. If the leads are in the form of printed lines on the surface of a dielectric substrate they may be gold or a combination of gold and copper. The dielectric in that case is a glass that is matched in expansion to both the substrate and printed lead.

An important feature of semiconductors used in hybrid assemblies is that they are only partially characterized or tested by their manufacturer because they exist only in the form of a chip. The packaged product that the manufacturer markets is more readily tested electrically and even thermally stressed because it is sealed and fits the test equipment fixtures. The bare chip, on the other hand, is quite fragile and has no leads attached that can be affixed to test connectors. In addition, it is quite tiny. This precludes handling the chip easily or even economically. For these reasons the bare chip is usually tested to some extent while it is in the semiconductor wafer. Testing done by probes does not fully characterize chip performance, and so the hybrid assembly that contains such chips must itself be qualification-tested the way that sealed semiconductor chips themselves would be tested.

Such testing is more costly than that done on the packaged semiconductors, but at the same time it can be more effective as a screening tool because it exercises the assembly as a whole. Hybrid assemblies that pass this screening test may have a greater probability of lasting and performing longer than discrete assemblies that do not undergo this screening.

4.4 Integrated Microwave Assemblies

Integrated microwave assemblies are electronic functions that physically are made from passive parts, discrete assemblies, and hybrid assemblies, all put into a closed box. They are larger than the other assemblies and usually have a microwave or rf section in the form of a hybrid assembly and a lower-frequency section which may be either a discrete assembly or a hybrid assembly. An assembly of parts at this level is usually large enough to be called a module. Modules are considered as entities in larger assemblies (radar subsystems, for example) that contain parts and electrical functions which logically and for the application belong together. This partitioning philosophy is applied to the system in order to separate functions and/or parts that grouped together would interfere with physical restrictions or electrical performance. Stated more positively, partitioning relegates to the appropriate locations the parts that function well together to enhance electrical performance. Modules have the physical characteristics required of independent assemblies, namely, connectors capable of carrying all signals, power, and ground into and out of the module, shielding (which is usually required for critical signals), and isolation of internal parts from stray signals. Mechanically, modules are also ruggedized for handling (plugging into higher-order assemblies, withstanding shock and vibration in service) and have provisions for cooling that interface with the system. Integrated microwave assemblies are especially important in radar systems because their functionality is usually proved before they are assembled into the subsystem. At this level of integration, modules can be assembled into the subsystem with confidence that the subsystem will function as required. Replacement of modules is straightforward since they are designed to plug into the rack or framework of which they are a component.

4.5 Monolithic Circuits

Monolithic circuits are chips that contain complete circuit functions and consist of semiconductor elements as well as all the passive components that comprise the circuit function. Monolithic microwave circuits are fabricated by growing and depositing active devices and passive circuit elements on a semiconductor wafer. The major technology differences between microwave monolithic circuits and conventional ICs are that gallium arsenide is the primary semiconductor used for microwaves, while silicon is used for lower frequencies, and that a relatively high percentage of the wafer area is used for passive circuitry such as microstrip lines or lumped elements, compared to a much lower fraction for passive circuitry on conventional circuits.

In this case a circuit function which may be made up of individual physical parts is reconfigured into a die or chip. The elements of the circuit which were larger parts are recreated as much smaller parts in a planar form and "printed" on the surface of the die. In the field of digital circuits, chips made from this construction are referred to as integrated circuits. The advantages of monolithic circuits are that ultimately the production cost is lower than for hybrids since the assembly labor content is minimal, the control of parasitics is excellent, and device characteristics in a circuit are matched because the devices are fabricated simultaneously in close proximity on the same wafer. The primary processes used for fabricating monolithic circuits are ion implantation in epitaxial layers grown on semi-insulating gallium arsenide wafers. Multilevel metallization is used for devices and passive components. Research work is under way on the use of other compound semiconductors, including indium phosphide and other materials and heterogeneous technology growth of silicon on gallium arsenide substrates and vice versa.

The term *monolithic circuit* came about because microwave engineers had referred to microwave hybrid circuits as integrated circuits prior to popular acceptance of that term as defining a circuit function being contained on a chip. These hybrids were called microwave integrated circuits (MICs). Consequently, as digital circuit functions began to be incorporated on single die, the microwave community realized that the technology could be adapted for use in forming components greatly reduced in size that would perform microwave functions. It should be noted that size reduction is not the only reason that integrated circuits (monolithic circuits) have gained such wide acceptance as they have. The cost to produce an integrated or monolithic circuit is very low after the initial engineering and tooling costs have been expended. When many circuits of the same type are demanded by the application or by the market, it often is more advantageous to undergo the costs of creating the artwork necessary to produce a monolithic circuit and processing the wafers of semiconductor material to produce the monolithic devices. The break-even point must be determined in each individual case through cost analysis and a tradeoff study of the alternatives.

4.6 Tradeoffs in Levels of Integration

The topic of tradeoffs is an especially important one for microwave packaging designers. The success of a product in terms of its producibility, cost, performance, affordability, and reliability is determined largely by the level of integration of its various components into a package that is functional while meeting all design requirements. While this may seem intuitive, it is an entirely different matter to make the specific decisions which will influence size, power dissipation, interconnections, device availability, manufacturability, cost to produce, and testability while still meeting performance requirements. A set of influencing factors must be chosen that are considered important and these must be prioritized so that the tradeoff conclusion reflects reality. An often overlooked feature is who will manufacture the assemblies. If the vendor community does not have the capability to meet the requirements, because of either equipment or other limitations, this may impact the level of integration decision. Certain vendors specialize in low-noise amplifiers, others in control devices, and expecting either to have the expertise to do both because of the level of integration desired may result in missed schedules, high costs, commitment of technical support to the vendor by the customer, etc. On the other hand, the benefits of a very high degree of integration can be low recurring costs to produce a product, compactness, low weight, and very likely, improved performance over distributed components. These desirable features must be balanced against design requirement, thermal dissipation, nonrecurring cost, time to effect integration, available packages, pinout capability, and testing. Thermal considerations are very important and circuits should be analyzed carefully to determine the degree of integration that is consistent with cost and performance.

System requirements set the groundwork for most of the decisions on tradeoffs. Typical system requirements might be availability and compatibility with weight, size, and cost goals. These requirements may dictate that the time to produce the product is minimal and thereby preclude the use of certain long lead time parts such as custom hybrids. The cost of items to be produced may influence the decision of what form they will take. If few items will be produced, there is little justification for expensive tooling and other engineering and design costs that would be incurred if a hybrid circuit or a custom monolithic circuit were to be specified. In that case, building the circuit from conventional components would be the correct choice. However, there are often more influences on the decision making process, so the previous case could be viewed as oversimplified. While cost may be a factor, it might be subordinated to another requirement such as size or weight or reliability. In those cases, the requirement that the complete function be enclosed in a maximum specified volume could be more important than cost in the packaging hierarchy of importance.

What has just been said refers more to narrowband radar circuits and systems than to broadband electronic warfare (EW) applications such as those that use an 8- to 18-GHz range. Specifically, the choice to use many components that are interconnected by microstrip connections on a large board may be valid at the lower frequencies. At the higher frequencies, however, the ground plane discontinuities normally encountered become troublesome because they are more pronounced. Maintaining proper microwave grounding for long runs is so difficult that the microstrip method may not be practical. Performance is usually achieved in these cases by using semirigid coaxial cable to interconnect individual shielded modules. If a high degree of electrical isolation is required between circuits, semirigid cable enhances this feature very well.

4.7 Component Selection and Specification

Component (hybrid or monolithic circuit) selection is a process that goes beyond electrical performance parameters. Depending on the target cost and reliability, a certain level of testing at the supplier is required to assure that the component will perform in the desired application and last as long as required. The military has addressed this issue through Military Specifications and Standards. They have generated these documents, which control components in conjunction with users and suppliers, so that the documentation represents realistic approaches to device specification and procurement. Full compliance to these documents may not be necessary in some military applications by virtue of individual contract requirements and the military's current trend to use commercial off the shelf (COTS) components whenever they can meet the requirements. This reversal of the requirement to use MIL-type components is changing the entire procurement picture. Although underway only since 1994, the move to commercial parts has picked up momentum and has affected the way OEMs and lower-order suppliers specify what they expect to see in the performance characteristics of the devices they are procuring. There is much debate at this time regarding the overall integrity of such components, and the debates will not be resolved any time soon. In commercial applications these documents need apply only to the extent the designer chooses. However, the principal test methods are contained in the following documents: MIL-M 38510, MIL-STD 883,

MIL-STD 750, and MIL-STD 202. Minimum part screening (testing) is highlighted in Table 4.1.

4.8 Designing Microwave Assemblies

4.8.1 Engineering

The role of the engineer in design is many-faceted. In large organizations, the responsibilities can be shared by individual groups of engineers that represent various functions, such as systems analysis, systems design and integration, components, electrical design, manufacturing, and reliability. Small organizations must accomplish the design with far fewer personnel who are skilled in several disciplines and therefore must be proficient enough to produce designs that will meet their customers' requirements. The discussion here is intended to assist designers through a systematic presentation of salient features of the design process for both high speed and digital circuitry. The procedure outlined here is typical; it describes activities that must take place and their interrelationship.

When a company decides to produce a certain microwave product, it has already studied the market for that product, decided how much of that market it can capture, what the price of the product must be, how manufacturing will be accomplished, what the specific detail parts will be, where those parts will come from, what specifications those parts must meet, how and where they will be tested, what the levels of assembly and the test strategy will be, and what all this will cost, including setting aside some funds for the unknown, the problems that will surely arise and plague the process somewhere along the line. These problems occur more frequently up front in the design process but can occur later as well. To illustrate this, consider the circuit (product) that had been developed in the laboratory and had performed well through engineering testing. All parts needed for this circuit had been proved to meet the specifications that govern them. Prototypes of the circuit were produced with parts from one vendor, and these were the ones that passed all testing. The drawing that described an individual part listed several vendors as qualified to supply that part. Procurement of the part was, for whatever reason, handled inappropriately and the actual parts may not meet the requirements of the drawing. If this happens when the production orders are being produced, the problem may not be discovered until a certain number of circuits have been manufactured. The suspect parts will have to be removed from assemblies and replaced. This type of occurrence is not infrequent and must be allowed for in schedule and financial planning. Quality improvements in processes can

TABLE 4.1 Minimum Part Screening

Test requirement	Hybrids	Passive components	Discrete devices	Integrated assemblies
Bond process control	883-2011	N/A	750-2037	883-2011
Internal visual	883-2017	Per agreement	750-2073	883-2017
Sealing	Seal*	Per agreement	Seal*	Seal*
Stabilization bake	883-1008	N/A	N/A	N/A
Temperature cycling at storage extremes	883-1010, 25 cycles	202-107, 25 cycles	202-107, 25 cycles	883-1010, 25 cycles
Mechanical shock	883-2002B	883-2002	883-2002	N/A
Burn-in	883-1015	883-2020B	750-1038, 1039	883-1015
Seal test	883-1014	202-112/A, B, D	883-1014	883-1014
External visual	883-2009	Per agreement	750-2074	883-2009

Numbers refer to the MIL-STD and the appropriate method and condition.
*Preseal bake at 150°C for 8 h minimum in a vacuum of ~1 torr.

and should strive toward eliminating such clumsy problems. Total quality is an improvement process that can be of great benefit along those lines. If a company has not investigated the use of total quality and its benefits, it will continue to pay for errors over and over.

For any product, the following processes should be followed as a minimum:

Evaluate the risk in using state-of-the-art processes and parts.

Develop cost vs. performance criteria and risk reduction methods.

Study areas of rf function integration and determine suppliers' capabilities in the processes that would be required to manufacture the component properly.

4.8.2 Partitioning circuits

Partitioning circuits in the microwave portion of the assembly is unique, or at least quite different from partitioning digital circuits. Microwave functions, being analog in nature, tend to require significantly different types of components for the accomplishment of different electrical tasks. Although there is a system interdependency among all components in a microwave system, these components often have a different form factor from each other. This can make the assembly of all of them appear incongruous unless an overall approach is taken to design a coherent and esthetic system. The system should meet the performance and design objectives while having a physical form that suggests that it was designed by one person, or at least a team of persons, who worked to the same rules.

4.8.3 Partitioning functions

The partitioning process begins with the designer dividing the entire system diagram or equivalent representation of major parts of the system into smaller blocks or functions that are deemed possible to produce, test, maintain, and fit into the space considered or allowed for the system. This activity requires knowledge of all the topics just listed so that serious complications do not arise later that are caused by invalid assumptions. Since all the required information may not be possessed by the designer alone, it is prudent that other engineering sources such as current literature, handbooks, and knowledgeable personnel be consulted to assist the designer. In large companies the depth of experienced personnel available for consultation and design review tasks will be greater than in small companies. However, literature sources can prove valuable. As time goes on, the literature containing required design information is growing rapidly. Excellent sources are the trade journals covering electronic packaging and components, vendor catalogs and supporting data, exhibits, seminars, professional society chapter meetings, and conferences. Meetings and conferences often provide an opportunity for dialog with other designers as well as papers on the topics of particular interest. This networking can lead to joint opportunities later such as teaming on projects or information transfer (the latter performed under nondisclosure agreements between companies).

Once the major components or functions are defined, it is necessary to verify that sources are available to construct them. This includes the vendor community as well as in-house capability. Much valuable information can be generated by a thorough discussion with potential suppliers of the functions. Although the degree of thoroughness is controlled by the amount of time and money available for this task, one should keep in mind that this step is critical to the success of the project. Good record keeping is important at this point since the generated data will prove very useful in planning and scheduling subsequent projects. The function partitioning defines the characteristics of the various functions and also the characteristics of the subfunctions contained within each larger functional block. Therefore, the subfunctions being grouped together must be studied for compatibility with each other in construction (materials, interconnection methods) and in electrical features (power requirements, noise susceptibility). Before proceeding with the design and detailed specification of the defined functional blocks or packages, it is important to hold a preliminary design review whether or not the customer dictates that the review take place. Those present at the design review should include marketing personnel to look after the customer's interests, system design personnel, electrical, mechanical, and manufacturing designers, and project and/or program managers. It is customary to include experienced designers not working on the project as well, to assure an objective overview. The purpose behind including what might be viewed as a large group to advise and comment is not to encumber the design process, but rather to enhance the process through avoidance of problems that could pose serious difficulties. The customer (if there is one) usually participates in the formal design review and either approves or disapproves of specifics presented during the review. If there is not a specific customer (for example, in the case of a new product being designed for a market) marketing and management responsible for new business must conduct the review to their satisfaction.

The preliminary review will produce action items to be attended to and perhaps will add a few tasks. These must be integrated into the project plan. The design process then proceeds, with attention to details. Some typical specifics are provided here as a guide. Systems requirements for environmental conditions, survivability, maintainability, performance, fabrication, and testability must be studied and incorporated into the designs with constant input from all designers responsible for specific features. This task management can be achieved only with painstaking attention to detail on everyone's part, particularly the project manager. Knowledge of the governing or applicable specifications is mandatory. The designs must be analyzed by appropriate personnel to determine the potential for meeting the specified requirements. This can include structural and thermal analyses, and circuit analyses as well. The output of such analyses should be considered in the design. Progress reviews should be held to assure conformance to the plan, which includes schedules and technical milestones. Schedules must contain sufficient measurable milestones since they are critical to successful project management.

Vendors were mentioned earlier because they can be a valuable asset in the design process as well as in the production stage of the project. Vendor review of detail designs is often a good way to check the realism of an approach. Vendor review and comment early in the design process can lead to cost-effective designs. A vendor who must survive by producing parts in as painless a manner as possible for the price quoted will be very careful to guide customers to design a producible and testable part. Vendor selection should be done with care and candor and with respect for business ethics. Vendor evaluation, which may take place before detailed designs are discussed, should include a quality, manufacturing, and purchasing department survey after engineering has determined technical competence. Further discussions with the vendor should hinge on that survey. Vendor past performance, resources including design and management personnel available for the task, and equipment must be considered before proceeding seriously with the potential supplier. At this point it can be helpful to generate a draft of a procurement specification for the functional package. Vendor review of that draft will produce suggestions and areas that require compromise. Vendor requests should be reviewed in turn by knowledgeable personnel in-house to represent the interests of the procuring company. This process will likely involve give-and-take several times before both parties agree on all remaining requirements. It is important to note that as the degree of complexity of a function increases, the number of potential suppliers decreases. This fact should be considered early in the design phase when it appears attractive to partition the system into a few functional blocks because they can be made into highly integrated assemblies. The advantages of that design approach could be outweighed by the dearth of suppliers who can produce the part. A tradeoff study could provide sufficient information for a decision.

4.8.4 Packaging

Packaging of the partitioned system is addressed by referring to the master block diagram. A block diagram shows the functions as they are broken down into subfunctions and sub-subfunctions, and so on as the situation requires. Specific packaging features and techniques should be chosen so that they are compatible with the circuits and devices being packaged. Wire bonding is used if unpackaged chips are placed in a hybrid package, however large that hybrid may be. Printed substrates of microwave-quality dielectric are used with both packaged and unpackaged chips as the interconnect medium. Ribbon bonding may be required where inductance is to be minimized or where electric current may be greater than can be handled by wire bonding. Special techniques may also be necessary, such as tape bonding (TAB, or tape automated bonding, a throwback to individually packaged digital devices where automation made the process economically feasible) and flip chip, which involves inverting a chip that has had bumps or balls of a compliant metal such as solder adhered to its bonding pads and attaching the chip to an interconnecting substrate. The significance of these techniques will become apparent later in this chapter. The first step in packaging a function after partitioning has been done is to perform an analysis that considers both dc and rf circuitry as they exist within the package. This analysis should assure that there is no mixing of dc with rf signals in the package. The analysis may suggest that a compartment be constructed within the package to provide shielding. The compartment must be made of metal, or the walls of the compartments, if they are a dielectric material, must be metallized. It is preferred that the metal not have very high conductivity so that the signals which radiate into the space above the top conductors on substrates are absorbed by the walls rather than reflected off the walls. Other physical features that must be considered are adequate grounding under substrates, especially in the vicinity of microstrip or stripline conductor launch points (locations of conductors on the substrates that mate with connectors or wire and ribbon bonds), length of transmission lines and ground paths, a potential source of ground discontinuities, and wire and ribbon bond lengths. The design may consist of several substrates within a package. That configuration requires a careful study to assure control of gaps between substrates and package walls as well as limitations on wire or ribbon bond lengths.

Available space. The allotted space or, more specifically, volume in which the electronic function will be placed should be broken into compartments that will be used for hardware other than interconnecting substrates. The items which will also be part of the package are cabling, access to mounting hardware and connectors, compartment walls between functions that must be electrically isolated, heat sinks or other cooling components, mounting hardware, and access. If mechanical isolation is required, consideration must be given to sway space. Other items that may be required are rf absorbing material, interconnects between isolation walls, grounding apparatus where dictated, and perhaps waveguide as well. When this allocation has been made, the remaining volume will probably seem quite restrictive. If the analysis shows that the required volume exceeds the allotted volume, it may be necessary to consider smaller components, another interconnect substrate material with a higher dielectric constant which will reduce conductor dimensions if dielectric thickness is kept the same, and smaller connectors. The consideration of these alternatives may suggest an acceptable volume. In all likelihood this relief may come with higher cost. Therefore, by studying the volume on a function-by-function basis, it may be possible that some volume can be taken from a relatively empty package and given to the overcrowded package if the size limitations of the affected packages permit such an exchange. This flexibility is greater in the case of packaging a large system such as an airborne radar than it is for the case of packaging a cellular phone system, but cost-effective alternatives are always available to the designer even in the most tightly packaged scenario.

Allowable weight. The weight of the product is an important factor in all electronic packages. This point becomes more obvious with handheld memo pads, cellular telephones, portable computers (laptops), and global positioning satellite (GPS) equipment. The list of available and emerging equipment is increasing constantly, and that increases the importance of weight in the product. In the recent past, equipment that was designed for airborne and spaceborne applications clearly required attention focused on the weight of each component, interconnect, and structure. Premiums were often paid for lower-weight materials for those applications. The situation of premium payments began to change with reductions in defense and space spending. Consequently as sophisticated commercial products become more and more commonplace, lightweight materials and technologies that previously could not be considered may actually be economical to use for interconnects, packages, and structures because of the increased sales volumes of those materials. Materials that are composites of individual materials (typically two or even three) can impart remarkable properties to the resultant combination that none of the individual materials possessed separately. Some more commonly used composites are graphite combined with epoxy, and aluminum blended with silicon carbide. These composites allow products to be designed with low weight and relatively high strength along with high thermal conductivity. Composites such as these allow the designer to achieve the elusive dream of a material with the density (weight) and heat conduction capability of aluminum and a tensile strength approaching that of steel. Composites are well known for their use in structural applications such as reinforced concrete and control surfaces on aircraft, and in early electrical applications such as glass-reinforced phenolic cases for housing small transformers. A printed wiring board is a form of composite and is an early example of composite usage in electronics. In a broad sense, a composite is a combination of materials which has properties that none of the constituent materials have by themselves. Materials are selected to be combined with each other to take advantage of one or more characteristic properties that the partner material possesses. The properties can be mechanical, such as tensile strength or thermal conductivity, and electrical, such as high resistivity or dielectric constant. A brittle material can be made less so by combining it with a softer one. A material's tensile strength can be improved greatly by combining it with a reinforcement which binds itself well to the weaker material, thereby imparting some of its strength to the combination.

Costly as these composites may be in the 1997 time frame, increased usage will certainly reduce that cost to allow these materials to compete realistically with conventional materials and technologies.

4.8.5 Materials

The materials which are to be used in microwave circuit construction must be selected first for their electrical characteristics and then for their mechanical properties. That is not to say that mechanical performance is secondary in importance to electrical performance, but rather that electrical requirements must be met before mechanical properties are considered. For example, if a dielectric material has adequate mechanical strength and stability at all anticipated temperatures but has a dielectric loss factor in excess of that which can be tolerated by the design under consideration, that material cannot be used and another one must be chosen that has adequate mechanical properties and meets electrical requirements as well. Consideration must also be given to the environments to which the circuit will be exposed. Performance at high- and low-temperature extremes under shock and vibration conditions (as in airborne or automotive installations) and corrosion resistance should be prime considerations.

Substrate materials. The wide variety of available substrate materials with good electrical characteristics can be appreciated by perusing microwave and analog packaging journals. The substrate materials range from inorganics such as alumina, beryllia, sapphire, and quartz to organics such as polyimide, polysulfone, polytetrafluoroethylene (PTFE), FR4 epoxy-glass, and aramid-reinforced BT epoxy, with many others in between. The properties of these materials should be studied to ascertain compatibility with the anticipated requirements. A list of typical properties can aid in selecting the most appropriate material for the application. In this day of access to the Internet, one can easily find the home page of a material manufacturer and download specific material designations and properties. The choice of the most appropriate material should depend on the allowable design parameters (that is, if the choice involves a material that requires physical support, consideration must be given to the effect of the weight and cost of that additional material). If the choice involves a departure from design goals that would result in unaffordable penalties in weight and/or cost, it may be prudent to consider materials with less attractive properties.

Attaching components to substrates. Electrical components may be attached to their supporting substrates by several methods. The primary method is with an adhesive that is compatible with the substrate and the component. Adhesives are used to provide physical support to the component and also in some instances to provide an electrical connection. Adhesives in this context generally are epoxies with fillers to give them specific properties such as electrical conductivity, improved strength, and in some cases, flexibility. When electrical connections are expected of the adhesive, its electrical conductivity should be verified through measurement. This point is important because the conductivity can be a limiting feature if it is too low. If the epoxy is to provide mechanical attachment and grounding as well, attention should be paid to the resistance of the interface between the adhesive and the component electrical connection metallization. That resistance can cause difficulties if a low-resistance path is required in that part of the circuit. Unwanted heat would be generated through I^2R losses at the interface, causing the component to be at a higher temperature than desired. In the case of chips mounted in this way, the bias voltage on the chip at the mounting surface would not meet the design value as well.

Another method of mounting is through solder. In this case, the component may be mounted only by its electrical connection pads.

The electrical and also the mechanical “connection” are provided by the solder. While this method is used frequently, it is not a recommended technique for all applications because mechanical shock and possibly vibration as well could disturb or destroy the electrical connections. It is always advisable to provide mechanical mounting and securing of components independent of electrical connections. Some electrical connections may be a bit tolerant of vibration and shock. Among them are wire and ribbon bonds, flip chip bonds, and TAB or cantilevered beam bonding. Components electrically connected to the surface of a printed wiring board must be secured properly. This can usually be accomplished through the use of an epoxy. Epoxies are often used to provide both electrical and mechanical functions. Some epoxies have high filler content composed of silver or gold particles which form an electrically conductive path. Although solders have been used for some time and continue to be used to provide electrical connections between components and their supporting printed wiring board, this method cannot apply in all cases. If only the solder and no wire or lead is used to make the connection and the component is mounted directly to the board, there is a risk that such connections will fracture with time and temperature excursions because the board and components will expand and contract at different rates. If the components are small, however, this method may suffice.

For bonding components or substrates to aluminum chassis, any conductive epoxy bead appearing exposed should be covered with an overcoating of another insulating epoxy which serves as a moisture barrier. This procedure need not be performed if the chassis will be hermetically sealed before it is removed from the assembly environment of low relative humidity. Except for mounting components, nonconductive epoxies are used in a different manner from conductive epoxies. They can be used to overcoat lines that cannot tolerate foreign material settling on them (other than the nonconducting overcoat, the effects of which have been factored into the performance of the circuit). These materials are also used to secure tuning elements (stubs on the substrate transmission lines joined by metal ribbons). Both of these approaches are best served by “low-loss” materials. Low-loss in this context is a relative term. Organic materials are not as a general rule low-loss from a microwave perspective. However, the loss tangent of this class of materials varies from 0.001 to 0.05, both values measured at 1 MHz. Higher-loss materials in this group have a purpose in microwave packaging. Control lines and dc lines often carry rf signals because of coupling to rf lines. Application of a high-loss epoxy to these lines provides an effective means of attenuating the rf signal without affecting the low-frequency control signal or dc being carried.

Solders used to connect components to chassis should be from the class which do not corrode if exposed to moisture. An analysis of the metallurgies involved will determine what if any electrochemical activity can be expected. Solders require fluxes unless the soldering operation is capable of being performed in an inert atmosphere or vacuum. Inert gas enhanced soldering can be a valuable process. In this process, the lead in the hot solder is prevented from oxidizing by a blanket of nitrogen. Consequently the solder is able to wet and adhere to the metallization of the parts and chassis as the case may be just as if flux were used.[1]

Fluxes are governed by the requirement that they do not contribute to corrosion in spite of their activity during the soldering operation. Many no-clean fluxes are available, and the designer should assure that the materials being used are not negatively affected by the flux left in place. Older-style fluxes that require cleaning by solvents and detergents after soldering are also available.

Tin-lead solders that have been used successfully for many years are SN60, SN62, and SN63 per Federal Specification QQ-S 571. Although there is considerable concern these days over the use of lead, up to 1996 there has been no nonlead solder which can be used as a drop-in substitute for tin-lead solder. To be considered a drop-in substitute a solder must have the same processing parameters (reflow temperature, wetting, strength, thermal cycling robustness, conductivity, and inertness, to name a few) and performance characteristics. Cost differentials are seldom mentioned in the quest to eliminate lead from solders.

4.8.6 Thermal design

Where analysis has shown that heat dissipation is an area of concern, special attention must be paid to component derating and dissipation. Worst case versus average power loss through heat dissipation must be compared. Component mounting materials must be evaluated for thermal conductivity. Their wetting to the component mounting surface and the chassis affects the total surface area transferring heat. Checking for voids under components and establishing processes to guarantee freedom from voids is essential. Process verification through x-ray, and infrared techniques that report device surface temperatures while the device is operating, are a few quality checking procedures that will help.

The junction temperature of semiconductor components should not exceed 125°C for microcircuits and 150°C for power devices. A surface finish of 63 μm and overall flatness of less than 0.2 mm will assure that thermal interfaces provide maximum heat transfer. Special cooling may be required for FET devices, especially for those in the first stage of low-noise amplifiers (LNAs). As a rule of thumb, one can expect about 0.07 dB lower noise for every 1°C drop in temperature. To achieve this, "unheated" air or thermoelectric devices are used. Crystal oscillators used as frequency standards for radars also require special thermal control. In those cases, however, the devices are placed into tightly controlled ovens that keep the temperature of the device constant. It is easier to heat rather than to cool when very tight control is required on the temperature of a device. The control temperature is usually set a few degrees above the highest expected operating temperature. A closed loop is required for this type of control. Several rules should be followed:

The sensor should be placed as close to the element as possible.

The thermal time constants of the element, the sensor, and the heater should be very nearly the same.

Design so that sufficient heat is delivered to the oven or element so rapid warmup (~3 min) is achieved and temperature can be maintained at the lowest environmental temperatures expected.

Use as much insulation and thermal isolation as necessary to keep heat loss from the oven to a minimum.

Heating elements should have as low inductance as is possible for the situation so that the surrounding circuitry is not affected by stray fields.

Mismatches in coefficient of thermal expansion (CTE) between joined materials can result in stress failures during temperature cycling. A few things to keep in mind when joining materials of different coefficients of expansion will minimize thermally induced failures:

1. Where solder is used to join two materials, the shear stress in the solder should not exceed the allowable levels of the solder. For tin-lead solder, 500 lb/in^2 at 125°C should be the upper limit of the shear stress.
2. Strain relief should be provided for all lead attachment.
3. Substrate boundaries of substrates should not be constrained; freedom of movement should be provided.
4. Substrate structures that are a composite of several materials, for instance, ceramic mounted to a molybdenum plate, should have as large a radius of curvature as possible under temperature cycling. Small radii cause exceedingly high levels of stress in the composite.

5. Where coaxial connector center conductor pins are soldered to the substrate conductor through a chassis or module wall, either a slip pin or rf connector should be used. Both hermetic and nonhermetic styles of this type of connector are available.

4.8.7 Vibration

Vibration can cause problems in two ways. First, there is the possible destruction or permanent damage to the circuit from the energy imparted by vibration, and second, the circuit may operate out of its desired parameters during vibration. Precautions must be taken to prevent either from happening.

Preventing circuit damage. For the first case, circuit elements should be mounted directly to a mechanically stiff substrate of printed wiring board assembly whose natural frequency of vibration is 200 Hz or higher. A 200-Hz fundamental frequency yields about 0.002 in displacement for each *g* of acceleration at a transmissibility of 10. By transmissibility here is meant amplification. This is a fairly common value for transmissibility. Lightweight stiffening techniques should be used to minimize distortion. An example would be ribs added to a chassis cover.

Provide strain relief for leads so that they do not become torn from a component or a substrate. Employ sufficient holddown techniques to circuit assemblies and devices by staking through the use of adhesive interfaces and screws to reduce relative motion between components and their mounting surfaces. Study component structures for weak spots. The use of solder for primary holddown should be avoided if the shear stress may exceed 500 lb/in^2 at 125°C. Avoid prestressing component leads by bending during soldering. Leads should be soldered into position with very little prestressing. If lead prebending is required by design, this should be accomplished through very accurate tooling so that no springback occurs. A typical mistake is made by assemblers who first push on a lead with a tool to force it into position, then solder it while holding the lead in the forced position until the solder solidifies. This sets up a residual stress which will add to the stresses encountered during vibration and temperature cycling as well.

Always use locking hardware, determine torque requirements, and assure they are followed. Signal wires and rf cables should be clamped or staked at approximately 1-in intervals to prevent fraying during vibration. Mount components so as to yield the minimum overturning moment, that is, the shortest standing profile during vibration. Mount large components so that the relative displacement between leads is very small. Avoid work hardening of leads, which may result in early failure due to fatigue.

Assuring circuit operation during vibration. Certain circuits are extremely sensitive to the effects of vibration. Among them are crystal oscillators for use in stable local oscillators and filter devices. Crystal oscillators in radars are particularly sensitive to vibration that will cause them to resonate in a manner that produces a frequency-modulated signal which can mask radar targets. To prevent this it is necessary to provide vibration isolation which will attenuate the input sufficiently. Depending on the sensitivity, it may take more than one stage of isolation to accomplish the desired reduction in acceleration to the frequency source. Other precautions should be taken. All resonator components, transformers, and other inductive components should be bonded to their respective assembly boards to minimize motion. Chassis covers should be located sufficiently far above inductive components to minimize capacitive effects and should be stiff enough and mounted with screws located close together to prevent motion relative to inductive components. All cables should be clamped and staked at 1-in intervals to prevent signal modulation. Use sufficient hardware to mount assemblies to the chassis to guarantee proper grounding. RF connectors should be selected based on their ability to perform with low-phase noise during vibration. Wires and cables connected to circuits mounted on vibration-isolated platforms should provide strain relief so that they do not interfere with the vibration isolation. In other words, do not allow the interconnections to restrain the assemblies at all. Sufficient sway should be provided. Finally, sensitive crystal oscillators often require that they be packaged in a housing made from high-density materials such as steel or tungsten, whose purpose is to attenuate deleterious acoustic inputs by their mass.

Mechanical shock. Mechanical shock is an excitation pulse of acceleration or force of short duration. In most cases, the shock pulse is defined as a half-sine pulse of 11 ms of amplitude 15 *g*. Prevention of damage due to shock inputs can be achieved by following the guidelines for prevention of damage from vibration. If vibration isolators are used, special techniques such as use of a snubber may be required to limit displacement of the vibration-isolated electronics.

4.9 Package Seal

Microwave circuits require seals to prevent the entry of undesirable liquids and gases as well as to provide electromagnetic interference (EMI) protection. The seal must withstand stresses associated with environmental screening. The most severe test of seals is temperature cycling. For this reason the TCEs of the materials used to provide a seal are important. The type of seal depends on the packaging of the electronics within and the environmental requirements. If the package contains exposed semiconductors, the seal must present a barrier to water vapor. The materials which can provide this type of seal are metals, ceramics, and glass.

4.9.1 Level of seal

A bolted-on cover will not impede moisture but will provide EMI protection to some degree. A package of this type may be enclosed within a larger package that is protected against the environment. A more durable seal is hermetic. A metal-to-metal hermetic seal can be achieved by soldering or welding. Solderable plating is required if solder is used. Soldering usually requires heating the entire assembly, although this may not damage the components within because they will not reach the reflow temperature of solder in most cases. Welding, on the other hand, delivers heat locally and is less likely to heat sensitive parts. Welding of aluminum packages can be done by laser, TIG (titanium inert gas), electron beam, and electrical resistance. In any event, the electrical connections are achieved by glass-to-metal seals. Certain features of glass-to-metal seals are prone to problems. The principal one is cracking of the glass meniscus on the lead. This is especially troublesome with rectangular leads because of the extremely small radius at the corners of the lead. Minor flexing of the lead can cause tiny flakes of the glass to be cracked away, leaving exposed a small amount of bare Kovar which was previously protected by the glass. This could lead to corrosion of the lead and eventual failure of the device. Another potential problem is the cracking of the glass bead, which can be either radial or circumferential. Cracks can appear to be so deep as to penetrate all the way to the inside of the package. Radial cracks are specifically forbidden by MIL-STD 883. The ultimate tests of whether a crack is detrimental are the leak tests of this MIL-STD. Correlation of visual criteria and leak testing is troublesome and often inconclusive. For these reasons the trend toward ceramic packages has become very strong, although Kovar enclosures continue to be used in both commercial and military microelectronics.

An O-ring or epoxy seal can be used whenever limited entry of water vapor is acceptable. These materials impede water entry, but eventually the moisture content of the air inside equalizes with the moisture content of the air outside. If the conditions under which dew forms do not occur or occur infrequently and do not last for long periods of time, these seals may be satisfactory. O-ring seals consume more space than may be available, because of the need for grooves and mounting hardware. EMI-RFI protection, if needed, must depend on additional seals. The main advantage is the ease of opening the package for repairs. Epoxy seals are relatively easy to apply, and the materials can be tailored for EMI-RFI properties. Rework is more cumbersome than with O-ring seals. If protection from water is essential, only a hermetic package will meet that requirement. Hermetic packages with an internal volume of more than a few tenths of a cubic centimeter usually have a cover that is attached by soldering or welding. The design should allow for removal of the cover and resealing of the package. For ease of assembly, the cover should be self-locating by using a step in the package or the cover.

Soldering. If they are to be soldered, aluminum packages and covers need to be plated with nickel and tin. Cover and package wall should allow for a solder fillet to reside all around the seal. When the cover is large, provisions should be made for support posts located in several places near the center to minimize flexing during temperature cycling and during pressurization associated with leak testing, which could stress and cause fatigue failure of the solder seal. These screw holes are themselves sealed by small covers located only over the screws. The underside of the cover should not be plated with tin in order to keep the solder confined to the seal.

Welding. No plating is required for welding, but the alloy used for the package must allow welding. In the case of aluminum, common alloys such as 6061 are not weldable because the joint cracks as it cools from shrinkage. Excessive shrinkage can be prevented by the use of an alloy that has high silicon content. Such an alloy is 4047, which has a silicon content of 12 percent. It is available as thin sheet and may be used in the weld area as a weldment. Even though during welding the silicon will diffuse into the 6061 package wall and cover, sufficient silicon will remain in the weld to prevent cracking. Care should be exercised if the cover is removed and replaced. It is advisable to use a new 4047 strip if a new cover is used. This is because the silicon will be depleted by the new 6061 cover. Another approach is to use a cover made form 4047 alloy. However, because of the high silicon content, machining is difficult and carbide tools are required.

A hermetic seal can be accomplished by using a YAG laser. This method offers two distinct advantages over resistance (electrical) welding processes: (1) Very low heat input is required, and it is applied locally without subjecting the entire package to high temperatures, and (2) no electric currents are induced in the package during welding.

4.10 Connectors

Connectors can be attached to the package by threaded fasteners, soldering, or welding. The disadvantages are that plating is required for soldering, and the heating that occurs during soldering of the connector to the package may damage heat-sensitive components within. Welding as a connector attach method localizes the heat, as was pointed out in Sec. 4.7.1. Knife-edge connectors may be used, provided seal integrity is fully tested. Two types of connectors are needed for microwave packages. DC and rf connectors are implemented with either glass and metal feedthrough construction or a separate connector assembly. The feedthroughs consist of a glass-to-metal seal at the center conductor (pin) and a metal ring around the glass that is soldered to the package wall. Electrical connections are made to the circuit contained within the package by means of these pins, which consist of a central conductor insulated from an outer metal shell by a bead of glass. The arrangement of conductor, glass, and shell is a delicate one by virtue of the glass, which must maintain a hermetic seal between itself and the shell as well as between itself and the central pin. The glass, whose coefficient of expansion is very close to that of the pin, is attached to the pin by a chemical bond between the oxides on the pin and the glass. In certain pin designs where the shell is not matched in thermal expansion to the glass, the shell and pin are assembled to the glass by heating all of them together until the glass flows. The shell contracts more rapidly than the glass, forming a compression seal between itself and the glass.The shell is sometimes called a ferrule. The ferrule material is, of course, different from the central pin material since it is desired to have the shell constrict more rapidly than the glass as they both cool. If the pin were the same material as the ferrule, it would reduce its diameter more rapidly than the glass, causing the bond between them to rupture. This kind of seal is reputed to be more achievable than a matched seal in which all elements of the feedthrough have approximately the same coefficient of expansion. The glass-to-metal seal is a compression seal with the center pin matched in coefficient of expansion to the glass. Important factors in this design are the clearance between the package and the connector, the type of solder to be used, and the material of the housing. A major drawback to using welding for the connector attach method had been that the connector frame, made from low-expansion nickel alloy, could not be directly welded to an aluminum package. This was overcome by first fabricating an adapter that was plated only in the area needed for soldering the connector into it, and then welding the adapter to the package. For aluminum, the alloy is

4047. These connectors are available now with the adapter already attached.

Microwave connectors are employed for use in cabled interconnections, for direct interconnection of packages, and for interconnection of plug-in supercomponent modules. Important electrical specifications for all microwave connector types are insertion loss, VSWR, and rf leakage. These and mechanical requirements are addressed in MIL-C 39012. Although adherence to the MIL specification may not be a requirement, it is an excellent source of information.

The most widely used connector for interconnection of microwave components is the SMA. Other similar types are available for circumstances where higher performance or frequencies above 18 GHz are required such as the APC-3.5 and K-connectors. For the installation of modular supercomponent assemblies into systems, microwave blind mate connectors have been developed. These are available from Sealectro as SMS and from Omni-Spectra as OSP connectors.

Semirigid coaxial cable is used most frequently to provide connection to the module connectors. Although termed semirigid, this cable is bent using tools. Semirigid cable allows some reforming by hand without significant changes in its electrical characteristics. This type of cable is used mainly because of its ease of use and small size. Most low-power cable of this type has a copper center conductor and copper outer skin. The intervening dielectric is usually a polytetrafluoroethylene because of its high temperature tolerance and low insertion loss. A widely used size is the 0.141-in-diameter cable commonly referred to as "141 cable." Although small in diameter, this cable is fairly rigid in small lengths. For that reason, the next smaller size, 0.086 in in diameter, is used for lengths less than 5 in. Both cables can be used through 20 GHz.

Flexible cable is available but has electrical characteristics inferior to those of semirigid cable. Flexible cable has higher loss, lower isolation, and higher VSWR; it creates noise spikes during vibration, is more difficult to assemble, and is more expensive. However, if the performance can be tolerated, the main disadvantage would be the increased cost over semirigid cable.

4.11 Reference

1. Mark Nowotarski and Stephen G. Konsowski, "Nitrogen-Based Fluxless Soldering," *Surface Mount Technology,* October 1990.

Chapter

5

Microwave Circuit Assembly and Development

This chapter is primarily concerned with the manufacture of microwave packages and hardware. Consequently, it deals with special assembly considerations, some pertinent and important features to consider for test purposes, and finally, the development process. The development process is a critical part of the manufacture of microwave circuits because during this phase problems surface that were not evident in the design phase. A methodical procedure will smooth out the bumps which inevitably occur as hardware is assembled and tested the first time. Breadboarding and prototyping are discussed here as procedures that still have a place in the product development cycle even though computer-assisted designs tend to diminish the magnitude of problems that will be encountered in this phase.

5.1 Discrete Components

Most transistors used in microwave circuits are packaged in planar fashion as opposed to the analog and digital counterparts which are packaged in TO cans. These planar parts have leads emanating from the package parallel to the surface of the PWB or substrate to which they are mounted. In the case of power transistors, the leads are in the form of ribbons. It is important that the site which will accept the transistor be flat and have a finish of no greater than 60 μm free of nicks and particles as well. If the transistor is mounted by having its flange bolted to the mounting surface, cracking of the transistor case can occur if the above requirements are not met. Transistor leads should be preformed to provide stress relief. This was discussed in Chap. 4. If the transistor is mounted on the surface supporting a substrate, the dimension of the distance from the conductor on the substrate to the bottom of the mounting surface must be controlled closely, or the transistor leads will be overstressed and will fail by cracking or by lifting the solder from the substrate.

If rf coils are used, check to ensure there is adequate attachment both at the ends of the coil and in the center body of the coil, to survive the loads imposed by vibration and expansion. A technique used frequently in such cases is to use a soft epoxy or RTV silicone rubber to permit relatively free unstressed thermal expansion. Axial leaded capacitors should be mounted to provide the required stress relief in the leads, and the body should be mounted with epoxy to the carrier or chassis to prevent excessive motion during vibration. Hybrids often require a good ground connection to the case as well as a good thermal path to the chassis. The preferred mounting method is to provide flanges so that mounting screws may be used. Clamps over the top of the package should not be used because they may crush the package, which is fragile, especially if it is ceramic. Passive chip components (resistors and capacitors) are usually attached by soldering. It should be determined and assured that the component has a barrier metallization such as nickel on the end caps to prevent scavenging of the end cap metallization during soldering. Chip capacitors are metallized almost exclusively by thick film pastes which though fired into the thin ceramic are attacked by the tin in the solder. The degree of damage (removal of original metallization) is a function of the temperature of the soldering iron and the time it remains at the site. An additional precaution should be taken to heat the chip capacitors slowly because of their susceptibility to cracking from thermal shock during soldering. Controlled temperature soldering irons are helpful, but a better method is infrared soldering in which the substrate passes under infrared heaters and all the joints are made simultaneously in a more controlled fashion. A drawback to this method is the price of such equipment. Small suppliers cannot generally afford to purchase it and must therefore rely on solder iron control and close monitoring of the process. Active chips (diodes and transistors) may be mounted by either solder or conductive epoxy. If a low and highly stable resistance is needed from the mounting material, conductive paste epoxy should be avoided. Materials are available in film form, however, that are loaded with silver and have demonstrated no loss in grounding or attachment strength after thermal cycling. When choosing the mounting material it is important to determine whether the material has in fact performed as described above. In the case of a repair, however, conductive epoxy may be the only feasible repair technique for replacing a die that has been removed. For high power dissipation, it may be necessary to verify the adequacy of the attachment medium with other inspection techniques or analysis (see Sec. 4.8.6).

5.2 Substrates

Substrates may be laminations of several materials such as PTFE clad on aluminum or ceramic soldered to a metal carrier. They may also be made up of multilayered PWB or ceramic. Extreme care should be taken to consider the different coefficients of expansion, especially where ceramics are concerned. Stresses transferred to the ceramic material could result in catastrophic failure due to cracking. Ceramic laminates may be reinforced through the use of a molybdenum carrier. This metal has high thermal conductivity (as opposed to the heavy, low thermal conductivity iron-nickel-cobalt alloy used in the early days of substrate integration). Both metals are closely matched in expansion coefficient to that of alumina. Where allowed by the design, a flexibilized thermally and electrically conductive epoxy may be the material to ruggedize the ceramic assembly. Epoxies are available that can survive many thermal cycles and maintain attachment to the substrate while providing good grounding characteristics.

5.3 Internal Interconnections

Various electrical connections need to be made within a microwave package, and the package must then be connected to other packages or to the system. Some of the internal interconnections can be incorporated into the substrate while the remaining ones such as component to substrate and substrate to package must be accomplished with ribbons, wires, or component leads. The usual methods for interconnection within hybrid assemblies containing thick or thin film metallized substrates are diffusion bonding, either thermocompression, thermosonic, or ultrasonic, and parallel gap welding.

The first to be described is thermocompression bonding. In this technique a gold wire previously formed to a fine diameter (0.001 in and less) and wound on a spool is fed through a tungsten carbide tip which is part of a wire bonding machine. The wire protrudes below the tip (called a capillary) and is formed into a ball by moving the capillary over a tiny hydrogen flame or causing a spark to jump to the wire and melt it into a ball. The ball is several wire diameters in size and is too large to be pulled up into the capillary hole, which is only about twice the wire diameter. This feature is important because as the capillary moves down vertically it pushes down on the gold ball and causes the gold wire to unreel from the spool. The spool tension is adjusted to be low enough to keep the wire from paying out too swiftly and becoming tangled while not exceeding the elastic limit of the very soft gold wire. Early capillaries were energized by heating alone, and this type of bond is called a thermocompression bond.

A design improvement which added ultrasonic energy to the thermal energy of the tip has been in place for some time. This type of bonding is called thermosonic bonding. The improvement allows tip temperatures to be lowered and thereby improves process yield. The process of attaching the wire to the bonding pads of the devices does not differ for either thermocompression or thermosonic bonding except in the application of ultrasonic energy for the latter.

Connections are made to the devices by causing the ball of the gold wire to be pressed onto the pads of the devices. These pads are almost always aluminum. The heat, and in the case of thermosonic bonding, heat and ultrasonic energy, cause the gold and aluminum to alloy, forming a metallurgical bond. Purity of the metallic constituents is a stringent requirement for a successful long-term bond. Typically the gold wire is 99.999 percent pure and the aluminum pad is 99.99 percent pure. Impurities can cause the formation of intermetallics which will, with elevated temperature, cause the bond joints to become embrittled and ultimately fail.

As just described, the ball of the gold wire is pressed onto the aluminum pad of the device to form the electrical connection. However, this bond is only one half of the connection. The other half is formed when the capillary is lifted from the bond on the device. At this point the capillary can be moved in any direction over the plane of the work area. This feature of ball bonding is especially important for producing wire bonds on substrates that contain large numbers of devices. The capillary is then moved to the desired bond pad on the supporting substrate or package and made to press the wire onto the pad. This time there is no ball under the capillary, but rather the wire emanating from the hole in the capillary and bent under a portion of the capillary bottom surface (parallel to the substrate). After the capillary makes the bond, which is called a stitch bond, it is raised up from the substrate. As this transpires, the wire is clamped above the capillary and held tightly. This causes the wire to separate from the flattened wire portion of the bond. A portion of the wire separated from the bond protrudes below the capillary and is made to form a ball, as indicated at the beginning of the cycle.

Ultrasonic bonding is the third form of wire bonding. This type uses aluminum wire instead of gold wire. Aluminum wire was introduced as an alternate to gold in the 1960s because many gold wire bonds were failing to remain attached to the aluminum pads on devices. The symptom of the failure was a condition known as "purple plague" wherein a small portion of the ball bond at the interface to the chip pad would become purple-colored. Not long after the purple material appeared, the bond would fail. The actual cause was quite elusive (it was a combination of causes which included contaminants in the wire introduced during wire production combined with high-temperature storage of the bonded devices), and while investigators struggled to find it, the alternate, aluminum ultrasonic bonding, emerged. In ultrasonic bonding, the tip is a grooved foot with the wire, which in this case is aluminum, passing under the foot and guided along the groove. The wire again is 0.001 in in diameter or larger. Whereas the aluminum bonding pads of devices are 99.99 percent pure, the wire contains about 1 percent silicon to improve its lubricity so that it does not drag on and become hung up on the extrusion die when it is being drawn into wire. This does not affect the quality of the wire bond. The reason thermocompression bonding cannot be used for aluminum wire bonding is that aluminum is an extremely active metal that forms oxides on its surface which do not get broken up sufficiently during thermocompression bonding and interfere with the formation of a diffusion bond between the wire and the pad. Ultrasonic energy, however, breaks some of these oxide bonds, but more importantly, the action of the tip scrubbing the wire back and forth opens up fresh unoxidized aluminum in both the wire and the pad and permits an aluminum-aluminum diffusion bond to form readily between the wire and pad. No ball is formed at the free end of the aluminum wire. A disadvantage to this type of wedge bond to wedge bond lies in the requirement that the wire be dressed almost in a direct line away from the chip bond. Bonding hybrid circuits in which wires must go from the chip to the substrate in any direction is severely restricted when this type of bonding is used. Ultrasonic bonds can be made closer together than ball bonds (0.003 as opposed to 0.004 in). This allows chip pads to be placed 33 percent closer, which on a device with hundreds of outputs saves a great deal of chip area that can be used instead for the active portion of the device.

Aluminum wire brings with it certain economies over gold wire. Although the cost is less than that of gold wire, the small difference can be significant when large amounts are used as in devices with hundreds of outputs or with power devices that carry large amounts of current requiring significantly increased diameters and multiple bonds to the same pad (it is not uncommon to use 0.020-in-diameter aluminum wire). Multiple bonds may serve the purpose for carrying high current in low-frequency and power circuits, but the excessive inductance associated with fine wires causes performance to become degraded in microwave circuits. In these instances relatively large

TABLE 5.1 Values of *a and* b for Common Ribbon Materials

	a	b
Annealed OFHC copper	0.2262	−0.4413
Cold-worked OFHC copper	0.0697	−0.2427
Pure nickel	0.516	−0.456
Gold	0.247	−0.404
Beryllium-copper (1.9% Be)	0.0778	−0.255

gold-plated copper ribbons are frequently used instead. A gold-plated annealed copper-nickel alloy is also used. Soldering may be performed, but it is a very delicate operation when solder is used with thick or thin film substrate metallizatons. Care must be taken to reduce the scavenging of gold from the plating, and cleanup must be particularly meticulous to make sure residues from the soldering operation do not remain on the substrate. Component interconnection by conductive epoxies is in its infancy and should not be used until the materials have demonstrated compatibility with the requirements.

When soft substrates are used, the metallization is usually rolled annealed copper that has been tin-plated. Therefore, soldering is the preferred method of interconnection. Parallel gap welding is used to some degree to connect the metallization on the soft substrate to a nonhermetic thin film metallized substrate. The substrate may contain ladder networks or surface acoustic wave (SAW) devices. Solder is used for stranded wire attachment for short runs where flexure is not a reliability problem. Stress relief should be provided in all connections to minimize stress in the solder joints. This is especially true when connecting materials of widely separated coefficients of expansion. A guide to establishing maximum loop radii (r, in) on interconnections as a function of ribbon material and number of thermal cycles (N) is given by the relation in Eq. (5.1). Table 5.1 relates the values of a and b for typical ribbon materials. The relation is valid for N up to 10^5. The numbers in the table are empirical, not theoretical.

$$r^2 < \frac{1}{150aN^b} \tag{5.1}$$

5.4 Substrate Selection

The choice of substrate dielectric material is made by the designer based on the critical electrical design parameters of the circuit, the ease of manufacturability, and required reliability. Mechanically, substrates can be classed as "hard" and "soft." Examples of hard substrates are quartz, sapphire, alumina, silicon carbide, and aluminum nitride. Examples of soft substrates are PTFE filled with alumina and other low-loss high-dielectric materials, epoxy-glass, and polyimides. Certain features are more achievable on hard substrates. Very fine lines (0.002 in separated by 0.001-in gap) such as required for Lange couplers may dictate hard substrates. If the circuit features are not as demanding, soft substrates may be the design choice. Soft substrates are easier to fabricate, and machining to size as well as cutting channels for transistors and other components can be done with ordinary cutting tools. Electrically, there is a quality that could drive the choice to hard substrate material. The loss tangent of hard substrates is an order of magnitude better (lower) than that of PTFE soft substrates and much lower than that of epoxy-glass and polyimide. Hard substrates are generally limited to 2- by 2-in size owing to breaking in processing and mounting.

The dielectric constant value is chosen based on the circuit density required by the application. Dielectric thickness is chosen based on adverse modes radiation, required circuit Q, and resonances. The effects of conductor thickness, surface roughness, surface resistivities, dielectric constant variations, thermal stability, moisture absorption, and loss tangent must be evaluated.

Most soft substrates use from 0.25 to 2 oz rolled or electrodeposited (ED) copper for the conductor. While ED copper adheres more strongly to the soft dielectric, the roughness of the surface against the dielectric leads to a higher-resistivity conductor. This makes the conductor more lossy. Rolled copper has a lower resistivity, but the adherence is less than that for ED copper. Electroless tin and gold track plating should be used for corrosion resistance. Finer tracks are achievable with the thinnest copper.

5.5 Grounding Techniques

The rf circuitry is generally grounded to the chassis or package case as applicable. For planar types of construction the grounding is normally done by means of a ground plane in contact with the chassis or case. The ground plane is usually a layer of copper under the rf circuitry separated by a dielectric medium. Soft substrates often have the dielectric bonded directly to an aluminum plate which can serve as part of the chassis. Ground planes can exist on the top surface of substrates as well, but they must be connected to the bottom ground plane.

Grounding difficulties in microwave circuits can exhibit themselves in circuit instabilities, poor flatness across the frequency band, mismatch discontinuities, active device failures, and interaction between circuit functions. Various techniques are used to solve these problems. Generally low-frequency and low-power circuits are not as susceptible to grounding problems. However, even these circuits can have problems if good design practice is not followed. A few rules may help.

Always use the shortest practical physical path to ground as allowed by the circuit. This assures that the circuit will be perturbed a minimum amount by any extra inductance in the ground path.

Use the minimum number of substrates (that must be connected) possible in the circuit.

Take care to ensure continuous ground planes between substrates and chassis interfaces.

Control interconnects to surface mount components and embedded components.

Several techniques exist for creating low-inductance rf ground paths from the top of the substrate to the bottom ground plane. When these locations are near substrate edges, wrap-around grounding using bond wires or straps is effective. Plated-through holes are a common grounding technique for epoxy-glass and polyimide multilayer boards. These are accomplished during the manufacture of the boards. Spring pins (roll pins) are used on soft substrates clad to aluminum carriers. The pin is inserted into a predrilled hole in the aluminum, establishing the ground. The other end of the pin protrudes a small amount above the soft substrate surface and is either soldered to a ground track or attached by a metal ribbon. Ribbons are used to assist in handling differential expansion between the pin and the aluminum. Such a pin should not be used if the aluminum cladding is less than 0.125 in. In this case, screws may be used instead of spring pins. A screw is substituted for a spring pin into a tapped hole in the aluminum and then soldered in place. Attachment to the ground on the top layer is accomplished in the same way as if it were a spring pin.

Cleanliness should not be ignored. Customer requirements aside, the performance can be negatively affected by solder balls left behind and flux residues that creep under components and even under substrate carriers. Alcohol is a good cleaner that was abandoned years ago because of fire concerns in favor of chlorofluorocarbons (CFCs), which presented no fire hazard at all. With the outlawing of CFCs, alternatives had to be used. This does not mean that there is no danger of fire with the use of alcohol, but proper venting, storage, and dispensing practice can allow the safe use of this material. Other commercial solvents are available from myriad sources that seek to fill the need of manufacturers who can no longer use CFCs. Any PWB or surface mount technical journal will reveal a list of suppliers who can furnish such solvents.

5.6 RF Gasketing

This is a method used to improve interface connections of all kinds. Gasketing is furnished as silver-loaded rubber, bellow springs, wavy washers, or corrugated copper springs. Achieving low rf leakage into the gap between a wall and a substrate requires that a positive contact be made between the edge of the carrier and the wall of the transition. This will suppress signals from causing wall currents that can couple to other circuits. Buttons of wire mesh are especially effective in this case. Miniature bellows have demonstrated their usefulness as well. These materials can provide the low-impedance path necessary to prevent arcing in the gap in high-power applications.

5.7 Transitions

Integrated assemblies require rf transitions between the various rf component subassemblies such as amplifiers and mixers. Lack of consideration of proper transition design will result in failure to achieve anticipated performance of the integrated assembly even though the subassembly tested to specification. Computer simulations are available that can analyze transitions. Transitions may be from one type of rf transmission medium such as coax (TEM) to waveguide (TE or TM), to stripline (TEM), and so forth. The VSWR required at the transition by both subassemblies to be interconnected must be known and the transition must be designed to accommodate that VSWR.

The input-output signal to an assembly is usually made through either waveguide or coaxial rf connectors. For connectors, the diameter of the center pin and dielectric is chosen so that the center pin will not short to the substrate metal carrier or be so close that excessive stray capacitance is generated. The rf connector is chosen for its VSWR and moding at the frequency of operation. The attachment of the center pin must be made to the microstrip in a manner that takes into account the differential thermal expansion. Floating center pins and soft copper ribbon from the pin to the substrate track have been used successfully to provide stress relief.

The total rf ground return path is the distance from the perimeter of the dielectric to the chassis floor, the distance from the chassis wall to the substrate carrier, and the thickness of the carrier. The gap between the chassis wall and track results in an impedance mismatch. The ground path length can be reduced by use of a knitted wire washer (commonly called a "fuzz button") or an elastomer that is conductive or beryllium-copper springs installed at the connector-to-substrate interface in the carrier so that they contact the chassis wall close to the connector dielectric. Another approach uses ribbon straps from the substrate ground to the chassis wall near the connector.

For substrate-to-substrate connections, the gap to ground path length should be designed to provide the VSWR required for no degradation of the interconnected circuits. Connection is made with either solid ribbon, mesh ribbon, or wire. The inductance can be tuned out with the use of open stubs of capacitance in microstrip circuitry. For channelized assemblies where substrates are separated by partitioning walls, the transition can be modeled to provide the required impedance and VSWR. The transition may be designed as a coaxial feedthrough. Wire bonds, ribbon bonds, and mesh bonds should be chosen with regard to rf impedances. Mechanical stresses should be calculated for temperature extremes to ensure reliability. In microstrip, ribbon and mesh are chosen so that their width approximates the width of the track on the substrate. The length, width, and height of the bonds will determine the impedance.

Wiring for power and ground must be routed within assemblies to prevent conducted and radiated susceptibility, or crosstalk failures. Shielding and filtering may be required. Ground loops need to be identified in design of the grounding for the signals and power, especially for low-noise devices.

5.8 Development Units

The term "breadboard" is seen in discussions of the early stages of the development of a product. A breadboard is an arrangement of components temporarily interconnected to determine the feasibility of some circuit. This constitutes a laboratory model, often wired together in a fashion that would not be considered deliverable to a customer. The breadboard provides an opportunity for an engineering demonstration of a technology principle, technique, or capability to be used in the upcoming product. Compliance with configuration requirements is not the function of a breadboard; hence it can be viewed as a working model that suggests feasibility of the design concept.

Since the effort in a breadboard program must be controlled, a critical circuit function should be identified as the purpose of the effort. If such a function is an integrated assembly, it is necessary to establish what is expected and how measurements will relate to the proof that the desired results have been achieved. The leading edge or at least critical function characteristic and parameter often bears the name of the breadboard effort. An example would be "extremely low noise amplifier." After the breadboard model has been tested sufficiently to convince the designers that the concept is feasible, work proceeds to the next stage, that of the prototype.

The prototype model allows the development of hardware that is capable of meeting the reliability and quality assurance requirements of the end product. After the prototype model has emerged from the breadboard stage, matching the critical function circuit into the next higher assemblies is the real test of the concept. While the prototype should have the form, fit, and function of the production hardware, it need not have its components screened (tested) to the level of that hardware. Each prototype item should prove it meets system requirements by passing a demonstration that clearly shows its success. Prototype demonstrations are usually the springboard into follow-on production. Evaluation of the prototype model should be directed to demonstrate some specific parameter of performance, producibility, reliability, or system compatibility. Tests must be conducted in accordance with a written and approved procedure, so that results are not left ambiguous. The test plan for prototypes should list methods, procedures, and test conditions. Any changes that have been included to meet producibility and manufacturability requirements must be included in the prototype so that they can be evaluated as well.

Sometimes production models are constructed after the prototype models have been evaluated. These items must be manufactured using all materials, processes, and quality assurance provisions including screening of full production hardware. The purpose of this effort is to evaluate the manufacturability of the part. The same rules apply to a supplier if the part is to be made out of house. These models may be the ones designated as usable for first-article inspection. Special screening or environmental testing may also be associated with these items. The evaluation of production models is very formal, and all results are reported. Corrective action, if necessary, is implemented immediately before production is permitted to begin. If the procedures leading up to this point have been conducted properly, that is, all test results recorded and changes made in parts, interconnects, assembly techniques, etc., implemented already, any problems should be minimal. A smooth transition to full production is a good indicator that the development process was carried out with the discipline, skill, and cooperation of all parties involved.

5.9 Specifications and Testing

Specifications govern the parts that will be used in deliverable product. Many parts are standard and meet government specifications. Although there is a strong trend toward the use of commercial parts, sometimes the only difference between a commercial part and one which meets the military specification is the degree to which the part has been screened. On the other hand, at times the parts that fail to meet the military specification may be sold as commercial parts, not as rejects, but as able to meet commercial use requirements. It is important to know what contractual requirements are placed on the parts to be used in a program before any part procurement can begin. This is a quickly changing environment, and it is necessary for designers to be fully aware of what requirements they must meet in part selection. Standard part selection should be a priority because it is usually the lowest-cost approach. Parts availability will be greater and the risk of depending on a sole source is diminished. Also, the military resists the creation of new requirements, especially if a standard part will suffice.

5.9.1 Purchased part drawing requirements

The procurement of parts often requires a purchased part drawing that specifies what requirements a specific part must meet. This procedure is used for nonstandard parts. The drawing follows a standard six-section format:

Section 1: Scope

Section 2: Applicable Documents

Section 3: Requirements

Section 4: Quality Assurance Provisions

Section 5: Preparation for Delivery

Section 6: Notes

Appendix: Nontechnical Requirements

Section contents

Section 1: Scope. This section simply defines the area the document covers.

Section 2: Applicable Documents. This section lists all the documents contained or used in the body of the specification.

Section 3: Requirements. This section describes the engineering requirements the part must meet. It also spells out what is specifically needed and specifies the bounds within which the manufacturer must operate.

Section 4: Quality Assurance Provisions. This section spells out the particular tests to be performed and test methods to be used to verify that the requirements listed in section 3 have been met.

Section 5: Preparation for Delivery. This section should include any special provisions such as electrostatic protection, air freight, storage, package marking, and standard carrier packaging.

Section 6: Notes. This section contains notes that may be helpful but does not list any requirements.

Appendix: The appendix should contain any nontechnical requirements the manufacturer should comply with before parts have been designated ready for acceptance testing.

5.10 Device Testing

Device testing is divided as follows: acceptance, sample, qualification, reliability, first article, and screening.

Acceptance testing (quality conformance part 1 or group A testing) is a level of testing intended primarily to verify that the part is the one required by the drawing.

Sample testing (quality conformance part 2 or group B testing) is done to verify that some critical parameter is being met or that no substantial changes have been made to the part. This test could be destructive.

Qualification testing is designed to evaluate manufacturers that wish to become alternate suppliers as an approved manufacturer. The procuring agency prepares and approves a qualification test procedure for this purpose.

Reliability testing is sometimes called a mini-qualification test. There is usually some environmental testing such as low-level vibration and temperature cycling.

First article approval may be required by contract in place of qualification testing. In this procedure, there is usually an in-process monitoring plan and an extended acceptance and qualification test.

Screening is not a test but a method of eliminating early failures. It should be severe enough to be effective but not so severe that it damages the parts in the sense that it causes premature failure. It is important to know what the customer requirements are for screening, both directly by specification and indirectly by reliability requirements. Thermal shock (several dozen cycles from −54 to +125°C nonoperating) is an effective tool. Burn-in is commonly performed as well. For semiconductors, current is made to flow in the forward and reverse directions while the device is at elevated temperature, usually 150°C.

Some specifications require screening, and when they are referenced, some level of screening is required. Although the requirement may not be stated and may not even apply in specific cases, the specifications commonly used for microwave components are listed here as a guide to assist in determining what to flow down to the supplier in the purchased part drawing.

Semiconductors: MIL-S 19500

Test methods: MIL-STD 750

Microcircuits: MIL-M 38510

Test methods: MIL-STD 883

Component test methods: MIL-STD 202

Environmental test methods: MIL-STD 810

Requirements, electronic equipment: MIL-E 5400

5.11 Supplier Testing

Most part testing is performed by the vendor, which includes screening, qualification, and quality conformance testing. The right to conduct similar tests in-house should be reserved and made a part of the agreement with the supplier. Qualification testing is primarily intended to verify that the design is capable of meeting the environmental requirements of the drawing. Life testing is often included to demonstrate part reliability. The sample size is usually limited, but the minimum size is two parts. Qualification testing is done one time on samples from the initial production hardware. Requalification is necessary when major changes to the design or manufacturing process are made. Qualifying a part as part of system qualification is sometimes done for economic reasons, but individual part qualification is better. To avoid any ambiguities and assure mutual understanding, a formal qualification test procedure is frequently written by the supplier and approved by the procuring agency. The following tests are usually included in the qualification and should be reviewed by the responsible product engineers as applicable: operating temperature, vibration, humidity, temperature cycling, mechanical shock, temperature-altitude, explosion, quality conformance, and screening. These tests are described briefly.

5.11.1 Operating temperature

The drawing should specify maximum and minimum operating temperatures. Common operating temperatures are −55 and 125°C, which should be specified as case temperatures. Any degradation allowable with temperature should be specified. A cold start should be simulated.

5.11.2 Vibration

In most cases this test will be performed with the part nonoperating. However, where phase noise can be expected during vibration, the parts should be tested operating. The same is true for devices such as oscillators which may be sensitive to vibration. The test will determine whether the package and the internal mounting of components and leads is capable of withstanding the input vibration level expected in end use.

5.11.3 Humidity

The purpose of humidity testing is to accelerate the effects of moisture on the part. The ability of metals to withstand corrosion is demonstrated by the test. Temperature cycling is used in some humidity tests to enhance the penetration of moisture into the part. Alternate drying and condensation accelerates corrosion processes. Moisture trapped inside the package during sealing of hermetic packages is of concern for circuits containing semiconductor chips. Test methods for determining the moisture level in parts per million have been developed, and limits based on reliability levels have been set in MIL-STD 883, Method 5008. Method 1018 of that MIL-STD measures the moisture of the package with a mass spectrometer after piercing the package in a vacuum chamber. Method 1013 measures leakage current during thermal cycling and looks for instabilities.

5.11.4 Temperature cycling

Temperature cycling subjects the parts to mechanical stresses that result from dimensional changes. The important parameters of this test are the temperature extremes, the soak times, and the transition time. Appropriate extremes are usually the part storage temperature requirements. The conditions can be extended a bit, but compatibility with the materials should be considered.

5.11.5 Mechanical shock

The purpose of mechanical shock in qualification testing is to ensure that the part can withstand shock that might result from rough handling. It can also be used as a screen to ensure the integrity of bonded joints. Qualification and screening levels should be consistent. The shock pulse is specified in terms of peak amplitude (*g*'s) and pulse duration. It may be either half-sine or sawtooth in shape.

5.11.6 Temperature-altitude

The temperature-altitude test is performed to verify the part's ability to withstand the effects of reduced atmospheric pressure encountered in high-altitude flight. This would not be applicable for hermetic devices except to test a cover seal. Temperature-altitude chambers are used for this test, and they range in size from a few cubic feet to dozens of cubic yards to accommodate large equipment.

5.11.7 Explosion

These are used to determine whether a part when operating will ignite an ambient explosive atmosphere. This test is performed in a chamber that can withstand such an explosion. The chamber is vented to allow combusted gases to escape. Under flight conditions, explosive atmospheres can exist (fuel leaks, etc.), so this test is important.

5.11.8 Quality conformance

These tests, intended to determine whether parts are in compliance with specified drawing requirements, are conducted on both a 100 percent and a sample basis. Group A testing is usually done as electrical testing on a sample basis after the parts have completed screening tests, which include final electrical testing on a 100 percent basis. For microwave parts, the final electrical tests and group A tests are synonymous and are performed on a 100 percent basis.

Group B tests are conducted on a sample basis to verify that assembly flaws in the device construction or the package are not passing undetected. This is accomplished by environmental testing which is between screening and qualification in severity. They also include long-term life tests for die-related failures. These tests are not considered destructive. Since the production of most microwave devices does not lend itself to lot formation, artificial lots should be formed with a reasonable size of samples.

5.11.9 Screening

Screening tests are designed to keep parts with assembly defects from being shipped. These parts usually fail during the early life of the equipment. The tests are considered nondestructive. Standard screening tests have been devised and documented as test methods in military standards such as MIL-STD 883, MIL-STD 750, and MIL-STD 202.

Chapter

6

Transmission Lines: Effects and Analysis

This chapter begins the discussion and analysis of transmission-line effects on high speed interconnections. Transmission-line effects may be referred to as *analog effects.*[1] They are an important part of signal integrity issues in high speed design applications.[2] This chapter describes the relationship of frequency, rise time, and line length. The physical and intrinsic components of transmission-line interconnects are described.

The intended applications are for digital interconnections implemented with at least two conductors separated by a dielectric. The conductors are often called *wires* in a cable, or may be printed circuits, ground planes, or shields. Although the low-frequency currents travel primarily in the conductors, high-frequency signals and transients travel in the dielectric. An important part of transmission-line design is to restrict the influence of the high-frequency signals and transients to the region near the intended conductors to avoid undesired coupling with other areas. Undesired coupling is often referred to as *crosstalk* or *radio-frequency interference* (rfi). Shielding and grounding are an integral part of a high-frequency transmission-line interconnection system to control undesired coupling for both interference emissions to other systems and susceptibility to emissions from other sources.

This chapter is intended to apply to binary digital signals transmitted at rates in the range from about 1 MHz up to 1 GHz, but the general principles may be extended to apply beyond this range as well as to other signals and other interconnection technologies. Typical applications for signal transmission at these rates are supported by the continuous development of silicon integrated circuit technology to increase the integration density and data rate capability. The increase in switching speed of advanced Schottky technology compared to earlier forms of logic devices made it necessary to consider the transmission-line effects for most applications of significant size, even at modest data rates. Many system engineers have discovered to their dismay that simply substituting a more recently manufactured part into an existing design may cause system malfunction. The newer part may carry the same generic part number or may be a pin-for-pin replacement for an obsolete part. The system does not operate properly even though the newer part is functionally equivalent and fully compliant to the older part specification, and the system clock rate remains unchanged. The engineer learns the hard lesson that the speed of the signal transition, called *rise time,* is often more important for system operation than the clock frequency.

The introduction of emitter coupled logic (ECL) led to very high speed logic that is designed to operate in a fully terminated transmission-line environment. ECL is a popular technology for large high speed supercomputers, signal processors, communications systems, and state-of-the-art instrumentation and test equipment applications. The relentless reduction in feature resolution has driven the technically simpler complementary metal oxide semiconductor (CMOS) transistor technology to dramatically denser levels of integration, as well as into the high speed range to replace advanced Schottky TTL and even rival the ECL technology of just several years ago in some applications. In addition to supporting advanced technology for industrial, military, and aerospace applications, CMOS integrated circuits supply most of the semiconductor technology for the wide range of electronics in the consumer marketplace, including personal computers and telephones, household appliances, automotive controls, compact disks, and digital television.

6.1 Basic Definitions

Transmission-line analysis is based on circuits that have "distributed" parameters. This means that the dimension of length (distance) separating circuit elements is an important characteristic affecting the performance of the electric circuit. The characteristics and signal levels are not the same at different locations along the same signal conductor. The distances separating such traditional components as resistors, capacitors, diodes, transistors, and integrated circuit chips influence their operation.

This section is primarily concerned with transmission lines with at least two conductors to carry binary (two-level) digital signals that may remain at a static level, or change state at up to a maximum data rate or clock frequency. A general digital interface includes data transmitted in simple *non-return-to-zero* (NRZ) format with a clock that is used to detect the data at a predetermined time in the bit interval between potential changes in the signal level. Other pulse coding techniques have been developed to provide characteristics such as elimination of the dc component, clock recovery from the data itself, and error detection.[3]

Although some digital systems are implemented with a clock which runs continuously at a fixed frequency, the general case includes clocks that may operate at various frequencies or may be halted. For example, typical high speed systems include built-in self-test (BIST) that requires that the clock be halted to switch between low-frequency interface to the test system, high-frequency self-test, and high-frequency system operation. Some high-frequency systems (for example, CMOS technology) may use clock switching to conserve power consumption during periods of reduced processing activity.

The wired transmission-line interconnection is a low-pass frequency function, with bandwidth from dc to beyond the maximum data rate or clock frequency. This section is generally not concerned with electrical power, telephone, or video distribution, or the use of a hollow waveguide for very high frequency energy transmission, optical fibers for transmission of light energy, or antennas to transmit or receive signals through free space (even though these applications use many of the same electromagnetic principles). Distance and transmission-line impedance may be considered to be similar to the influence that such parameters as resistance, voltage, and temperature may have on ordinary "lumped" circuit performance.

The distance that becomes important in analyzing circuit performance is not an absolute value but becomes more significant as the frequency becomes higher relative to the distance. "High speed" is a relative term whose meaning may vary for different applications and may change with technology. However, the following are rough order-of-magnitude (ROM) guidelines for considering transmission-line effects in wired interconnection networks:

- 1 ns or less transition times
- 10 cm (4 in) or more line lengths
- 1 MHz or greater clock frequency or data rates

It may be useful to combine these into a high speed figure of merit, where shorter rise times, longer line lengths, and higher frequency increase the concern for transmission-line effects. This figure of merit is expressed as follows:

$$\text{Figure of merit} = \frac{\text{frequency (MHz)} * \text{length (cm)}}{t_{\text{rise}}\,\text{(ns)}} \tag{6.1}$$

This figure of merit does not have a critical value below which transmission-line effects can always be ignored, but values above about 10 to 100 usually indicate that the interconnection design should be concerned with controlling transmission-line effects on signal integrity.

6.1.1 Rise time, overshoot, and undershoot

Although data rate or clock frequency is the term most commonly used to refer to the speed at which circuits perform, it is usually not the clock frequency that is most critical for digital circuit analysis of signal integrity. Maximum clock frequency itself is not a common cause of ringing and signal distortion. However, increasing clock frequency does reduce the time available for signals to approach steady levels. Therefore, transmission-line effects and waveshape distortion should be considered in the analysis of interconnection systems operated at high data rates.

The *rise time,* or the rate at which digital signals change from one state to another, determines the highest-frequency components of digital signals. The signal transition time is often referred to as *rise time,* regardless of whether the signal transition is in the positive or negative direction. Likewise, *overshoot* is defined as a transient signal level beyond the steady-state level in the same direction as the signal transition of either polarity. For example, if the steady-state low level is ground, a signal level below ground on a negative transition is overshoot, sometimes further defined as *negative overshoot* to clarify the polarity.

Similarly, *undershoot* is defined as a transient signal level between the steady-state levels. Undershoot is usually the most severe concern because it is a signal reversal in the opposite direction to the signal transition. However, the term undershoot may also apply to an intermediate signal level (or step) before reflections raise the signal level toward a steady-state level. In either case, undershoot represents a signal that is less in magnitude than the final static signal levels, regardless of whether the level is more or less positive. Undershoot is less positive for the positive-going transition but less negative for the negative-going transition. Because an undershoot level is closer to the switching threshold than the static signal level, it has risk that it will not be received as the intended logic state which was generated by the signal source.

Overshoot does not usually cause any direct adverse signal performance. Reasonable overshoot is actually beneficial for reducing the transition time through the threshold region and increasing the noise margin of the signal, especially during the time that other signal transitions may induce transient noise. However, excessive overshoot can result in adverse performance or serious malfunction such as overstress or latchup of sneak paths. If semiconductor junctions such as transistors or diodes are driven into heavy conduction during overshoot, slow recovery time may impact subsequent speed response. Heavy reverse recovery currents may result in a serious noise pulse when recovery results in a sudden change in current. Excessive overshoot is typically followed by undershoot, so limiting overshoot is often an effective means to control undesired undershoot.

Actual signals tend to have rounded corners at the beginning of signal transitions and approach signal levels asymptotically, so the definition of rise time and signal levels is somewhat ambiguous. Therefore, the rise time of actual signals is defined at less than the full 0 to 100 percent excursion between final steady-state levels, although the full excursion is used for most computer simulations of signal transitions. The 10 to 90 percent levels have been a popular rise time definition, especially for charging a resistor-capacitor circuit. However, the 20 to 80 percent levels are more practical and consistent for high speed signals that tend to have severe departure from linear waveform segments, especially when bandwidth limits and/or skin effects are significant.

Examples of the 20 to 80 percent rise time, overshoot, and undershoot are illustrated in the two complementary waveforms of Fig. 6.1.

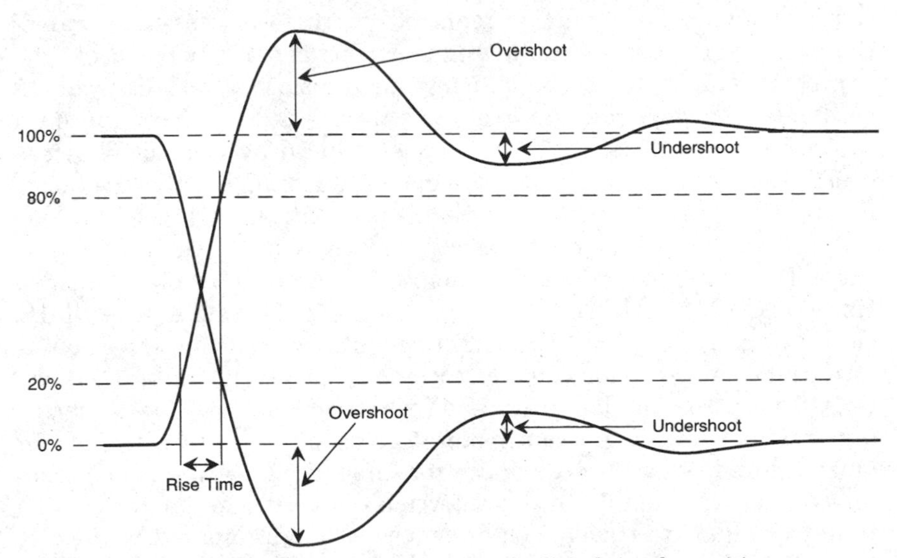

Figure 6.1 Rise time, overshoot, and undershoot of a digital signal transition.

These waveforms illustrate underdamped ringing, so there is initial overshoot, followed by reduced levels of undershoot and overshoot as the ringing converges toward the (two) static signal levels. It should be noted that rise time and the overshoot and undershoot levels are with respect to the steady-state signal levels, not the unloaded or open circuit driver signal levels or any of the transient levels. Therefore, the defined rise time may be significantly less than 60 percent of the peak excursion of a signal with severe overshoot and ringing. It should be noted that the actual waveshape of a signal transition may be significantly different from that illustrated in this figure. For example, lines with very little high frequency loss will result in a waveform that tends to have discrete steps. Signals on lines with no load termination may have an additional step at locations that are not near the far end of the line.

6.1.2 Rise time and frequency

Rise time and frequency are usually somewhat related, because the rise time must be less than the period of the frequency. When complex data processing must be performed within a clock period, the rise time for the data signals should be much less than the clock period. For example, if the rise time is to be less than 10 percent of the clock period, the rise time requires at least about 10 times the frequency bandwidth as the clock frequency. For a technology example, if CMOS circuits operating at 3-V power supply are capable of 1-ns rise time to a high-impedance load, that rise time is 10 percent of a 100-MHz clock rate. However, even if the clock rate is less than 1 MHz, the same frequency bandwidth is required for the 1-ns rise time transition. The average power of the frequency components per unit of time is proportional to the rate at which transitions occur, but the frequency bandwidth per transition remains unaffected by clock and data rates. For example, if the frequency is reduced by a factor of n, the *average* power is reduced by the same factor of n, but the frequency bandwidth to process each transition is unaffected.

A 1-ns rise time is compatible with distances of less than a few centimeters (1 or 2 in) in common dielectric materials before transmission-line effects may become more significant. Therefore, a 100-MHz clock rate may be achievable without requiring transmission-line terminations (or time for line reflections to settle) for single silicon CMOS chips or small multiple chip hybrid packages with circuit dimensions within a few centimeters. However, circuit dimensions for large printed circuit boards and interconnections between them usually exceed this, as do interconnections between major units. The time for signals to travel such distances is long compared to the rise time. Therefore, these interface signals must include the capability to drive the transmission-line impedance, even if the receiving circuits are high impedance. Critical signals such as clocks must be terminated to maintain proper waveshape, and any unterminated lines must allow adequate time for their signals to settle after line ringing. Typically, interfaces traveling these longer distances are limited to clock rates nearer to 10 MHz, even though the input rise times may still be only a few nanoseconds.

6.1.3 Line length and frequency

When longer (e.g., 1 to 10 m or greater) wired interfaces of clock and data lines are operated above about 10 MHz, both the driving and receiving ends of the lines must be terminated to control both signal and crosstalk ringing. The time for the signal transition to travel long distances is long compared to the entire clock interval. The required termination at the receiving end of a line always consumes power, which adversely impacts the power source, thermal dissipation, and reliability. However, it should be recognized that even an unterminated (open) long line consumes power during the time that a signal transition travels the two-way round trip propagation delay time. For example, if the receiver is a high-impedance CMOS input without a parallel termination and the line length is not long relative to the clock interval, the line consumes power each time the signal changes state and the transition travels from the source to the load and reflects back to the source. Lines exactly terminated at the far end consume power at a constant level. Lines not terminated at the far end consume power at a level that depends on the line length and the rate at which signal transitions are launched on the line.

The data rates of very long lines are limited by the rise time degradation of wired lines. The rise time tends to increase in proportion to line length. The rise time degradation effect is similar to bandwidth limiting and is primarily caused by skin-effect losses and dielectric dispersion. Skin-effect losses attenuate the higher-frequency components more than the dc resistance. This results in a gradual approach to steady-state levels, even for lines that are perfectly terminated.

The waveshape is somewhat similar to *RC* charging, although underdamped terminations can partially compensate for these losses. The gradual approach to the steady levels tends to be more linear than the exponential approach for an *RC* charging circuit. The signal level gradually approaches the steady-state levels as the lower-frequency components finally increase the signal level.[4]

Distortion on a long line results not only directly from resistive losses but also because there is continuous reflection back toward the source. Energy from the leading edge of a traveling wave such as a digital transition is reflected back toward the source, giving the later parts of the waveform a "tail." However, a steady-state sine-wave train is also subject to the same distortion, but the waveshape remains sinusoidal. When energy is removed from the leading edge and added to the trailing edge of each sine wave, it appears that the resulting sine-wave train travels more slowly than the leading edge of a digital transition. This result is part of the reason that lower frequencies travel more slowly than high frequencies on the same line.[5] Another part of the reason for reduced propagation velocity at lower frequency is that as more of the low-frequency wave penetrates farther into the conductor, not only is the net resistance lower, but the dielectric constant inside the conductor is higher.

A fast rise time digital transition is rich in wide bandwidth frequency components. Unless all the frequency components of the transient waveform are transmitted at the same speed and with the same attenuation, they arrive at the far end of a long line with different phase and amplitude relationships. These nonuniformly altered frequency components combine to result in a different waveform at the far end. The observed propagation delay of a line depends on the waveshape transmitted and the sensitivity of the receiver.[4]

Dielectric dispersion means that the different frequency components of a transient travel at different rates, so they arrive at the load with a different time relationship than when generated at the driver. Typically, the higher-frequency components of transients travel faster, which results in rounding of corners of the waveform and slower rise time. Mixed dielectrics with different dielectric constants may allow parallel paths to propagate portions of the signals at different rates, which may exaggerate the frequency dispersion and waveform distortion.

6.2 Technology Trends

The advancement of electronics technology continues at a rapid pace. The following is a brief reflection on a few highlights of the past. Possibly an appropriate starting point is Augustin-Jean Fresnel, an early nineteenth century French mathematician who challenged the prevailing corpuscular theory that light was comprised of particles that traveled through the ether and were absorbed by the eye. Fresnel developed the mathematics based on the idea that light traveled as transverse electromagnetic (TEM) waves at different speeds, depending on the dielectric material.

Lee De Forest developed the triode vacuum tube in 1906. The introduction of the control grid provided amplification that began the electronics revolution.[6] The control grid is comparable to the gate of field effect transistors and analogous to the base of bipolar transistors. The

ENIAC computer, developed at the University of Pennsylvania in 1946, is generally regarded as the first large electronic digital computer. It contained about 17,000 digital vacuum tubes and weighed 30 tons. Engineers of that time dreamed of computers that would weigh "only" one ton! The Whirlwind computer was developed for the U.S. Air Force about 1950. Whirlwind operators developed a preventive maintenance technique of voltage margin stress testing to improve operating reliability. The plate voltage supply for the vacuum tubes could be lowered, so that weaker tubes would fail during test. Replacing the weaker tubes before a run would reduce the chance that tubes would fail during the calculations. Voltage margin or speed margin stress testing is also used to improve reliability by detecting weaker elements such as memory units or transistors, or to verify that designs have adequate performance margins.

The development of the transistor at Bell Laboratories in 1948 began the solid-state electronics revolution about four decades after De Forest developed vacuum tubes with a control grid. By 1960, Fairchild Semiconductor, Texas Instruments, and Westinghouse Electric had developed the technology to isolate multiple bipolar transistors and diodes on a single "chip" of silicon. Westinghouse developed a product line called "molecular circuits" which included a single flip flop, or 1 to 3 diode-transistor logic (DTL) gates per device. Texas Instruments developed logic devices that became the transistor-transistor logic (TTL) product line. Motorola developed the Motorola emitter coupled logic (MECL) product line that paced the industry for high speed logic performance.

Seymour Cray was a pioneer in the supercomputer industry. His early computer work stretches back to the Navy Tactical Data System (NTDS) Unit computers and the Univac 1103. The NTDS computer weighed just over 1 ton. Seymour Cray joined with William Norris of Westinghouse Electric to found Control Data Corporation in 1957, where he pioneered the first all-transistor 1604 computer. Cray continued supercomputer development at Control Data with the CD6600 and CD7600, then founded Cray Research in 1972 which produced the Cray supercomputers that led the industry in processing performance throughout the decade of the 1980s.[7] Cray later founded Cray Computer Corporation and developed dense packaging technology and bus architectures for very high speed performance. However, a market for the gallium arsenide (GaAs) processing modules did not result.

Much of the incredible increase in digital processing capability in the past decades has been achieved with the dramatic and continuing increase in the capability to integrate more memory and processing density into smaller spaces without a corresponding increase in power consumption. Increasing the semiconductor circuit density permits more complex functions to be interconnected within a small space that can be operated at high speed without the adverse impact of transmission-line effects. The density of more complex functions with parallel processing and pipelined stages has increased dramatically, while the clock frequencies within a single chip or small multichip hybrid have risen less rapidly.

The first large digital computers of the decade of the 1950s (which rapidly became obsolete museum displays) gave way to the silicon transistor computers and digital signal processors (DSP) with magnetic core memory of the 1960s. These were superseded by the early integrated circuit processors of the 1970s, although large systems still required the space of a large room and many kilowatts of power. These were superseded in turn by the integration of large CMOS memory and processors on single chips of the 1980s. The decade of the 1990s has achieved submicron resolution technology which is capable of putting millions of transistors and millions of bits of memory on a single silicon chip, which in turn is capable of operating reliability at clock rates of over 100 MHz. It is interesting to note that the finer resolution of deep submicron photolithography is limited by the wavelength of the light used to define the circuit features.

However, the clock frequencies for data transfer between larger units have risen very slowly over these decades. Although the transistors for driving the signals have improved performance, the speed of signal propagation remains limited by the speed of light, which is about 3×10^8 m/s in empty space. The speed of TEM propagation is about 50 percent of that in typical printed circuit board material and wire insulation. The speed of light is a significant constraint on the performance of technology—from the resolution of integrated circuit features to the operating speed of interconnections between large system units.

6.3 Transient Currents

The dynamic transmission-line impedance is seldom above the 50 to 100 Ω typical of most high speed wire lines. Edge speeds of 1 ns rise time mean that transmission-line effects become significant for line lengths of only several centimeters (a few inches). Although power supply voltage has dropped to 3 V for many of the submicron CMOS outputs, it still requires 60 mA to drive a full rail-to-rail signal into a 50-Ω line. Even ECL and other lower voltage interfaces that only switch approximately 1-V signals require 20 mA of current to drive the signal transient into a 50-Ω line.

Even when the driver transistors can switch the transmission-line current in 1 ns, the transients induced into power and ground systems and the package leads are significant. For example, if the power and ground package pins each have about 10 nH of self-inductance, one output switching 20 mA in 1 ns will cause up to 400 mV of transient drop in the power supply voltage to the driver. If multiple outputs drawing power from common power and ground pins attempt to switch simultaneously, the power supply voltage will collapse to a level that limits the transient current to all drivers. Therefore, some lines switching will change the signal timing of other lines or may corrupt signal lines that should have remained static or even upset the logic state of memory elements sharing the power supply voltage. The design of high speed transmission-line interfaces requires consideration of the electrical effects at the ends of the line, as well as the line itself.

6.4 Transmission-Line Components

A transmission line will be considered to be a dielectric between at least two conductors that are intended to transfer a dynamic signal from a source at one location to at least one other load at another location. The source is often called a *driver* (power systems often call the source a *generator*), and the load is often called a *receiver.* For digital applications, the transmission line is ordinarily intended to transfer information, which is often called *data.* Other information such as clocks or format controls may also be transferred with the data on separate transmission lines or may be encoded in the data and recovered by processing of the received data. Decoding a clock frequency from encoded data is called *clock recovery* and typically uses a *phase locked loop* (PLL). A PLL may also be used to maintain synchronous timing between units when the data rates are high compared to the length of the line, or for fault tolerance or noise immunity.

6.4.1 Hollow waveguides, free space radiation, and optical fibers

All transmission lines can be considered as waveguides that direct signal energy from the source to the intended load or receiver(s). The common usage of the term *waveguide* means a hollow waveguide designed for high-frequency propagation in the dielectric inside the waveguide. Round waveguide can be considered as analogous to coax, with the center conductor replaced with the capacitance of the dielectric. The more common rectangular waveguide can be considered as two flat conductors representing the two broader sides, but the shorter sides are spaced far enough away so that at high frequency they are inductive.[8] Rectangular waveguide can be considered as somewhat analogous to a stripline conductor between ground planes. The equivalent two-terminal pair circuit of round waveguide has capacitance in series with the line inductance, so it does not pass low frequencies. Equivalently, the sidewalls of rectangular waveguide result in inductance in parallel with the dielectric, so low frequencies are shorted and do not propagate.

Of course, high-frequency energy may be radiated throughout the dielectric of free space. In this case the source is commonly called a *transmitter,* and the signal is coupled to the dielectric with an antenna. A receiving antenna captures some of the energy in the dielectric for detection of the signal. The energy density is rapidly reduced for radiation into free space, approximately as the distance cubed for radiation from a point source, because the volume of a sphere increases as the cube of radius, or about as the distance squared for a directional antenna beam. Microwave signal transmission is typically appropriate for long-distance line-of-sight links at frequencies over about 1 GHz.

A hollow waveguide of dimensions appropriate for the transmitted frequency will contain nearly all of the signal energy, so the signal can be delivered directly to the desired load without much loss. Furthermore, waveguides also isolate external signals, so the desired signal is less corrupted by undesired signals. The transfer of the desired signal from the source to the receiver is more efficient than for signals that are transmitted through free space. The division between coax and hollow waveguides for signal transmission is determined primarily by the physical size for the frequency to be transmitted. A hollow waveguide is capable of transmitting higher power levels at less attenuation than coax cable of comparable dimensions. Hollow waveguides must be large enough so that the dimensions are comparable to the full wavelength of the frequency to be transmitted. Therefore, they usually become physically impractical for frequencies below about 1 GHz because the free space wavelength exceeds 30 cm (1 ft). Conversely, the smaller size required for TEM propagation in coax and shielded twisted pairs, as well as the higher attenuation, results in a practical limit below a few GHz.[8]

Very small fibers designed to transmit energy at frequencies near the visible light spectrum are called optical fibers, or the interconnect network is referred to as fiber optic. Fiber optic transmission lines offer the capability to transmit much more information in less space over longer distances than wire conductors. Furthermore, fiber optic cables can be less vulnerable to crosstalk and other electromagnetic interference.

Hollow waveguides, microwave radiation, and fiber optics are typical technologies for transmission of power and signals at very high frequency. Unlike ordinary two-conductor transmission lines, they do not transmit dc and low frequencies. Therefore, arbitrary digital data must be encoded to maintain a minimum activity level or may modulate a carrier frequency within the frequency range of the line. However, the scope of these chapters is limited to transmission lines with two signal conductors that provide interconnection networks at frequencies from dc up to the practical limit of about 1 GHz.

6.4.2 Coax and twinax

A single conductor inside a round waveguide is commonly known as coax; the outer conductor is called a shield. The initial signal transient still travels at the speed of light in the dielectric between the inner conductor and the shield, but steady-state current can flow in the two conductors after the transient has established a new voltage level in the dielectric between the conductors. The two conductors carry the current loop that can transfer a dc level directly to the receiver.

A pair of conductors may be enclosed in a shield, with differential signals of opposite polarity on the two inner conductors. Part of the initial signal transient establishes the new voltage level between the two inner conductors; part of the signal is established between each of the two inner conductors and the shield. Transmission lines with two conductors inside a shield may be called shielded pairs or twinax. The load or termination resistor may be connected between the two inner conductors so that very little static current flows in the shield. Transient currents circulate in the shield as the rise time of the transition travels along the line. The two conductors are usually twisted to reduce coupling between other signals as well as to maintain constant line-to-line spacing so that dynamic line impedance is controlled within reasonable tolerances. Twisted pairs within a shield are commonly called shielded twisted pairs, which may be designated as an STP line.

6.4.3 Open lines

Conductors do not need to be surrounded by a shield to transfer signals from a source to a load. Two individual conductors reasonably close to each other (one may be a ground plane) can transfer signals at frequencies from dc up to a maximum that depends on the spacing between the two conductors. The impedance between these two conductors will be lower than free space, so most of the signal energy remains captured between the conductors, which guide most of the signal from the source to the load. However, some of the transient signal (particularly the higher-frequency components of faster transients) may radiate away from the conductors or may couple into nearby conductors. Similarly, other transient signals may couple more undesired noise into the signal than if the conductors were fully shielded. Signals that are transmitted on conductors between ground planes (called printed circuit stripline) are partially shielded. The shielding is very effective if the conductor height from the ground planes is much less than the spacing to adjacent conductors and the distance to the edge of the continuous ground planes.

6.4.4 Basic transmission lines

A basic transmission line consists of a single line connecting one source to one load as illustrated in Fig. 6.2. Actual transmission systems may be highly complex, with such characteristics as multiple sources which must be properly timed (or *synchronized*) to avoid conflicts, often called *contention.*

The transmission line may contain components of different impedance, and branches or stubs to multiple loads, and may be subject to coupling to other lines. A more complex transmission line with multiple receivers and coupling to another line is illustrated in Fig. 6.3. Some interface circuits may be capable of both transmitting and receiving signals; these are often called *transceivers,* or bidirectional (sometimes shortened to "bi-di," pronounced "bye-dye") interfaces. An ordinary transmission line includes both the capacitance and inductance of the dielectric, as well as lossy components such as resistance and leakage. External components may add lumped capacitance, resistive loads, termination resistors, or diode clamps that affect transmission-line operation.

When a transient signal arrives at the load end of a line, the load resistance may not match the line impedance. The response to this

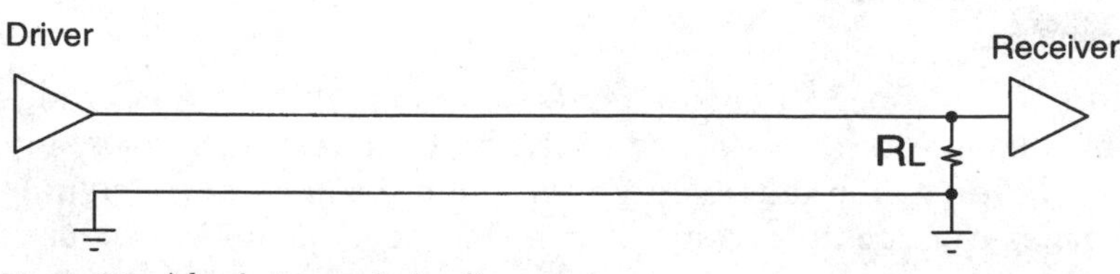

Figure 6.2 A basic transmission line.

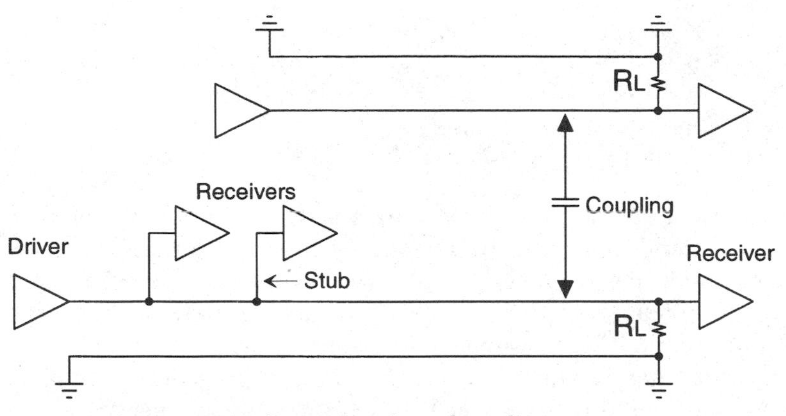

Figure 6.3 A transmission line with stubs and coupling.

mismatch travels back to the source, also at the speed of light in the dielectric. Unless the source is matched to the line impedance, further adjustment iterations occur until the source and load conditions are compatible. It is possible that the iterations result in conditions that adversely affect operation. For example, a power surge in a large interconnected power distribution system may result in shutdown if transient conditions exceed acceptable limits. Overshoot on unterminated digital lines may result in signal levels that degrade operation or may combine with subsequent driven signals to exceed maximum ratings.

6.5 Basic Equations

The basic equation for the characteristics of a two-conductor transmission line that propagates dynamic signal in the dielectric applies to the simple mode called transverse electromagnetic (TEM). TEM essentially means that both the electric and magnetic fields are perpendicular (transverse) to the direction of travel. Therefore, there is no electric or magnetic field component in the direction of travel. Of course, to the extent that there are finite losses along the line, there will also be a small electric field component in the direction of travel.

TEM is the mode by which electromagnetic energy is transmitted through empty space at the speed of light. Other dielectric materials transmit energy at lower speed. Higher-order modes may exist for a particular conductor geometry (for example, a hollow rectangular waveguide) that will transmit energy only at limited frequencies and at lower speeds (the group velocity) than TEM propagation in empty space. Higher-order modes must be avoided for digital signal transmission so that the signal is received with as little loss and wave-

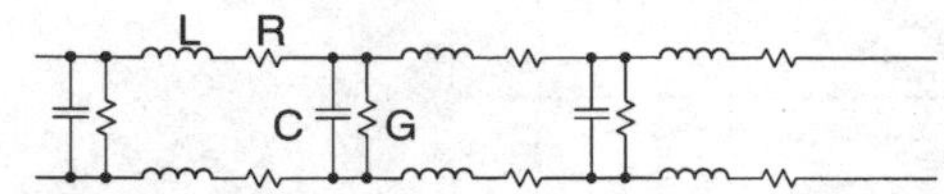

Figure 6.4 Elements of a two-conductor transmission line.

shape distortion as possible. This is ordinarily accomplished for two-conductor transmission lines by keeping the separation between the conductors much less than a quarter wavelength for the significant frequency components of the signals being transmitted so that TEM propagation dominates. However, digital signal transitions may contain some very high frequency components that may not propagate uniformly along the line or may be coupled into nearby lines.

The basic two-conductor transmission line is considered to have series and shunt impedance elements as illustrated in Fig. 6.4. The combined series impedance is designated as Z with no further designation. The shunt impedance is ordinarily defined as the inverse of impedance, called admittance, and is designated as Y. These are distributed parameters, defined per some convenient unit of length, such as inches, feet, or miles in English units; centimeters, meters, or kilometers in the metric system.

In general, $Z = r + j\omega l$ is the series impedance, and $Y = g + j\omega c$ is the shunt admittance, where r, l, g, and c are line resistance, inductance, conductance, and capacitance per unit of length, $j = \sqrt{-1}$, and ω represents radian frequency, where $\omega = 2\pi f$ and f is ordinary frequency. The series resistance r primarily represents the series resistance of the conductor, including skin resistance at the frequency components of the signal transition. The frequency is assumed to be high enough that essentially all of the current flows on the outer surface of the conductors, so the inductance and capacitance internal to the conductors can be ignored for transient analysis. Therefore, the l, g, and c terms are associated only with the dielectric in the space between the conductors.

The inductance l and capacitance c are intrinsic characteristics of the electric permittivity and magnetic permeability of the dielectric (l is not the inductance of the conductor, which is normally nonmagnetic). For most uniform dielectric materials, l and c are reciprocal parameters, which means that their product is a constant that defines the minimum propagation delay through the dielectric. The presence of closely spaced conductors reduces the inductance and increases the capacitance of the dielectric compared to free space.

g is the shunt conductance, which represents the dielectric leakage between the signal conductors. For usual conductor spacing and dielectric materials at frequencies used for digital data transmission, the leakage (which may include any radiation from open lines) is so low that g can be ignored compared to $j\omega c$ at sufficiently high frequency. However, it is important to consider if any dielectric absorption bands may exist within the signal bandwidth.

The resistance of ordinary conductors is not zero, and increases at higher frequencies due to *skin effect.* However, many common transmission lines have sufficient conductor size so that the resistance is low compared to $j\omega l$ at reasonably high frequency. Although both inductive reactance of the dielectric and conductor skin-effect resistance increase with higher frequency, the skin effect increases the high-frequency resistance of a circular conductor only about as $\sqrt{f}$. Therefore, inductive reactance dominates resistance at higher frequency, at least until very high frequency cutoff effects interfere with TEM propagation.

6.5.1 The dynamic line impedance

The characteristic dynamic impedance of a uniform transmission line is designated Z_0 and is defined as the ratio of a change in voltage to the change in current. The basic equations are

$$Z_0 = \sqrt{\frac{Z}{Y}} = \sqrt{\frac{r + j\omega l}{g + j\omega c}} \tag{6.2}$$

Low-frequency impedance. At very *low* frequency, so that reactance terms vanish compared with r and g, Z_0 approaches $\sqrt{r/g}$. If g times the length of the line is insignificant compared to the load impedance, then only any series resistance is considered for dc analysis. The steady-state static output signal amplitude is

$$V_{\text{out}} = \frac{V_{\text{in}} * R_L}{R_L + R_S + R_0}$$

where R_L is the load resistance, R_S is the total source resistance, and R_0 is r times the length of the line for both conductors.

Short line lumped capacitance. For a short line with no termination at the far end, if both r and g can be ignored, capacitance dominates. The transient analysis of a short line can be analyzed as if the total line capacitance C_0 of c times the length of the line (plus any lumped capacitance on the line) is a lumped capacitance.[4] This is often a more convenient method to analyze short line stubs as their equivalent capacitance rather than analysis of the reflections when the stub length is shorter than the signal rise time.

High-frequency transmission-line impedance. At sufficiently *high* frequency so that r is small compared to $j\omega l$ and g can be ignored, then Z_0 tends to approach $\sqrt{l/c}$. This is the ordinary form used for transient analysis. However, this common form is only an approximation. Although this common form indicates that Z_0 is purely real and independent of frequency, practical lines with series resistance will have complex impedance that will tend to have a capacitance component. Furthermore, many digital data transmission lines have short line stubs and lumped capacitance that further increase the effective line capacitance. High-frequency skin-effect losses will have a significant effect on signal waveshape, primarily increasing the rise time as a function of line length.

6.5.2 Transmission-line propagation

The basic equations for the propagation characteristic P_0 of a transmission line are

$$P_0 = \sqrt{Z*Y} = \sqrt{(r + j\omega l)(g + j\omega c)} \tag{6.3}$$

The propagation characteristic consists of two parts, both the imaginary propagation delay part and a real attenuation part due to losses.

When r and g are not significant at sufficiently high frequency, the propagation delay approaches a minimum of

$$t_{pd} = \sqrt{l*c} \tag{6.4}$$

Note that propagation delay is per the unit of length of l and c. The propagation delay may be slightly greater for lines with losses, but losses primarily affect signal rise time and amplitude. Although the propagation delay may be measured at some threshold level for the receiving device, the intrinsic delay of the above equation is only to the beginning of the arrival of the signal. Increased rise time will also add to the effective delay of a practical transmission line due to the rise time delay until the receiver detects the signal. The propagation delay of the receiver is in addition to the intrinsic and rise time delay of the line.

6.5.3 Line attenuation

The attenuation of the signal amplitude as it travels along the transmission line is approximately

$$a = \frac{r*c + l*g}{2\sqrt{l*c}} = \frac{r*c + l*g}{2t_{pd}} \tag{6.5}$$

When g can be ignored because the dielectric losses are insignificant, the attenuation equation reduces to the convenient form of

$$a = \frac{r}{2Z_0} \tag{6.6}$$

Note that this is attenuation per unit of length of r and that skin effect increases r for the higher-frequency components, which limits signal rise time. For example, when the total loop resistance is equal to Z_0, 50 percent of the initial signal amplitude is lost. Also note that r is the series resistance of both conductors (the loop resistance) per unit length, not just the resistance of a single conductor. However, for some applications, such as a single wire over a ground plane, or inside a coaxial shield, the resistance of the single wire may be much greater than the ground or shield. However, a common ground may share return currents from other signals. It should also be noted that when differential signals are transmitted on twisted pairs, the twist interval increases the electrical length of the wires compared to the physical length.

Skin effect concentrates high-frequency current (rapidly changing currents flowing in the same direction in the same conductor) toward the outer surface of the conductors, including crowding toward the opposite edges of thin conductors. This concentration of current density increases the resistance losses, especially for very high frequency where the surface roughness and conductivity, as well as contact between multiple strands of stranded wires, affect resistance.

There is also a similar effect for closely spaced conductors, called *proximity effect,* which concentrates the current toward another conductor carrying current in the opposite direction. Proximity effect is present even at low frequency, including dc. There is little distinction between proximity and skin effects in coaxial cable because both concentrate current toward the outer surface of the inner conductor, which is equally spaced to the shield. Thin conductors such as printed circuits and shields are affected less by proximity effect, due to their higher proportion of surface area to cross section. Stranded wires and braided shields have additional high-frequency resistance due to the relatively poor contacts between individual strands along the surface (in essence, a very rough surface). This resistance may increase with time if the surface conductivity of the individual strands deteriorates to higher resistance as compounds such as oxides or salts of the conductor form on the surface.

6.6 References

1. Zaid Ayoub, "Beware of Analog Effects in PC-Board Conductors of Fast Digital Systems," *EDN,* Jan. 18, 1996, p. 115.

2. James E. Buchanan, *Signal and Power Integrity in Digital Systems: TTL, CMOS & BiCMOS,* McGraw-Hill, New York, 1996.
3. *INTERFACE: Line Drivers and Receivers Databook,* National Semiconductor Corporation, Santa Clara, Calif., 1992 ed., sec. 6, Application Note 808.
4. Richard A. Matick, *Transmission Lines for Digital and Communication Networks,* McGraw-Hill, New York, 1969, chap. 5.
5. Hugh H. Skilling, *Electric Transmission Lines,* McGraw-Hill, New York, 1951, chap. 14.
6. Jacob Millman, *Vacuum-Tube and Semiconductor Electronics,* McGraw-Hill, New York, 1958, chap. 7, p. 156.
7. C. H. Chen (ed.), *Computer Engineering Handbook,* McGraw-Hill, New York, 1992, chap. 17.2.
8. Hugh H. Skilling, *Electric Transmission Lines,* McGraw-Hill, New York, 1951, chap. 13.

Chapter

7

Ideal Transmission Lines

This chapter deals with the analysis of transmission lines that are ideal in the sense that losses and distortion can be ignored. This does not mean that losses and distortion in practical high-speed digital transmission lines can be ignored but that the analysis is more reasonable. Losses and distortion significantly affect the waveshape of high speed digital signals on long lines, but the basic propagation of the fundamental information is nearly equivalent to propagation on lossless lines. It will be shown later that an ideal line by itself is not ideal to the extent that it is highly desirable for the transmission of information. Losses such as termination resistors or clamping diodes must be added to transmission lines to absorb energy to limit excessive current and waveshape distortion caused by ringing on the line.

There are limits to the rate at which information can be reliably transmitted over a given line, but these limits cannot be determined by the equations for ideal lines. However, many high speed interconnections are operated over lines that are so short that losses are not significant. The signal integrity issues of ground bounce, power droop, overshoot, ringing, crosstalk, and line terminations are often of major concern on shorter high speed transmission lines.

This chapter provides the basic equations for common forms of transmission lines and describes the virtual ground effect, both for wires near a ground plane and for differential pairs. Control of interference from adjacent signals, called crosstalk, is discussed for twisted pairs and flat flex cables or ribbon cables. An example of the application of a linear analysis of the impedance equations is included to provide sensitivity analysis to changes or variations in the parameters affecting transmission-line impedance. The conditions for transmission-line propagation are summarized.

7.1 Ideal Lines

If a transmission line meets the special relationship $r*c = l*g$, it is called an *ideal line.* This relationship may be rewritten as $Z_0 = \sqrt{r/g}$, which is also the dc impedance. When this relationship is true (at least over the frequency bandwidth of interest), the following results apply:

- Z_0 is purely real and independent of frequency.
- Propagation delay t_{pd} is minimum and independent of frequency.
- Attenuation $a = \sqrt{r*g}$ is minimum and independent of frequency.

An ideal line transmits all the frequency components of a signal waveform alike, so that the driven waveform is preserved without distortion at the receiver, even if attenuated.

7.1.1 Lossless line as an ideal line

A lossless line is a special case of an ideal line; because if both r and g are zero, there is no attenuation and no distortion. It is common to refer to lossless lines as ideal lines.

Practical lines, of course, are not lossless. However, ignoring losses simplifies the analysis of transmission-line propagation. Lossless analysis provides a reasonable approximation to transmission-line impedance and propagation delay but does not adequately predict waveshape for most applications. Lossless delay is a common method used to simulate the performance of circuits with transmission lines.

Closely spaced lines (which are necessary to propagate high frequencies) with low leakage are not exactly ideal. They have $rc >> lg$, and it is not usually practical to add inductance to balance the intrinsic line capacitance. However, ferrite beads (often called *shield beads*) may be added to critical lines such as digital clock distribution to control waveshape. In this case, the ferrite beads also provide lossy attenuation of the higher-frequency components to reduce ringing.

Practical transmission lines have finite resistance, skin effects, and lumped capacitance loading. Skin effect increases resistance and slightly reduces inductance for higher-frequency components. In thin rectangular conductors (such as printed circuits), skin effect crowds the current transients to the conductor edges, increasing the capacitance fringe fields. Dielectric absorption losses and radiation effects may become significant at very high frequencies.

Even though practical lines are not ideal, most transmission lines used for digital data transfer are short enough that losses are so low that the ideal line parameters are adequate to analyze the basic impedance and delay characteristics. Loss of amplitude is usually insignificant for shorter lines but may be estimated for longer lines if the dielectric losses are ignored. Rise time and other waveshape characteristics of the signal at the output of a transmission line are not as easily analyzed with mathematical models.

Exact analysis of the transient step response of either the dielectric or the conductor is difficult, but the results for most ordinary applications can be considered in the design analysis. Observation and measurement of the operating performance of an initial model should be performed to verify adequate performance margins. Corrective action is needed if there is not adequate performance margin to assure reliable operation, including allowances for production variability and the operational environment.

7.1.2 Low-loss approximations

The ideal line approximations for low-loss line equations are the following:

$$Z_0 = \sqrt{\frac{l}{c}} \qquad t_{pd} = \sqrt{l * c} \tag{7.1}$$

The propagation delay is a function of the dielectric and is about 3.33 ns/m in free space, or just slightly over 1 ns/ft. The velocity (or speed) of propagation is the inverse of propagation delay, or about 3×10^8 m/s, which is just under 1 ft per 10^{-9} s (ns) in free space.

The velocity in air is essentially unchanged from free space. Most transmission lines for digital data transmission use some other dielectric material, which often also serves as part of the mechanical structure. Examples of typical dielectrics include the insulation that separates wires in a cable bundle or the inner conductor from the shield of coax, the glass-epoxy (or other material) that separates conductors from ground planes in a printed circuit board, the ceramic substrate of hybrid packages, and the oxide layers that separate conductive (usually metal) interconnections of integrated circuits. Some dielectrics integrate as much air (or other inert gas with a low dielectric) as practical into the dielectric to reduce the effective dielectric constant, which increases propagation velocity. Examples include expanded dielectrics with air cells (also excellent thermal insulation) and air bridges in very high frequency integrated circuits.

7.1.3 Propagation as a function of relative dielectric constant

It is convenient to describe a dielectric by the electric permittivity relative to free space, called the relative dielectric constant. This is so common that the shorter term "dielectric constant" is generally understood to be the dielectric constant relative to empty space. The relative dielectric constant is designated as ε_r, so the impedance Z_0 decreases and propagation delay t_{pd} increases as $\sqrt{\varepsilon_r}$. Therefore, the more common form for the minimum propagation delay of an ideal line is

$$t_{pd} = 1.017\sqrt{\varepsilon_r}\ \text{ns/ft}$$

or (7.2)

$$t_{pd} = 3.33\sqrt{\varepsilon_r}\ \text{ns/m}$$

The total propagation delay for a length of transmission line is the propagation per unit length times the length, which is designated as $T_0 = t_{pd} * \text{length}$. Similarly, the total capacitance for a length of line is designated as $C_0 = c * \text{length}$. These are related as $T_0 = Z_0 * C_0$. The relationship to the common time constant of an RC circuit as $T = R * C$ may be noted. For applications where the capacitance of a length of line must be determined, it is conveniently computed as $C_0 = T_0/Z_0$. Similarly, for inductance $L_0 = T_0 * Z_0$, where both T_0 and Z_0 are easily determined by a time-domain reflectometer (TDR) measurement. A TDR instrument is convenient for accurate measurements, but a reasonably terminated source and a mismatched load (an open load is appropriate) can be used for TDR observations of the reflections returning to the source, using an ordinary oscilloscope.

Propagation delay is determined by observing the delay from the beginning of the launched wave until the beginning of the return of the reflected wave from the mismatched far end. The exact time of the beginning of the reflected wave is subject to some measurement uncertainty,[1] but this is not significant for ordinary measurement accuracy. The time is ordinarily defined to be the linear intercept of the more linear portion of the reflected transition with the level before the reflected wave arrives.

Line impedance may be measured either at the far end of the line or at the source. Line impedance may be determined experimentally at the far end by selecting a resistance value that most nearly eliminates reflection, or calculated from the observed reflection amplitude with a known termination resistor. This method may not be practical for some cases because it depends on the measurement accuracy of small values which may not be significant compared to line losses.

Likewise, line impedance may be determined at the source by selecting a source resistance that most nearly launches an initial signal level that is 50 percent of the static no-load signal level. Similarly, line impedance may be calculated from the measured initial signal level with a known source resistance as a fraction of the unknown line impedance. A laboratory time-domain reflectometer measures line impedance by comparison of the reflection at the interface of the line being tested to a known standard value, typically 50 Ω.

7.2 Line Impedance Equations

The dynamic impedance of a transmission line is determined by the ratio of the intrinsic inductance to the capacitance. The impedance of empty space to a TEM wave is about 377 Ω. However, the presence of conductors reduces the inductance and increases the capacitance (in inverse proportion, so that the $l * c$ product remains constant). This significantly reduces the impedance of most two-conductor lines. Typical values for the impedance of digital signal transmission lines range from about 40 to 100 Ω.

The dynamic impedance of two-conductor transmission lines is primarily determined by the relative dielectric constant and the ratio of spacing between conductors to the conductor size. The equations for some common types of line geometry in a uniform dielectric are given below. The round wire equations are based on high-frequency analysis, where the skin effect concentrates the current at the outer conductor surface but ignores the conductor resistance. The impedance reduces to zero when the conductors touch at $D = 2H$ (this is when

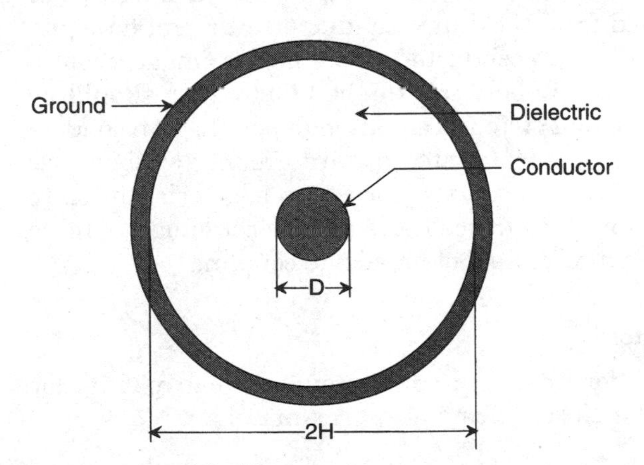

Figure 7.1 Coaxial cable geometry.

the distance H from the ground or shield is equal to the radius of the round wire). The microstrip and stripline equations for thin conductors with ground planes are empirical approximations (derived from measurements) that apply over a reasonable range of frequency and H/W ratio. H is the height of the center of a round conductor from a ground plane or the coax shield, or the height of a thin conductor above the ground plane. D is the round wire diameter. W and T are the width and thickness of thin conductors ($T < W$). The geometry and dimensions are illustrated in Figs. 7.1 through 7.4.
Figure 7.1, coaxial cable:

$$Z_0 = \frac{60}{\sqrt{\varepsilon_r}} \ln \frac{2H}{D} \tag{7.3}$$

Figure 7.2, round wire over ground:

$$Z_0 = \frac{60}{\sqrt{\varepsilon_r}} \ln \frac{4H(K_p)}{D} \tag{7.4}$$

Figure 7.3, buried microstrip over ground:

$$Z_0 = \frac{60}{\sqrt{\varepsilon_r}} \ln \frac{5.98H}{0.8W + T} \tag{7.5}$$

Figure 7.4, stripline between ground planes:

$$Z_0 = \frac{60}{\sqrt{\varepsilon_r}} \ln \frac{3.80(H + 0.5T)}{0.8W + T} \tag{7.6}$$

Figure 7.5 illustrates that two-conductor impedance in the absence of ground can be derived from the single conductor over ground equations because the fields for two conductors are a mirror image about a *virtual ground plane* midway between the conductors. It should be noted that the microstrip and stripline equations apply to thin conductors ($T < W$) with the broadside oriented toward ground (or the other thin conductor); they are *not appropriate* for a thin edge orientation to ground or other conductors. The fringe fields of edge coupling are more complex than the more uniform field of broadside coupling.

7.2.1 The proximity factor

The proximity factor K_p for round wire over ground is nearly unity for a large H/D ratio but for close spacing is approximately

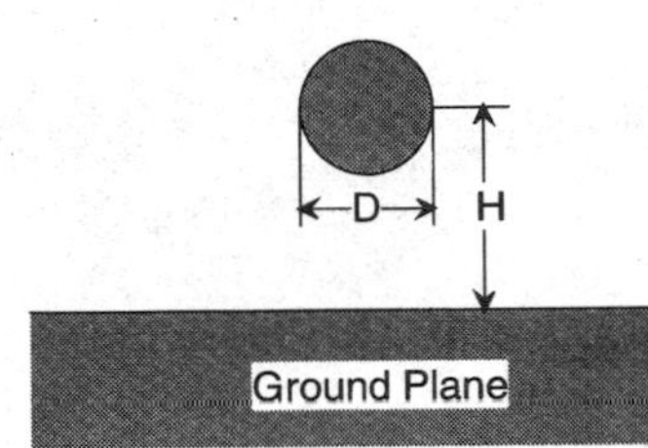

Figure 7.2 Round wire over ground geometry.

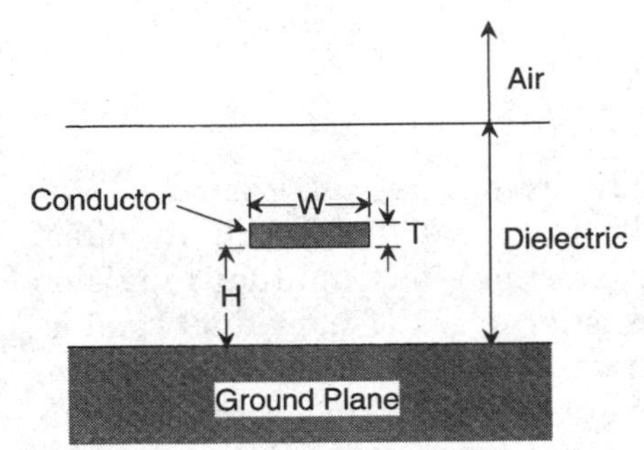

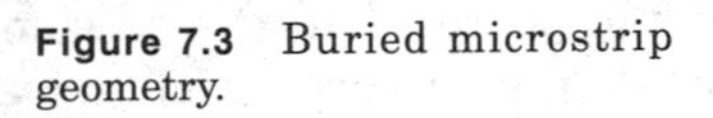
Figure 7.3 Buried microstrip geometry.

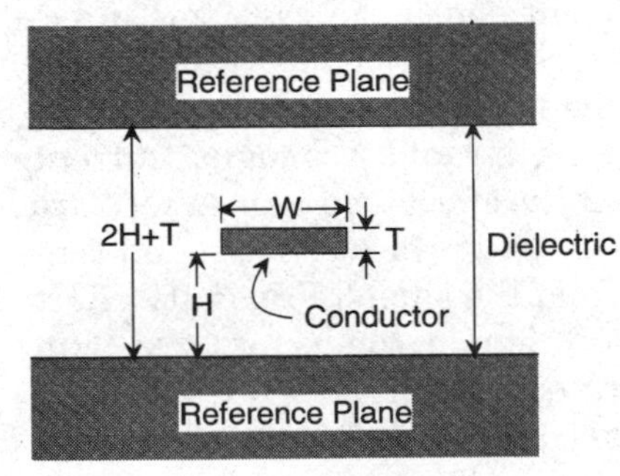

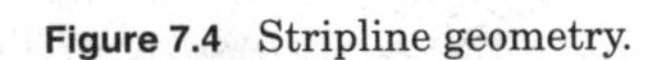
Figure 7.4 Stripline geometry.

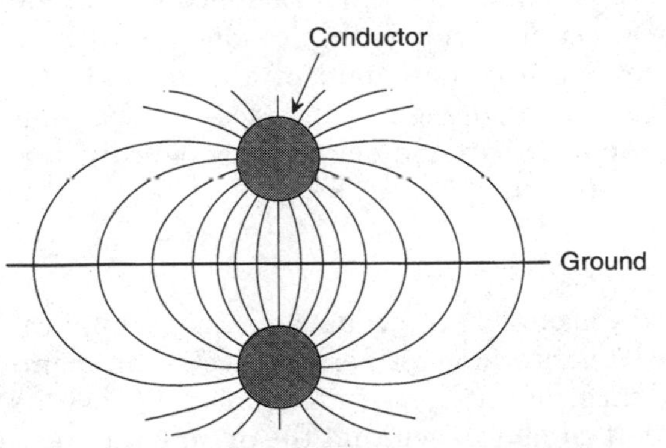

Figure 7.5 Two-conductor symmetry to virtual ground.

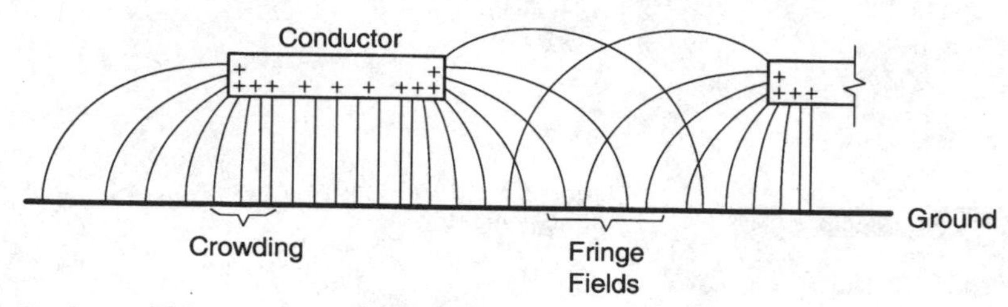

Figure 7.6 Thin conductor skin effect, crowding, and fringe fields.

$$K_p = \frac{1}{2} + \frac{\sqrt{4H^2 - D^2}}{4H} \tag{7.7}$$

K_p reduces to 1/2 when the round wire touches the ground at $D = 2H$, so then $\ln(1) = 0$, which results in $Z_0 = 0$. The proximity effect results from the same mechanism as skin effect; mutual repulsion drives like currents to the extreme outer edges of individual conductors, while mutual attraction drives unlike currents toward other conductors carrying current in the opposite direction. This crowds the current in round wires toward the side nearest a ground plane (or the return conductor). This is illustrated in Fig. 7.6 for thin conductors over a ground plane. + represents current density in the same direction. Current density in the opposite direction (not illustrated) is in the ground plane. Note that "current" here means high-frequency current, or for the common digital signal case, a rapidly changing current when the signal changes state. Proximity effect and skin effect are indistinguishable for coax because the entire surface of the round center conductor is at the same distance from the shield. Proximity effect is not ordinarily considered for thin rectangular conductors, but skin effect does drive the currents toward the edges of thin conductors.

As a rough approximation for ordinary proportions, most of the current in the ground plane under a thin conductor is contained in a region 3 times the width of the conductor (or one conductor width on either side of the conductor). A similar rule is that conductors with an open space between them of 3 times the height of a thin conductor above the ground plane will not have significant crosstalk because there is not significant overlap of the ground plane currents or the corresponding fringe fields in the dielectric.

7.2.2 Practical impedance

About 50 Ω is a typical and reasonable impedance for many applications and is a common instrumentation standard for test equipment, as well as the load impedance for high speed technologies such as emitter coupled logic (ECL). It can be shown that the optimum impedance for a coax cable of a given size with a relative dielectric constant near unity varies from about 75 Ω for minimum attenuation to about 30 Ω for maximum power transmission before dielectric breakdown occurs.[2] An optimum exists because a very small conductor size (*D* for coax, *W* for printed circuits) for a given *H* increases series resistance faster than Z_0, which increases losses faster than the transmitted signal amplitude for the same current. Conversely, a very large conductor size reduces Z_0 toward zero faster than resistance increases, so the signal amplitude is reduced faster than the losses for the same current. Therefore, losses are higher at both high and low Z_0 extremes, with an intermediate region where the ratio of losses to signal amplitude are near a minimum and not very sensitive to small changes in Z_0.

Of course, if *H* can be increased, the power capacity may be increased significantly because the voltage breakdown is much greater. High-voltage ac and dc power distribution over open lines on towers is an example of the use of very large spacing dimensions, but these are not suitable for high-frequency data transmission lines. The conductor diameter cannot be increased to the size that would minimize losses because the cost of the conductors would be prohibitive, and the weight of wires that can be supported by the towers is limited. It may be noted that for narrowband transmission of high frequencies (typically a modulated carrier), a hollow waveguide will transmit more power at the same frequency than a coax cable. However, hollow waveguides, microwave transmission links, and fiber optic lines cannot transmit low frequencies (or dc), so a minimum frequency or data activity must be maintained.

An H/W ratio of about unity results in nearly 50 Ω impedance for a single microstrip line over ground, or stripline between ground planes, for common printed circuit dielectrics. This is a typical compromise for printed circuits for digital applications. Because the impedance does not increase very fast for increasing *H,* board thickness would need to be increased significantly for small increases in Z_0 to reduce signal currents slightly. Conversely, spacing conductors much closer together reduces Z_0 more rapidly. Differential lines can have a line-to-line impedance that is nearly double the individual line-to-ground impedance if the individual conductors are spaced farther apart than the individual spacing to ground.

Although most digital data transmission lines are not long enough to be concerned for series resistance losses in the line conductors, the losses in the signal drivers and the power supply increase with signal current. Conductor width *W* for printed circuits is usually limited to the minimum that can be consistently produced for the available technology and performance requirements. Therefore, the usual compromise is to make the dielectric space *H* large enough to avoid high signal currents (to limit transient currents as well as the size and weight of the power supply) but small enough to limit the size and weight of the printed circuit board and the chassis and support structure. Larger-diameter holes interfere with conductor routing between pin fields. Board thickness is also limited by manufacturability for the number of layers and the diameter of plated through holes. Increasing board thickness increases the ratio of board thickness to hole diameter (referred to as *aspect ratio*), which limits the manufacturing ability to produce reliable plated through holes for the vias that interconnect between layers.

Equations (7.3) to (7.6) apply for reasonable dimensions, at frequencies that are high enough that skin effect concentrates currents at the edges of conductors but low enough that the dimensions are small compared to the quarter wavelength, and for only two conductors in a uniform dielectric, and ignore losses in either the dielectric or the conductor. Obviously, these conditions do not apply exactly for many common signal applications. For example, digital signals contain a very wide spectrum of frequency components. The lower-frequency components (including the dc component that maintains static levels) travel farther toward the interior of the conductor, and at lower velocity. The higher-frequency components of a signal transition experience higher losses and tend to radiate energy into adjacent space instead of being concentrated between the conductors. Losses and lumped capacitance loading are common causes of waveshape distortion in a uniform dielectric.

7.2.3 Mixed dielectric lines

The dielectric of high-performance lines is often not perfectly uniform or may be composed of more than one material. This is called a *mixed dielectric* and may result in waveshape *distortion* because portions of the signal travel in different dielectric materials (including the lower frequencies that travel inside the conductor), so they travel at different velocities and arrive at the destination at different times. The combined signals may have different phase relationships, which result in a different waveshape than that transmitted from the source. For example, microstrip on the surface of a PC board may have a coating layer above the board that has different dielectric properties than the PC board, and air with a lower dielectric above the coating. Another example is coax or twinax (twisted pair inside a shield, called *shielded twisted pair*) constructed with different dielectrics to provide desired electrical and mechanical properties. Air provides a nearly ideal dielectric for signal propagation, but mechanical support is required to maintain the conductor spacing despite

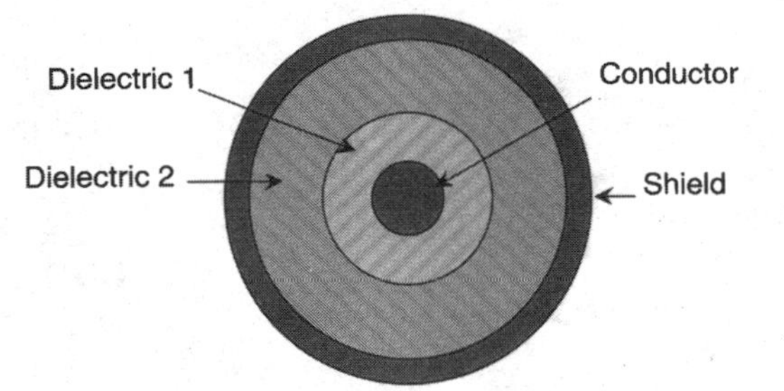

Figure 7.7 Complex mixed dielectric.

stresses such as bending or crushing. An example of a complex mixed dielectric designed for high-performance coax is shown in Fig. 7.7. The conductor is silver-plated copper to maintain low skin-effect resistance despite oxidation and minimize the opportunity for corrosion. The primary dielectric surrounding the conductor is an open-cell porous (called *expanded*) material to reduce the effective dielectric constant. The primary dielectric is surrounded by a solid material to provide mechanical strength for crush resistance for the softer porous material. The outer shield surrounds the outer dielectric to shield the signal on the inner conductor and provide the return current path.

7.3 Differential Pair Lines

A pair of wires, including the mixed dielectrics inside the shield as illustrated above, may be used to transmit differential signals. The wires are twisted as pairs and shielded to reduce interference (called *crosstalk*) and provide impedance control. The spaces between the wire pairs and the shield are also filled with air. When this cable construction is used in applications where the air pressure is subject to change (as in aircraft or space), it must be ensured that the cells in the porous dielectric are open and all spaces will vent when air pressure is reduced and refill when repressurized. The dielectric constant of air is so nearly that of free space (a vacuum) that there is no significant change in electrical transmission properties whether the empty spaces are filled with air or evacuated.

An example of the electric fields in a shielded twisted pair is illustrated in Fig. 7.8. The insulation surrounding each wire may be a complex mixed dielectric as in Fig. 7.7. It may be difficult to analyze the line-to-line impedance accurately by the basic equations defined above. However, the round wire over ground basic equation can be used for a reasonable preliminary estimate of one-half of the impedance between the lines. We can recall that there is a *virtual ground plane* at a line equally distant from the two conductors, as in Fig. 7.5.

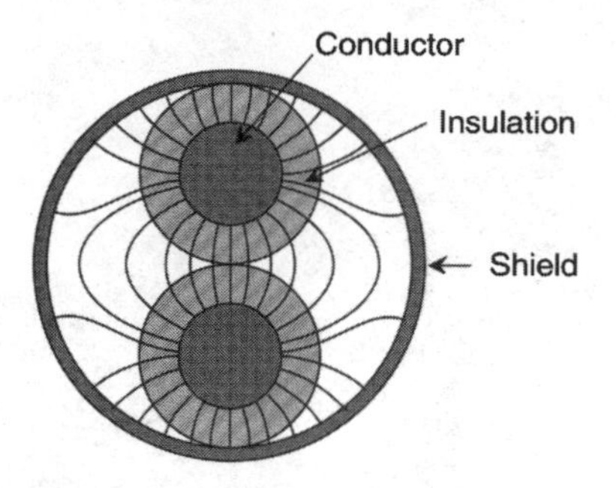

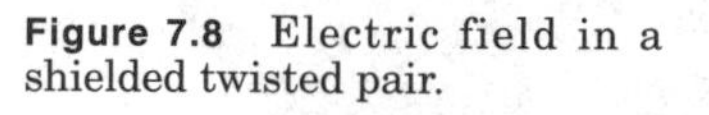

Figure 7.8 Electric field in a shielded twisted pair.

Therefore, if we substitute *one-half* of the distance between centers of the two conductors into the round wire equation, we have an upper bound estimate of one-half of the line-to-line impedance of the pair in the absence of the shield.

The presence of the shield reduces the net impedance to less than twice the round conductor to ground estimate. However, it should be observed that the minimum shield to conductor spacing is the same as the spacing to the virtual ground between the conductors. Therefore, the coaxial cable equation is a lower-bound estimate of the individual line-to-ground impedance, which is one-half of the line-to-line impedance. The line impedance between the two grounds is likely to be closer to the coax value than the wire over one ground plane. However, these two equations may be used to estimate the possible range of line-to-line impedance for shielded twisted pairs operated as a balanced differential transmission line.

Rule of thumb for a lower bound for shielded twisted pair differential impedance: double Eq. (7.3) (coax), with $2H$ equal to the outer insulation diameter of each wire.

The cable vendor for most cables purchased for high speed applications will have characterized the actual impedance in accordance with the particular manufacturing processes used in the cable fabrication. Major vendors of high speed cables and printed circuit boards should have access to computer software models that can estimate such transmission-line parameters as impedance, crosstalk, and propagation velocity consistent with their manufacturing processes.

Inspection of Eqs. (7.3) and (7.4) for coaxial cable and round wire over ground reveals that they are of similar form. They differ by ln 2 times the proximity factor. Since the proximity factor reduces to 0.5 as $2H$ approaches D, both equations approach the same limit of zero. At a more typical H to D ratio, $2K_p$ approaches 2, so the round wire over ground impedance is greater. For example, at $H = D$, the wire over a single ground plane impedance is about 88 percent greater

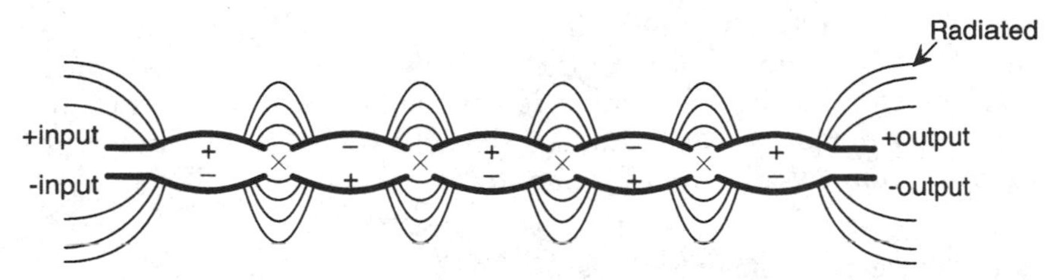

Figure 7.9 Cross section of twisted pair illustrating electric fields.

than the coax impedance. A similar comparison of the buried microstrip and stripline between ground plane equations indicates that the presence of the second ground plane reduces the impedance by somewhat less than a factor of 2 in the logarithm. It should be noted that the presence of any other conductors also reduces the impedance, even when there is no shield or ground plane. However, the shield reduces the impedance of twisted pairs in a consistent (even if difficult to analyze) manner and reduces interference (crosstalk) between signals.

A primary purpose for twisting wire pairs is that the positive and negative electric field polarities rotate as the signal travels along the twisted pairs. This means that the fields intercepting any fixed point at a sufficient distance away from the pairs nearly cancel because the positive and negative polarities approximately cancel. Figure 7.9 illustrates a cross section through a twisted pair line, with electric field lines between alternate turns of the wire pairs. Each line is labeled + or − at each turn to indicate that each polarity alternates. The external field is approximately canceled by the alternate turns along the length of the line. Similarly, any external field impinging on multiple turns of the twisted pair line should couple almost equally to both wires as *common mode* crosstalk. When the differential receiver is more sensitive to differential signals than common mode signals, it will reject the common mode and properly detect only the differential signal. However, the common mode coupling and common mode rejection of the receiver are not perfect and not without limit. It remains important to limit the common mode interference to less than both the common mode rejection capability as well as the common mode to differential mode conversion that will interfere with proper detection of the differential signal. As indicated in the figure, twisted pair lines are often most vulnerable at the ends of the cable, where there is no continuation of the alternate polarities and the pairs are typically run straight for interfacing to the connector.

An often overlooked characteristic of differential lines is that a typical differential receiver does not terminate common mode crosstalk even if it perfectly terminates differential signals. This means that only the driver can terminate (or at least dampen) common mode interference. The driver termination (usually called *source* or *back termination*) must be well balanced for both lines, as any unbalance will cause common mode conversion to differential crosstalk.

7.3.1 Parallel pair differential lines

Another common configuration for differential pair transmission lines is parallel pairs in printed circuits. The parallel lines do not have the cancellation properties of twisted pairs but may be more economical to construct for many applications. Either the buried microstrip or stripline configuration may apply, depending on whether only one or both ground planes are in close proximity. Additional grounded lines (usually called *guard lines*) may be placed adjacent to the signal lines to reduce crosstalk. A stripline pair is illustrated in Fig. 7.10. The $3H$ spacing between the lines is not essential but is recommended to reduce the mutual coupling between the differential pair lines. When the mutual coupling between lines is limited, the line-to-line impedance is primarily determined by the impedance of each line to reference plane(s). The reference planes may both be electrical ground, or one of the reference layers may be a dc power plane well decoupled to the ground.

The parallel pair configuration is used for several common interfaces. Thin, flexible printed circuits may be called *flex cable*. These are often used between assemblies that move with respect to each other or must be connected and/or disconnected within a limited space. A popular flat cable is often called *ribbon cable*. Ribbon cable is often an economical molded construction, which may include a metal foil layer to act as the shield. A foil shielded ribbon cable is illustrated in Fig. 7.11. The illustration includes ground wires adjacent to every signal wire (every third conductor) because the foil shield is not adequate as the signal return for high speed signal lines.

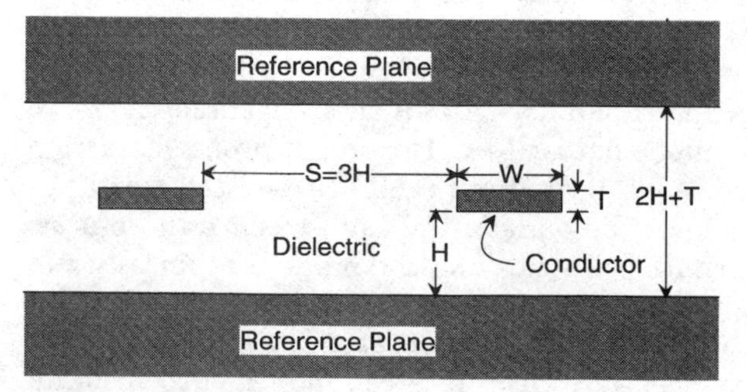

Figure 7.10 Stripline pair.

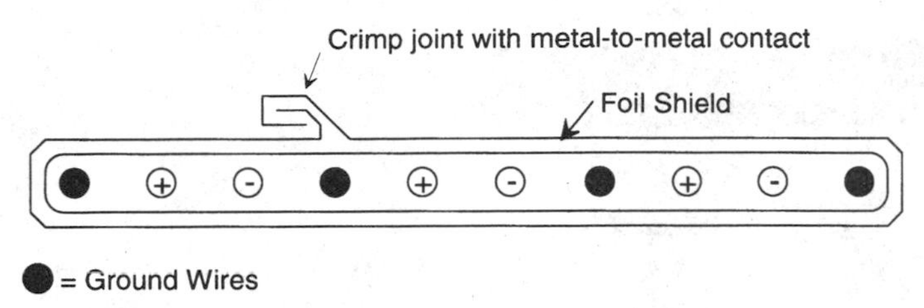

Figure 7.11 Ribbon cable with foil shield.

Many of the flat cables can be mass terminated by connectors that pierce the insulation and make physical contact with the individual conductors. This economical method of contacting the wires with the connector is called *insulation displacement connection* (IDC). Connectors may be placed at intermediate points along the flat cable, not only at the ends. Flat cables are not limited to parallel wires; groups of individual wires or cables can be physically attached together in various forms, including flat cables. One form is flat woven cable, which is tied together by the same technique as rug weaving. For example, the individual wires may include twisted pairs, shielded twisted pairs, and coaxial cables. These cables remain reasonably flexible and can be formed into various shapes and terminated to multiple connectors (sometimes called the *octopus cable form*) to replace several individual cables. Another form of flat cable is twisted pairs that are molded together. Often these have straight (untwisted) sections at intervals that allow for IDC connectors. These cables have most of the advantages of twisted pairs to control crosstalk and reduce external interference, while offering the convenience and economy of flat cables.

Low-performance unshielded ribbon cables may include only ground wires to provide the return conductor for single-ended signals. A ground wire every third wire provides a ground return adjacent to each signal wire, but crosstalk may be significant. Low-frequency data bus signals may be combined on several adjacent wires, with only a few ground return wires, but this configuration does not provide a reasonable transmission-line environment. Therefore, the data rate must be limited so that the cable length is not significant, because crosstalk and ringing of both the signal and ground lines will be significant until transient effects have subsided. Unshielded cables are also vulnerable to external interference and are notorious for radiating energy from digital signals with fast rise times into sensitive receivers.

As was noted for Fig. 7.10, if the space between conductors is at least 3 times the conductor width, there is little overlap of the fringe

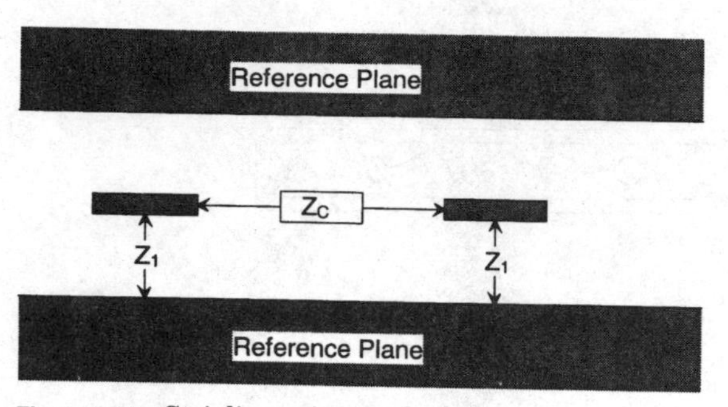

Figure 7.12 Stripline pair impedance.

fields (less than about 10 percent) and therefore relatively low coupling between lines, which results in high line-to-line impedance. If there is a reasonable space between the two signal conductors (needed to avoid a low Z_0), the impedance to ground for each conductor is less than the line-to-line impedance. The thin conductor over ground equation (for stripline or microstrip as appropriate) can be used to estimate the impedance of each line to ground. The net impedance of the differential pair will be less than twice the single conductor to ground impedance because there is some common impedance between the lines. This is illustrated in Fig. 7.12. Z_1 is the impedance from each conductor to the reference plane(s), and Z_C is the mutual coupling impedance between the lines. The greater the space between the conductors, the higher the common impedance between the lines, and then the line-to-line impedance approaches twice the single conductor to ground impedance.

An issue which should be noted for any guard lines added to reduce crosstalk is that they should be connected to the signal ground planes at sufficiently short intervals. Generally, the interval between connections to ground planes should be short compared to the distance that a signal travels during the signal rise time, so that they appear as short line stubs. These short line stubs are capable of conducting the circulating dynamic currents that flow in the ground plane and adjacent guard lines during the signal rise time. These circulating currents are induced in the ground plane and adjacent lines as a result of the changing currents that flow through the dielectric during the signal rise time.

7.4 Impedance Analysis

The equations for transmission-line impedance are not in linear form but include ratios, square root, and the logarithm of the ratio of dimensions. Analysis will indicate that for common ranges of the parameters determining line impedance, the square root and logarithm functions reduce the sensitivity to tolerances of these parameters. This means that, for example, a 10 percent change in a parameter that affects impedance will result in less than a 10 percent change in the impedance value.

A convenient analysis technique to transform the more complex line impedance equations into linear form is called *analysis of binary combinations* (ABC). This technique is based on the design of experiments (DOE) analysis.[3] This analysis is ordinarily used to analyze test results or to analyze the sensitivity of design performance to various design and environmental parameters. This DOE technique is based on the statistical analysis of variance (ANOV) and is the linear form of least squares line fitting, also called *linear regression.*[4] However, the analysis is also appropriate for fitting a linear model over a limited range of the parameters of a nonlinear function. The linear coefficients essentially determine the sensitivity of the function for a range of the parameter, as measured at the endpoints of the range. However, the sensitivity is not simply measured at two points, with all other parameters at some fixed level, as in *one at a time* analysis. ABC analysis also varies all the other parameters simultaneously. Therefore, one-half of all the data points are taken with each parameter at one of the endpoints, with the other half of all the data points with that parameter at the other endpoint. Analysis of k variables at two levels each is called a 2^k factorial design.[5] The analysis is structured so that the sensitivity for each of the parameters is the average difference between each half of the data points at each of the endpoints, while the other parameters are also varied over their endpoints. Therefore, this is *combinational analysis.*

An analysis of the sensitivity of line impedance to the four parameters in the buried microstrip equation was performed as an example. The selected center condition was the following:

$$H = 5 \qquad W = 5 \qquad T = 1.0 \qquad E = 5 \rightarrow Z_0\,(\text{center}) = 48\ \Omega \tag{7.8}$$

7.4.1 Results of the analysis

Each of the four parameters was varied to limits ± 20 percent of the center value, which is ±0.2 for T and ±1 for H, W, and E. The results for traditional DOE analysis would be the average change in Z_0 from the average Z_0 for all 16 combinations. The following sensitivity results are normalized to the average Z_0 and the percentage change of the parameter (± 20 percent for this case). The normalized results (rounded to the nearest percent) for the average sensitivity for this range of conditions are

$$\frac{Z_0}{H} = +57\% \qquad \frac{Z_0}{W} = -45\% \qquad \frac{Z_0}{T} = -12\% \qquad \frac{Z_0}{E} = -50\% \tag{7.9}$$

The average over the 16 combinations is only about 1 percent greater than the calculated center value. The estimated standard deviation of this linear model is less than ± 2 percent, so the linear model is reasonable to estimate line impedance over this range of parameters within a 3σ accuracy of about 5 percent.

The linear model predicts a maximum impedance of about 65 Ω to a minimum of 33 Ω over this range of values. The calculated result for all 16 combinations over this range of parameters was a range of a maximum of 66 Ω to a minimum of just under 34 Ω, very nearly identical to the linear model prediction. The calculated results at the extremes are a little greater than the model, and the average is a little greater than the center value, indicating that the Z_0 result is a little more sensitive at extremes that increase the impedance. This indicates that the shape of the Z_0 response function tends to be concave upward, with higher sensitivity to parameter variations at the higher impedance values.

The calculated range of impedance is an increase of about 36 percent to a decrease of about 30 percent from the average impedance value when all parameters are at their worst combination ± 20 percent from their center values. The ratio of the maximum to the minimum impedance is nearly 2 to 1. This is a range of impedance that would be considered reasonable for most practical applications. The close agreement of the average and the calculated center value also indicates that interpolation of the results for less than a 20 percent change in the parameters should be reasonably accurate. The normalized response sensitivity coefficients of Eqs. (7.9) are for a full 100 percent change in each parameter, so the predicted Z_0 response is scaled by the percentage change in each parameter. For example, the −45 percent normalized sensitivity to conductor width W means that for each 1 percent increase in W, the Z_0 impedance is reduced by about 0.45 percent, or about a 4.5 percent reduction for a 10 percent increase in W. The analysis indicated that this sensitivity should be reasonably accurate for a parameter range of up to ± 20 percent, with a slight reduction in accuracy significantly beyond ± 20 percent.

The worst case predicted tolerance for all parameters is the sum of the magnitudes of the normalized sensitivity coefficients times the tolerance for each parameter. For example, if the dielectric height H and dielectric constant E are controlled to within ± 10 percent and the other two parameters are controlled to within ± 20 percent, and if all four parameters go to the worst limits simultaneously, then the worst case predicted Z_0 tolerance is 5.7 + 9 + 2.4 + 5 = ± 22.1 percent.

Of course, all of the parameters at the worst case combination of extremes *simultaneously* is highly unlikely in actual production if the individual parameters are controlled to be near the nominal, with rejects at the extremes very rare. The results will tend to cluster more closely about the average because it is likely that the effect of some of the contributions will be small, and some will tend to counteract others. For example, consider the *tolerance* on the parameters to be a distribution of the process controls for the dielectric constant, thickness, width, and height parameters. If *the tolerances on* these are independent, the standard deviation of the result is the square root of the sum of the squares (RSS) of the response to each parameter because the variance (the square of the standard deviation) of the sum of *independent* distributions is additive for normal distributions.

For the example above, the sum of squares of a ± 10 percent random tolerance on H and E and a ± 20 percent random tolerance on W and T is about 144, so the square root is a ± 12 percent tolerance on the Z_0 impedance. This is just over one-half of the calculated 22 percent worst case Z_0 tolerance for the same tolerances on the parameters. This is about as expected, because the standard deviation of the sum of n equal independent effects increases as $\sqrt{n}$, and n is 4 in this case. The reduction for this case is a little less than $\sqrt{n}$ because the magnitudes of the individual effects are not quite equal. RSS is not much lower than worst case if just one parameter dominates the tolerance. The analyst is also cautioned that the individual parameter values should be reasonably independent and tend to cluster near the nominal value for an RSS combination to be valid.

If the tolerances on the individual parameters are equal, the normalized tolerance on the Z_0 impedance may be calculated directly as either the worst case sum of the magnitudes or the RSS of the percentage sensitivities from Eqs. (7.9). The RSS is about 90 percent of the tolerance on the individual parameters. For example, this means that a random combination of no more than a ± 10 percent change in all parameters usually tends to result in no more than a ± 9 percent tolerance on Z_0 impedance. However, the worst case sum is about 16.5 percent for a ± 10 percent change in all parameters. The worst case limits for Z_0 impedance would be from a maximum of 56 Ω to a minimum of 40 Ω.

The average effect indicates that the average over all 16 combinations is 48.55 Ω. This is just slightly higher than the calculated impedance at the center values for all the parameters, indicating that the impedance response function is slightly concave upward. An example of a slightly concave (but monotonic increasing) function in

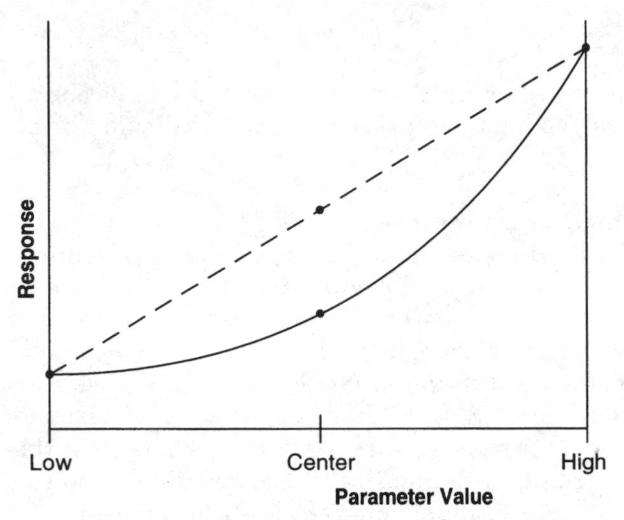

Figure 7.13 Concave monotonic increasing function.

one parameter is illustrated in Fig. 7.13. This figure indicates that the response function is more sensitive to changes in the parameter at the high value than at the low value. A monotonic concave function will be more sensitive at one of the extremes. This higher sensitivity means more variability to tolerances of the parameters. A basic objective of robust designs is to reduce the variability of the response. Therefore, a concave response indicates the possibility that the design may be shifted toward less variability. Conversely, a strictly linear response has the same response sensitivity throughout the range of analysis, so the principal method for reducing the variability of the response is to control the tolerance of the input parameters.

If curvature is significant, extrapolation beyond the range of analysis may be vulnerable to increasing departure from the linear model. Concave functions that are not monotonic may have a maximum or minimum inside the range analyzed. An internal maximum may mean that "nominal is the best," but the binary analysis will determine which extreme is the worst to support worst case analysis. The response for values inside the range analyzed may be better than predicted by the linear model, and the calculated center value will be better than the average of the combinations at the extremes.

7.4.2 Nonlinear interactions of parameters

Interaction is the name given to effects where the sensitivity to one parameter is not similar when other parameters are at different levels. Figure 7.14 illustrates a response that has no interaction. This means

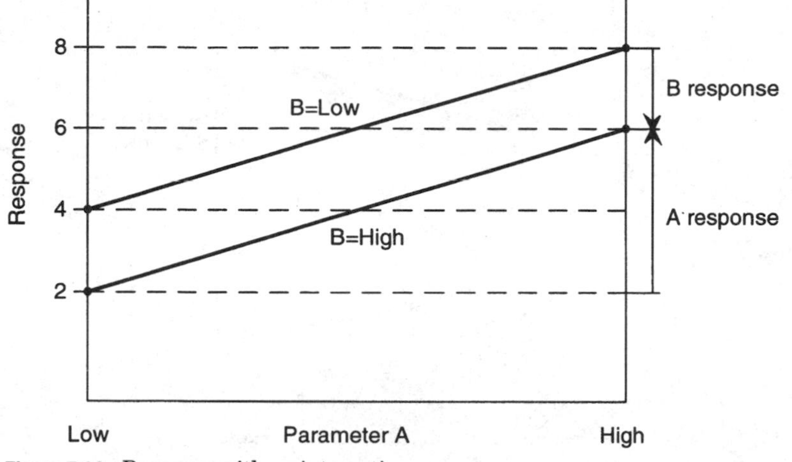

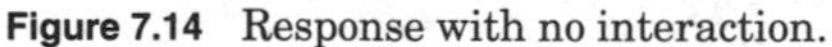
Figure 7.14 Response with no interaction.

that the response for a change from low to high in parameter A is essentially the same regardless of whether parameter B is high or low. If there is a two-way interaction, the average sensitivity of one parameter will be different when another parameter is high than when it is low.

A common type of interaction is a reinforcing interaction, where one level of another parameter increases the sensitivity of the first parameter. The response illustrated in Fig. 7.15 shows that the response to parameter A is greater when B is low than when B is high. An example from medicine is a drug interaction, in which some effect of one medicine may depend on whether another medicine (or condition)

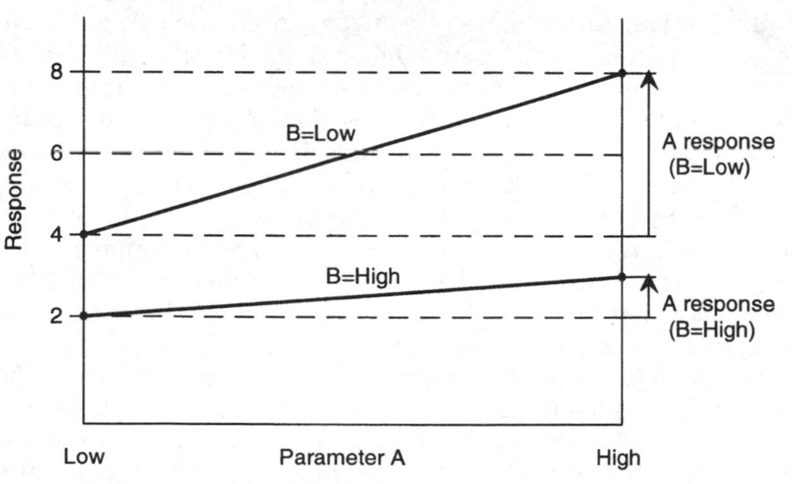

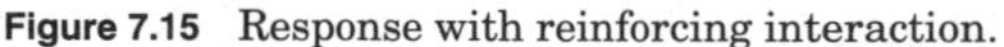
Figure 7.15 Response with reinforcing interaction.

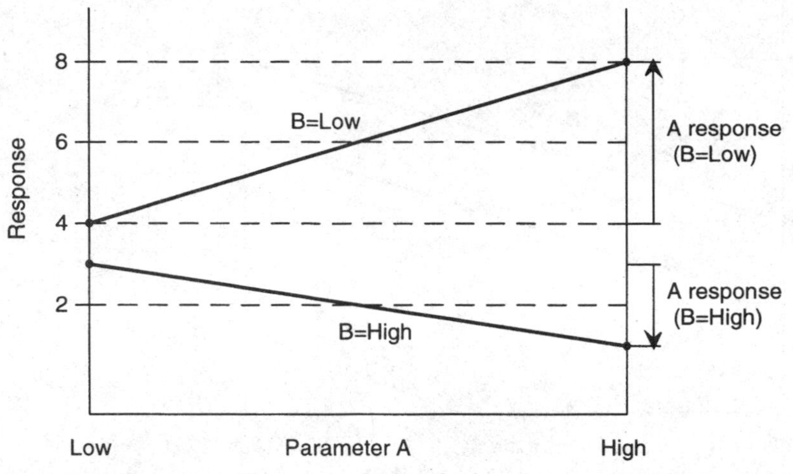

Figure 7.16 Response with contrary interaction.

is present. Another example, from digital logic, is that the output of the AND gate (the carry output of a half-adder) will respond to a logic HIGH on one input only if the other input is also a HIGH.

A more unusual interaction example is a contrary interaction, where the polarity of the response is opposite, depending on the level of the other parameter. Figure 7.16 illustrates a contrary interaction. This example shows that the response to the parameter A is positive when B is low but negative when B is high. A contrary response is rather unusual in natural functions, but it can be very elusive because the *average* response may be near zero. However, the actual response may be very sensitive to different combinations of two (or more) parameters. An example of contrary interaction from digital logic is an EXCLUSIVE-OR gate (the sum output of a half-adder). The output will go to a HIGH in response to one HIGH input if the other input is LOW. However, the response is the opposite (the output will go to a LOW in response to the same HIGH input) if the other input is HIGH. An EXCLUSIVE-OR gate may be considered a conditional inverter, because the state of one of the inputs determines whether the output is the true or the inverse of the other input.

If one at a time analysis had been done for parameter A with B set low, the conclusion for this illustration would have been that the response would be minimum for parameter A low. The analysis for parameter B would conclude (properly in this illustration) that the response would be minimum for parameter B high. This would seem to indicate that the minimum response would be when parameter A is low and B is high. However, the correct combination for a minimum response is parameters A and B both high.

It should be noted that the magnitude and polarity of the responses shown in these examples are arbitrary. The response function (the microstrip impedance for this analysis) may either increase or decrease as an individual parameter is changed from LOW to HIGH, or the response may not change at all, or the change of the response function may depend on the value of another parameter, as in the interaction examples.

The results show that microstrip impedance is relatively more sensitive to the height H above the ground than the other parameters. All other parameters being held constant, the 20 percent change in height changed the impedance by an average of about 5.5 Ω. This is about 11 percent of the average impedance, or only about 57 percent of the 20 percent change in height. For an example of interpolation to a 10 percent increase in H, the impedance would increase less than 6 percent. Figure 7.17 shows all 16 results (8 for the higher height, 8 for the lower height) of the experiment, with the line showing the average effect for the 20 percent change in height. The figure also shows as a histogram on each axis the range and relative distribution of the 8 results at each level of dielectric height.

The sensitivity to the relative dielectric constant is almost exactly 50 percent, as would be expected for the square root of a parameter reasonably close to unity. For most construction methods, the dielectric constant can by held rather constant, so its contribution to impedance tolerance will be reasonably small. For an example of ± 10 percent variation in dielectric constant only, the line impedance would vary about ± 5 percent.

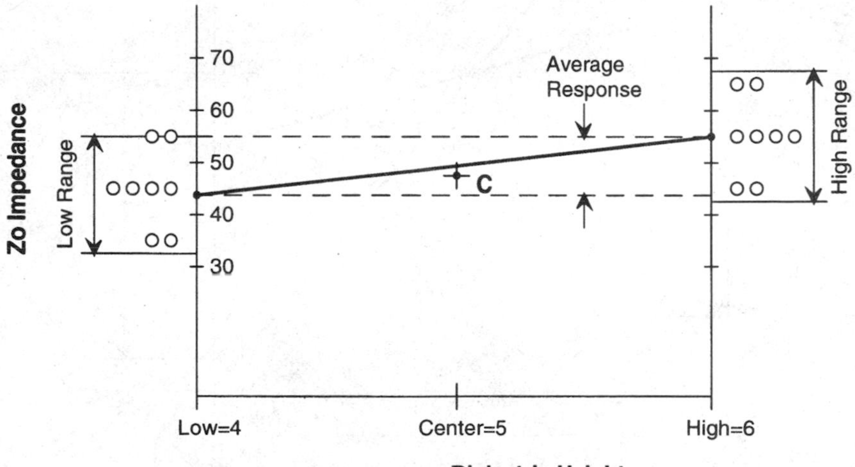

Figure 7.17 Results of impedance experiment for dielectric height.

The sensitivity to line width W is slightly less, at about -45 percent, so a 20 percent increase in W results in about 9 percent reduction in impedance. The width of fine printed circuit lines is often one of the most difficult dimensions to produce accurately, so it may be a significant contribution to impedance variability. Furthermore, many production processes result in nonuniform edges that also increase variability and high-frequency losses due to the skin-effect current crowding at the line edges. The average sensitivity (in ohms) to dielectric constant, line width, thickness, and the calculated center effect is shown in Fig. 7.18 as an overlay of the individual effects.

The sensitivity to line thickness is essentially insignificant by itself, so it can usually be ignored as a direct contribution to impedance variability, or assumed to be a constant. However, the tolerance on line width is typically proportional to thickness. Therefore, it is usually appropriate to use the minimum thickness practical to improve line width accuracy. Line width was shown above to be important in determining microstrip line impedance.

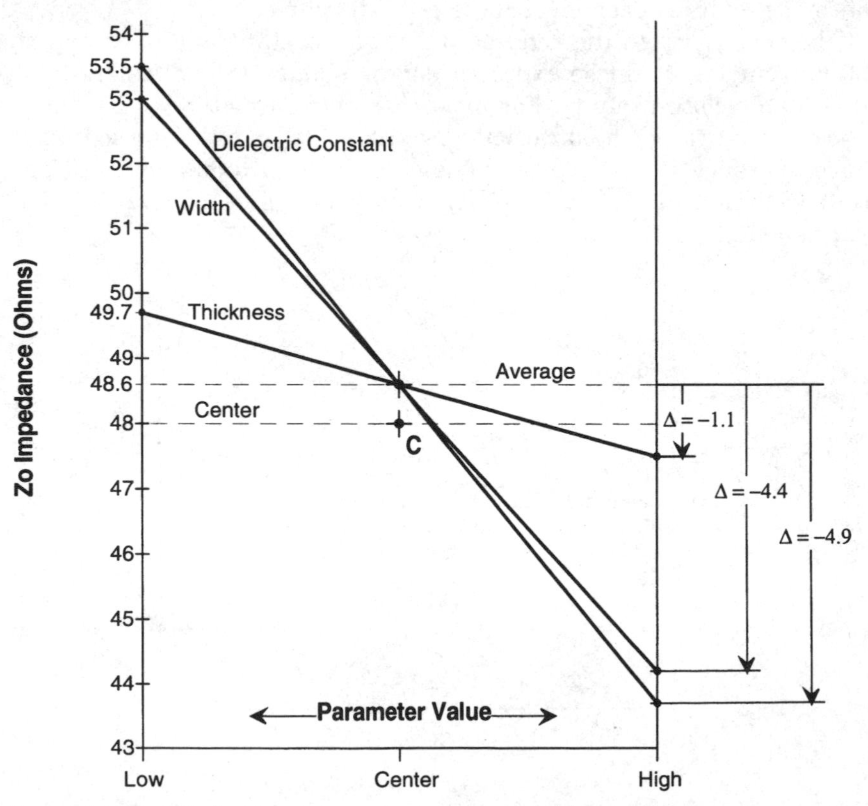

Figure 7.18 Results for dielectric constant, width, and thickness.

7.5 Summary of Conditions for Transmission-Line Propagation

The $l * c$ product that determines high-frequency propagation velocity is only a function of the dielectric. The l and c are reciprocal parameters for most uniform lines. This means that high-frequency electromagnetic waves in the same dielectric travel at almost the same velocity, independent of conductor geometry, including radiation in free space without conductors. However, losses can cause distortion, which appears as a slight reduction in propagation velocity (as well as increased attenuation) at different frequencies, or for the frequency components of complex waveforms. The maximum velocity in a dielectric is often called the "speed of light," because light is just one frequency range of electromagnetic energy. Light energy itself can be efficiently transmitted through optical fibers that direct most of the energy to travel to the receiver to minimize interference and radiation losses.

Conductors such as wires are used to guide lower frequencies (down to dc) of electrical energy to travel from the source to a receiver. The electrical energy travels as a transient electromagnetic wave in the dielectric between the conductors. Some of the energy may be lost owing to finite resistance in the conductors, and some may be lost to coupling to other conductors or radiation into the dielectric surrounding the conductors. Electrical signals are subject to interference from other electromagnetic energy that may travel in the dielectric, or other currents that may flow in the conductors.

7.5.1 The velocity of propagation

The speed of light in free space is nearly 3×10^8 m/s, or almost 1 ft (12 in = 30.5 cm)/ns. Therefore, the full wavelength of a 1-GHz sine wave traveling in free space is nearly 30 cm. The quarter wavelength is about 7.5 cm, or nearly 3 in. The propagation delay (the inverse of velocity and wavelength) is proportional to the square root of dielectric constant. For example, if the dielectric constant relative to free space is in the range of 4 to 9, the propagation delay per unit length increases 2 to 3 times, and the wavelength is reduced to one-half to one-third of the free space wavelength.

7.5.2 Transient propagation in the dielectric

Most high-frequency and transient propagation travels in the dielectric, not in the conductors. However, the presence of conductors increases shunt capacitance, with a corresponding decrease in inductance. Therefore, high-frequency impedance between conductors is

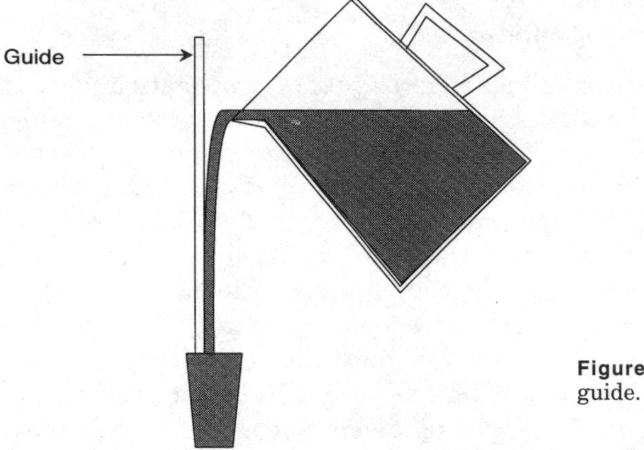

Figure 7.19 Water follows a guide.

reduced compared to free space, so that most of the signal energy travels along the dielectric between two conductors, taking the "path of lower resistance" rather than radiating in other directions. The conductors act as an open "waveguide," so that most of the signal energy travels along the desired path to the receiver at the destination and does not interfere with other signals. A physical analogy for fluid flow is that the surface tension of many fluids results in a tendency to follow a surface. For example, water will tend to follow a rod (such as a long screwdriver shaft), so the rod can be used to direct water being poured into a rather small filler, as illustrated in Fig. 7.19.

Shielded cables (including coaxial cables) and hollow waveguides completely surround the dielectric with a conductor, so that very little of the signal energy escapes. The analogy for fluids is a pipe or tube that contains the fluid. However, to the extent that the shield is not a perfectly zero resistance electrical conductor, surface currents result in voltages that will generate external fields. Electrical shields are like a pipe that lets a little of the fluid seep out of the walls. However, like pipes carrying a fluid, the most serious risk for leaks is often at the connections. Terminating a shield and connecting it to a shielded enclosure is an important part of preventing signal interference.

7.5.3 Conductors carry the dc loop current

Conductors do carry the dc loop that flows behind a traveling transient wavefront to maintain static levels. Lower-frequency currents do penetrate into conductors but travel at a much lower propagation velocity (the dielectric constant inside conductors is much higher), distorting the trailing (or later) portion of the waveshape. The low-

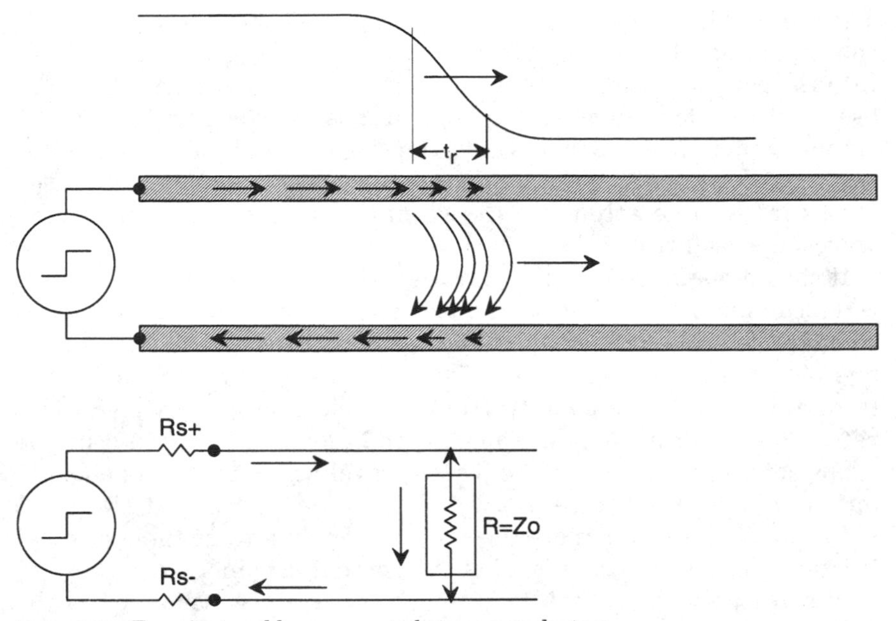

Figure 7.20 Transient and loop currents between conductors.

frequency currents that travel more slowly inside the conductors also encounter lower resistance, so there is less loss than the skin-effect loss encountered by the high-frequency transients of the leading edge of the initial transition. This means that if enough settling time is allowed, the signal level will gradually approach a final level limited only by the dc resistance of the conductors. However, the final static signal level may never be achieved at higher clock and data rates.

Figure 7.20 illustrates transient propagation and the loop current that follows behind the transient wave. The current flows as displacement current through the dielectric during the time that the signal transition (the signal *rise time*) travels along the space between the conductors. The figure illustrates that the displacement current density in the dielectric increases to a maximum during the central portion of the transition, then declines toward zero as the signal voltage approaches the steady-state level. The ratio of the change in voltage to the change in current is the dynamic impedance of the dielectric, which is designated Z_0. This can be considered as a resistance between the conductors during the time that the signal is traveling in the dielectric.

The conductors carry the loop current behind the transition after the initial signal level is achieved. It should be noted that the current density that is illustrated as rapidly increasing "in" the conductor actually flows on the extreme outer edges of the conductor surfaces.

Therefore, there may be a significant increase in line resistance near the transition, due to the skin effect. The current gradually diffuses into the conductor until the steady-state current eventually flows at essentially uniform density. Although simple transmission-line analysis may ignore the losses and consider the dynamic impedance to be resistive, the actual impedance does include reactive components. The losses in the line attenuate the signal amplitude and distort the waveshape as it travels.

It should be understood that arrows indicating current flow "in" the conductors do not imply that the electron flow rate in a conductor is comparable to the rate at which the transient effect travels. The average net rate at which electrons flow in a good conductor is so slow (typically far less than 1 m/s)[5] that it is more properly called "drift rate." The water analogy is appropriate here; that is, when a pebble drops onto a pond, the surface waves travel away from the pebble at an observable, rapid rate. It is primarily the waves that travel; so very few of the water molecules travel very far from their original location. The rise in water level at the shore line is insignificant.

A more appropriate analogy in which there is also some flow of the water may be dropping the pebble, or a fish jumping on a deep tidal estuary or bay. The rate at which the surface waves travel away from the pebble or the fish is much greater than the net flow rate of the entire estuary with the rise and fall of the tides. The pond, or the bay, or a water pipe is full of an enormous number of water molecules; the net rate at which the molecules flow is usually very small compared to the speed at which a disturbance travels. Likewise, a good electrical conductor is full of an enormous number of electrons, but the net rate at which they flow is insignificant compared to the rate at which an electrical transient travels.

7.5.4 Conductor geometry guides the transient wave

The conductor geometry must be designed to properly guide the traveling wave to the desired destination without interference to or from any adjacent signals. As was shown above, the spacing between conductors (or the height above ground) is important in maintaining the proper transmission-line impedance. It is important that transmission lines that use a common ground have a continuous ground reference along the line. The common ground does not need to be a dc ground potential; multiple power and ground planes can each be used as the common ground reference. However, if a signal line changes reference planes along the line, it is important that the different reference planes be coupled together with a good-quality high-frequency capacitor at essentially the same location that the line changes reference planes. The total capacitance needs to be large enough that the voltage across the capacitor remains nearly constant during signal transients. Therefore, signal lines should not switch between different reference planes that cannot be coupled together at the same location.

7.5.5 The conductor spacing must be small for propagation

The spacing between the two conductors guiding the transient signal must also be small enough so that the highest frequencies needed will propagate along the intended transmission line. The spacing must be much less than the quarter wavelength of the highest frequency needed to adequately transmit the signal waveshape. Very long lines will need additional frequency bandwidth to allow for high-frequency attenuation. The highest frequency needed is determined primarily by the signal rise time of digital signals, not just the highest clock frequency or data rate. A reasonable "rule of thumb" is that digital signals usually require significant energy at the third harmonic of the transition time (often called *rise time*). Therefore, it is reasonable to require that the frequency capability of the transmission line be at least the inverse of the transition time. For example, if the transition time (measured between the 20 to 80 percent levels) is about 1 ns, the required bandwidth for transmission is at least 1 GHz. Digital signals with sharp transitions to a linear rise time may contain some energy at much higher frequency components, but those may be lost during propagation due to skin-effect resistance or radiation into the surrounding dielectric. Therefore, the rise time at the far end of a transmission line may be slower, and the waveshape may be distorted compared to the signal source.

7.5.6 The spacing to other conductors must be larger

The spacing to adjacent conductors must be kept large enough to avoid unwanted coupling between other signals. Unwanted coupling between lines is often called *crosstalk*. This terminology originated with telephone lines; if there is sufficient coupling between adjacent telephone lines, the people on one circuit can hear a conversation on an adjacent circuit. Since space is often limited in complex systems with large numbers of signals, one of the most effective methods for limiting crosstalk is to keep the space between the signal conductors (or the height above a ground plane) relatively small compared to the space between adjacent conductors. Close spacing between signal conductors also increases high-frequency bandwidth. However, closer spacing reduces the dynamic impedance, increasing transient currents and power dissipation, so the conductor size (diameter for round wires, width for flat conductors) must be small enough to achieve reasonable transmission-line impedance.

Although small conductor size increases losses, this is usually not significant except for very long lines. Caution must be used to ensure that close spacing between conductors does not approach the dielectric breakdown rating of the dielectric for the voltages that might occur. Twisted pairs, shielded pairs, or coax reduce crosstalk between different signals, even with relatively close spacing. The length of wire conductors for high speed data transmission is limited by losses which will reduce the signal levels so that they cannot be reliably recovered by a receiver. Longer-distance high speed data is typically transmitted by other methods, such as microwave or fiber optics.

7.6 References

1. Richard A. Matick, *Transmission Lines for Digital and Communication Networks,* McGraw-Hill, New York, 1969.
2. Hugh H. Skilling, *Electric Transmission Lines,* McGraw-Hill, New York, 1951.
3. C. H. Chen (ed.), *Computer Engineering Handbook,* McGraw-Hill, New York, 1992, chap. 6.6.
4. George E. P. Box, William G. Hunter, and J. Stuart Hunter, *Statistics for Experimenters,* Wiley, New York, 1957.
5. Douglas C. Montgomery, *Design and Analysis of Experiments,* 3d ed., Wiley, New York, 1991.

Chapter

8

Transmission-Line Interconnections: Loads and Ends

The equations for transmission-line propagation are based on uniform conditions. However, lines of useful interest have at least two ends, one where the signal is generated and another where the signal is received. The source and loads on an electrical transmission line may not be perfectly matched to the line impedance. Many lines have lumped loading, branches, and changes in characteristics when the signal travels through connectors. Interconnections typically consist of multiple lines, which interact or interfere with transmission of the signal. This chapter discusses various transmission-line characteristics, including line termination techniques, line loading factors, rise time, crosstalk, shielding, and grounding.

8.1 An Analogy to Aircraft Flight

An analogy of transmission of electrical signals to an aircraft flight is that the principles of flight apply along the route, a route must be followed that travels from the origin to the destination, and interference with other aircraft must be avoided. However, takeoff and landing are also important for safe arrival at the destination. These compare to the principles of the propagation of an electronic signal along conductors that guide the signal from the source to the load and avoid interference with other signals. The takeoff and landing are also essential events of the flight that are critical to the objective of transferring the passengers and/or cargo from the origin to the destination. The aircraft must be loaded with enough fuel to complete the flight and must be capable of takeoff and then safely landing at the destination.

There are special cases where delivery occurs along the route; in some cases, some of the passengers may jump out as parachutists and cargo or bombs are dropped at selected locations along the route. An analogy to electric transmission lines may be that some lines have multiple taps where the signals are received. A general conclusion is that the ends of the aircraft flight include the primary activities; the journey along the way is only the means to accomplish the objectives at the ends. The ends of an electric transmission line must be reasonably compatible with (but not necessarily exactly terminated to) the line characteristics for the signal to be properly propagated along the transmission line and accurately recovered at the receiver(s).

8.2 The Ends of the Transmission Line

The source and destination ends of a simple two-ended line are discontinuities to the transmission line. Various other conditions are discontinuities in otherwise uniform transmission lines. Examples include branches, stubs, changes in dimensions of the line geometry, transitions between reference planes, transitions through connectors, and discrete components such as load capacitors and termination resistors. Any discontinuity in a uniform transmission line may cause a reflection of the traveling wave. The original wave that arrives at a discontinuity is called the *incident* wave. The portion of the incident wave that is reflected back toward the source is called the *reflected* wave. The resulting signal level at the discontinuity is the sum of the incident and reflected waves. It may be a change in both current and voltage (but, of course, no increase in energy). For example, the ratio of voltage to current in the uniform transmission line is defined by the line impedance Z_0; if the end of the line is terminated in a resistance, the resistor determines the ratio of voltage to current at the end of the line. For the case of a perfect termination resistor that exactly matches the line impedance, there is no change in the voltage and current, and no reflection.

When there is a very high resistance or a nearly open circuit at the end of a line, the voltage must increase, the current must drop to nearly zero, and most of the energy must be reflected back toward the source. This is approximately the case for many unterminated digital receiving devices (often just an input to a logic device); most of these inputs are very high input impedance and respond to voltage levels without absorbing much static energy. The high input impedance also allows several devices to be connected in parallel without significant static loading, although they may add capacitance loading. The increase in voltage as the traveling signal arrives at the open end of a line typically results in *overshoot* beyond the static signal level. This overshoot by itself is not usually a problem because it results in a faster transition through the threshold region of the receiver. However, recovery from the overshoot can degrade speed and signal levels.

The reflected energy from an unterminated end of a line travels back toward the source, which is also a line discontinuity. If the source impedance exactly matches the line impedance, the reflected energy is absorbed by the source without further reflections. This is called *source termination.* Most sources of digital voltage signals are naturally low impedance. A resistance must be added in series with the source to match the line impedance, so it is often called *series termination.* Termination at the source is sometimes called *back termination,* referring to its location at the source end rather than the far end of the line. Source termination is usually required for termination of common mode interference on long differential lines, because the high common mode impedance of most differential receivers reflects any common mode crosstalk.

8.2.1 Standing waves

Source termination alone may be sufficient for most simple two-ended lines, especially when the frequency is low enough so that the reflection from the open end of the line can return and establish steady-state levels before the next signal is launched. Certain lines can be operated at higher rates, but the interaction of incident and reflected waves must be considered. The reflected waves may combine with newer incident waves to form *standing waves* at periodic locations along a long line. This ratio of maximum to minimum signal levels at high frequency is often referred to as the *voltage standing wave ratio* (VSWR). If standing waves are to be controlled on long, very high frequency lines, the reflections must be controlled by a load termination. When both source termination and load termination are used, the line is called *double terminated.* Double terminated lines can control both common mode crosstalk reflections and signal reflections. Double terminated lines can effectively eliminate signal ringing, even on shorter lines, because the termination does not need to be precise. Less accuracy required for termination resistors may allow the termination resistors to be integrated into the driver and receiver circuits.

8.2.2 Source termination

Source termination only offers the advantages of higher signal levels and less power dissipation for static signal levels. High-impedance loads can be added or removed without concern for providing a termination at the new end of the line. Lines with multiple sources may use source termination because the series terminations do not affect drivers disabled to the high-impedance condition. Therefore, drivers may be added or removed without concern for a termination at the end of the line.

A driver with a source impedance of exactly the line impedance will launch a signal that is one-half of the steady-state open circuit voltage. If the receiving end is essentially open circuit, the voltage will double at the receiver, so that the steady-state level is reached as soon as the initial signal arrives at the receiver. Practical systems prefer a source impedance less than the line impedance, so that larger signal levels are achieved, with a small amount of ringing allowed. The driver will launch an initial signal level that is greater than one-half of the steady-state signal level. If the initial signal level is large enough to cross the threshold of receivers along the line, they will switch to the new state as soon as the initial signal arrives. When the initial signal is large enough to assure switching, this is called *initial wave switching*. All receivers along the line receive a sufficient signal to switch as soon as the initial signal arrives from the source, rather than needing to allow enough time for the reflection to return from the far end of the line.

A line with adequate resistance to limit ringing to acceptable levels is referred to as terminated, even though the resistance does not precisely match the line impedance. When the resistance is not closely matched to the line impedance, it may be more technically correct to refer to the line as *dampened* rather than terminated. However, it is common to refer to any separate resistance added to a transmission line to limit ringing or standing waves as a *termination* resistor.

When the transmission line is adequately terminated so that the ringing is dampened within limits acceptable to the receiver(s) on the line, the signal level will be adequate at all points on the line after the first reflection returns to the source termination. This condition is referred to as *incident wave switching*. The distinction is that *initial* wave switching establishes and maintains an adequate signal along the line after the first incident wave travels toward the far end; *incident* wave switching establishes and maintains an adequate signal along the line after the first reflection returns to the source. The two conditions are equivalent for a single load at the end of the line where the incident and reflected waves coincide but are different for multiple loads nearer the source.

Incident wave switching requires control of ringing so that adequate signal levels are maintained after the first reflection. Source termination only is sufficient to provide incident wave switching, but the initial signal level may not be adequate. Initial wave switching requires both an adequate initial signal level and control of ringing. Some degree of load termination and/or overshoot clamping is typically required for initial wave switching.

8.3 Line Discontinuities

When a traveling wave arrives at a discontinuity that changes impedance, the voltage and current must change to accommodate the new impedance level. Part of the energy must be reflected to establish the new signal level on the line on which the incident wave arrived. If the new impedance is another transmission line, a modified incident wave continues along the new line. If the new impedance is a branch into two (or more) lines, new incident waves continue along each new line. A traveling wave that arrives at a discontinuity that changes impedance always causes a reflection of part of the energy in the incident wave.

When the impedance change is to a higher impedance, the voltage must increase and part of the current is reversed in the reflected wave. For example, if the discontinuity is an open circuit which accepts no current, the voltage doubles, so the reflected wave is equal in voltage to the incident wave, but all of the current is reversed; the reflected energy is equal to the incident energy.

If the impedance change is to a lower impedance, the current must increase and part of the voltage is reversed in the reflected wave. For example, if the discontinuity is a branch to two new lines (each less than double the impedance of the incident line), the voltage is reduced when the signal arrives at the branch. The reduced voltage signal travels along each branch, while the reflected wave is an increase in current with a voltage reversal. For the case of all lines of equal impedance, the voltage at the branch is reduced by one-third, so that a one-third reduction in voltage (about 11 percent of the energy) is reflected, and two-thirds of the voltage (about 44 percent of the energy) travels along each branch. However, for the case of each branch exactly double the impedance of the incident line, there is no change in voltage. This configuration, called a *hybrid divider,* is an ideal method for sending transmission-line signals to separate loads. An exactly terminated hybrid divider is illustrated in Fig. 8.1. One-half of the current (50 percent of the energy) travels along each branch, and there is no reflection.

Although it may not be practical to achieve exactly double the impedance for ordinary digital signal lines, it is desirable to make each branch a higher impedance. Even if it is not practical to double the dynamic impedance of the transmission line in the branch, each

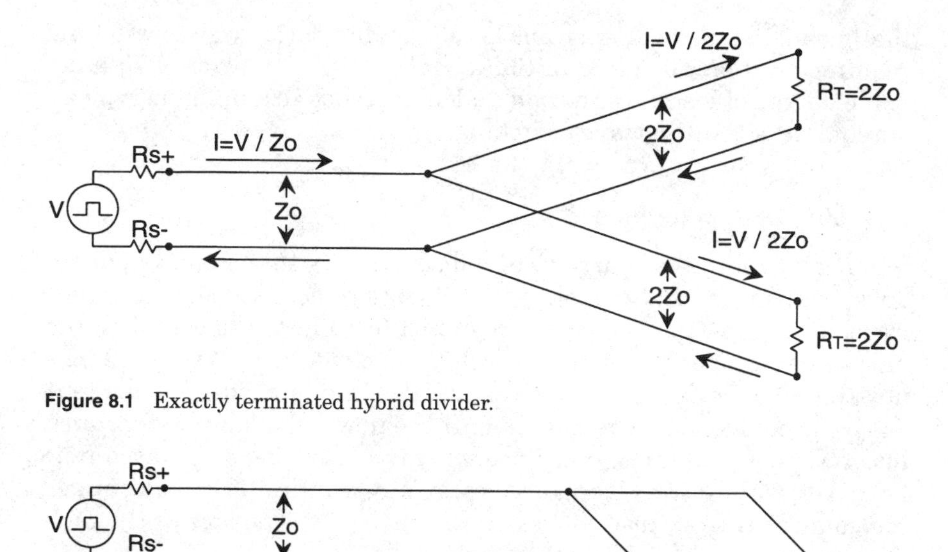

Figure 8.1 Exactly terminated hybrid divider.

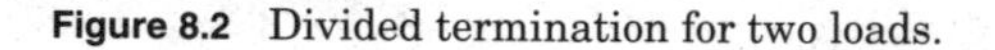

Figure 8.2 Divided termination for two loads.

branch may be terminated with double the appropriate load termination for a single line. The waveshape may be acceptable, especially if the length of the branches can be kept reasonably low compared to the signal rise time. Two separate loads, each with an appropriate termination, are illustrated in Fig. 8.2. This may be convenient when the receiver load termination is located on units (e.g., board assemblies) which may be removable, so that both of the units are interchangeable with equal terminations. In addition, this may be appropriate for applications where one of the termination loads may be missing in certain circumstances. For example, if the two units represent parallel redundancy, acceptable operation may be maintained with only one unit functional, or when either one is removed for test or maintenance. It should be emphasized that the appropriate termination may be significantly greater than $2 * Z_0$ at each load.

8.3.1 Short lines

When the line length is short enough compared to the quarter wavelength of the highest-frequency component of the signals to be transmitted, reflections that would otherwise occur are absorbed by the signal source during the signal rise time. It is convenient to assume that any lumped capacitance such as loads, line stubs, or mutual coupling can be added to the intrinsic line capacitance of short lines. This assumption is also reasonable for lines with multiple capacitance loads that are spaced no more than a quarter wavelength apart. Note that the equivalent wavelength for a single digital step of finite rise time is approximately the distance that the signal travels during the signal rise time. Therefore, the equivalent wavelength for a single traveling step is the distance between the leading and trailing ends of the rise time. The quarter wavelength of a 1-ns rise time edge is a few centimeters for most high speed dielectrics.

8.3.2 The loading factor

The intrinsic capacitance of a line is c per unit of length, so the total intrinsic line capacitance is $C_0 = c *$ length. As was shown previously, $c = t_{pd}/Z_0$, so $C_0 = t_{pd} *$ length$/Z_0 = T_0/Z_0$. Therefore, the total line capacitance of a line loaded with additional capacitance is $C = C_0 + C_L$, or $C = C_0(1 + C_L/C_0)$. The line inductance remains essentially unchanged, so the total line capacitance is no longer the reciprocal of line inductance. This means that the propagation velocity of a loaded line is slower than the maximum velocity determined by the dielectric constant of the line, because the $L_0 * C$ product is increased. The propagation delay of a line section with lumped capacitance loading is approximately

$$T_0' = \sqrt{L_0 * (C_0 + C_L)} = T_0 \sqrt{1 + \frac{C_L}{C_0}} \tag{8.1}$$

The factor $\sqrt{1 + C_L/C_0}$ is called the *loading factor* and will be designated K_L. Therefore, the loaded propagation delay is $T_0' = T_0 * K_L$. The line impedance is reduced, so the loaded impedance $Z_0' = Z_0/K_L$. It is often more convenient to estimate propagation delay and impedance than intrinsic capacitance, so $C_0 = T_0/Z_0$ can be substituted into the loading factor. Therefore, an equivalent form for the loading factor is the following:

$$K_L = \sqrt{1 + \frac{C_L * Z_0}{T_0}} \tag{8.2}$$

8.3.3 Cutoff frequency and rise time

The cutoff frequency for a short section of a transmission line with lumped inductance and capacitance is the same as for a two-terminal pair low pass filter with total series inductance L and capacitance C per section, as follows:[1,2]

$$F_c = \frac{1}{\pi \sqrt{L * C}} \tag{8.3}$$

Therefore, the cutoff frequency of a short transmission-line section with lumped loading is approximately

$$F_c = \frac{1}{\pi * T_0 * K_L} \tag{8.4}$$

One of the consequences of the presence of cutoff frequency for a transmission line is that input impedance to the line is not real for frequencies above the cutoff frequency. Therefore, the frequency components of a complex waveform that are above the cutoff frequency are rejected by reactive attenuation and reflected to the source. These high-frequency components do not pass through the transmission-line section.

The rise time for an exponential waveform is about 2.2 time constants between the 10 to 90 percent levels (as illustrated in Fig. 9.2), or $2.2/2\pi * F_c$ for the step response of an *RC* network. This is commonly expressed as $t_{rise} = 0.35/F_c$.[3] The estimated minimum rise time is determined by the cutoff frequency as follows:

$$t_{rise}(\text{min}) = \frac{2.2}{2\pi * F_c} = 1.1\, T_0 * K_L \approx T_0' \tag{8.5}$$

This means that the estimated rise time out of a short line section with any lumped loading may exceed the one-way loaded propagation delay, even without considering losses such as skin effect. The loaded propagation delay (and therefore the rise time) is greater than the intrinsic delay by the lumped loading factor K_L. The lumped loading factor of a line segment was determined previously from the ratio of lumped capacitance to the total intrinsic capacitance ($C_0 = T_0/Z_0$) of the line segment. Although some very long uniform lines with low loss may achieve faster rise time, the one-way delay is usually a reasonable estimate of the output rise time of lines, especially if any discontinuities such as connectors or high-frequency skin-effect losses are significant. Of course, when the input rise time is much greater than the line delay, the rise time increase through the line may be negligible. Similarly, if the intrinsic line capacitance is much greater than the lumped load capacitance, the lumped loading factor is nearly unity.

8.3.4 Rise time of multiple line segments

Transmission lines may consist of several line segments, each of which tends to have discontinuities at the segment boundaries.

Typical discontinuities which add lumped capacitance are connectors or load points. The timing analysis should include any alternatives such as any test connections or extender boards that may be used. Although not analytically exact, a reasonable estimate of the output rise time of line segments with lumped loading is the RSS (root sum of squares) combination of the rise times of the input and each of the line segments.[4] The line should be reasonably terminated to limit overshoot, all segments should have essentially the same impedance, and the propagation delay through each discontinuity should be short compared to the rise time, so that reflection effects can be reasonably ignored. This estimate may be shown as follows for the input rise time and n line segments:

$$t_{\text{rise}}(\text{out}) = \sqrt{[t_{\text{rise}}(\text{in})]^2 + [T_0'(1)]^2 + [T_0'(2)]^2 + \cdots + [T_0'(n)]^2} \quad (8.6)$$

8.4 Crosstalk

The basic transmission-line equations have been described above for a single line, but practical networks often consist of numerous lines which may be in close proximity. Adjacent lines affect the transmission-line impedance, depending on the signal conditions on the other lines and the distance to the other lines relative to the ground. The typical condition is that the other line does not have an active signal transition at the same time as the transition on the primary line. In this case, the additional capacitance to the other line (called *mutual coupling capacitance*) reduces the line impedance in a manner that is similar to additional ground capacitance. Some of the signal is coupled into the other line, which is called *crosstalk,* a term derived from telephone networks, where faint conversations might be heard from other lines.

Crosstalk actually consists of both distributed capacitance and inductive coupling of approximately equal magnitudes.[5] Unlike lumped capacitance, which affects both impedance and propagation delay, distributed mutual coupling affects impedance, but not propagation delay. Although it may be easier to visualize the electric field of mutual capacitance between adjacent lines, it should be understood that there is also an equivalent magnetic coupling between the adjacent lines. The magnetic coupling is important to understanding the difference between the crosstalk at the two ends of the coupled line. When signal energy travels in one direction on the primary line, the portion of the signal coupled into the adjacent line travels in both directions on the adjacent line.

It is reasonably obvious that mutual capacitance coupling is in phase with the signal on the primary line at both ends. However, the coupled lines are essentially the primary and secondary of a transformer due to mutual inductance coupling, although both sides of this transformer have only a single "turn." Therefore, the two ends of the inductively coupled secondary have opposite signal polarities.

The inductively coupled crosstalk that travels backward toward the source of the primary signal has the same polarity as the primary signal. Therefore, the total backward crosstalk is the sum of the inductance and capacitance components. The forward crosstalk is the difference of the (usually nearly equal) inductive and capacitance components. The forward crosstalk travels along with the primary signal, so the duration of forward crosstalk is only about the rise time of the primary signal. However, the duration of the backward crosstalk is the two-way propagation delay of the coupled length. Since backward crosstalk is usually of greater amplitude and longer duration than forward crosstalk, it is the greatest concern for crosstalk. Crosstalk is usually defined in terms of the proportion of the primary signal that can be coupled to the adjacent line as a backward crosstalk voltage ratio, which is designated as K_B.

8.4.1 The short-line scaling factor

The crosstalk voltage is reduced in proportion to the coupled length when the time for the signal transition on the primary line to travel through the coupled section is less than the rise time. This reduction is called the short-line scaling factor and is designated as K_S. When the short-line scaling factor is applied to mutual coupling, the length is only the mutually coupled length; although when applied to line reflections it is the length over which the reflections occur. The short-line scaling factor for mutual coupling is the following:

$$K_S = \frac{\text{coupled length} * 2t_{pd}}{t_{\text{rise}}} \quad \text{if less than 1}$$
$$K_S = 1 \quad \text{otherwise} \tag{8.7}$$

The coupled length at which K_S just reaches 1 is called the *critical length.* Above this length the crosstalk amplitude is fully developed and any further increase in the coupled length only increases the duration of the backward crosstalk, but not the amplitude. Below the critical length, the crosstalk amplitude is proportional to the coupled length. Faster rise time does not increase the crosstalk amplitude for long coupled length but reduces the coupled length at which full amplitude is reached. Nothing "critical" or dangerous happens at the critical length; it is just the length at which the sensitivity to length changes, as illustrated in Fig. 8.3.

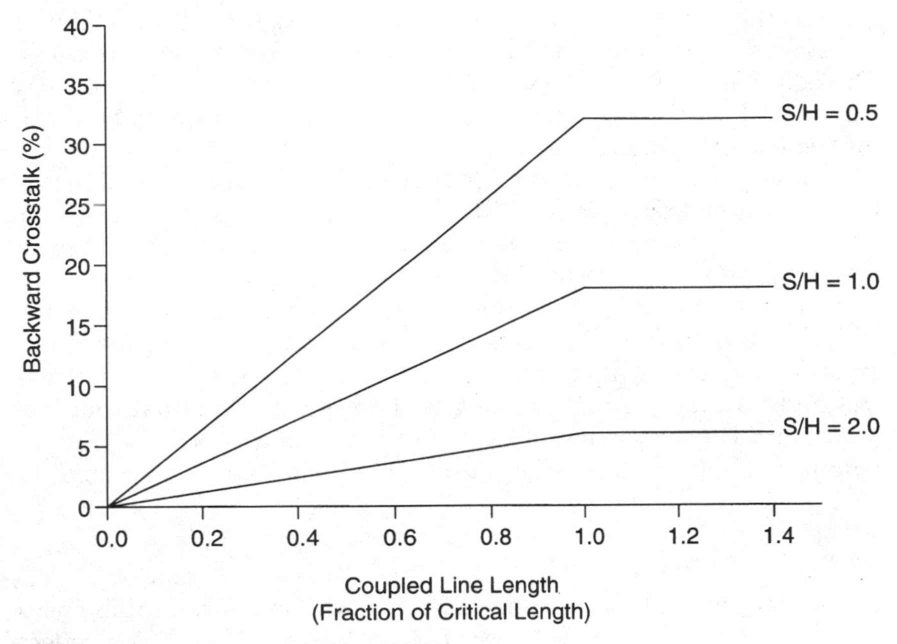

Figure 8.3 Crosstalk vs. coupled length.

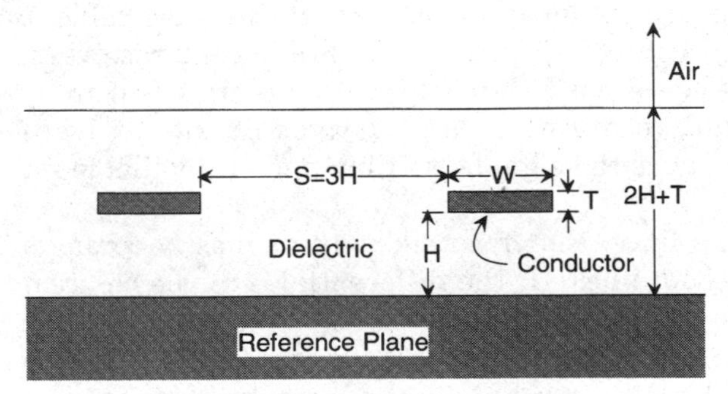

Figure 8.4 Buried microstrip crosstalk geometry.

The data[6] is for a range of the ratio of spacing between conductors to the dielectric height above a ground plane for a typical buried microstrip printed circuit configuration as shown in Fig. 8.4. The crosstalk data shows that for an S/H ratio of at least 2.0, the backward crosstalk for a single line should be less than about 5 percent. This is the basis for the recommendation of an S/H ratio of 3:1 to reduce mutual coupling to acceptable levels for critical signals. The 3:1 S/H ratio is also recommended for parallel differential pairs because reducing mutual coupling between the pairs means that the line-to-line impedance approaches double the line-to-ground impedance of the individual conductors. Reference 6 provides calculated crosstalk data for other conditions and printed circuit configurations. The additional reference plane of dual stripline and stripline reduces the mutual coupling compared to buried microstrip, especially at the higher spacing-to-height ratios.

It may be noted that because the critical length is proportional to rise time, crosstalk increases more rapidly with coupled length for faster rise times. Therefore, a network is considered more vulnerable to crosstalk at faster rise times, even if the coupled lengths are shorter. Networks with faster rise time will tend to have more lines with crosstalk, even if the worst crosstalk on lines with long coupling is not greater. More care will need to be taken in routing signal lines to control mutual coupling and the length of lines where mutual coupling does occur. Higher speed drivers and receivers may need better terminations to limit crosstalk reflections, as well as for signal reflections.

8.4.2 Reflections of crosstalk

Crosstalk is also subject to reflections, but with significant differences from signals. Since crosstalk travels in both directions, it can reflect from both drivers and receivers. If adjacent lines carry signals in the same direction (this is typical for data buses and interface cables), forward crosstalk might interfere directly with adjacent receivers. However, this may not be serious because forward crosstalk is usually low amplitude, is short in duration, and arrives at the adjacent receiver at the same time as the signals, so time may be available for data signals to settle.

For many differential lines, the crosstalk may be mostly common mode (nearly equal in both lines). If the differential receiver has reasonable common mode range, the receiver should reject most of the forward crosstalk. However, differential receivers are typically terminated line-to-line and have very high impedance to the receiver ground. This is usually necessary to achieve high common mode rejection and to avoid ground loop currents. Therefore, the common mode crosstalk is reflected from the receiver even if the receiver is well terminated for differential signals. If the two differential lines are not equally well terminated with respect to ground at the driver, some of the common mode reflection will be reflected from the driver, and some of that may be converted to differential crosstalk and re-reflected to the receiver. This reflected crosstalk arrives at the receiver later than the initial crosstalk, so less settling time is available. Reflected crosstalk on long lines may even interfere with data in later bit interval times, resulting in intersymbol interference.

Backward crosstalk traveling toward an adjacent driver may seem to be of less concern because it should not upset a driver (which is true). However, unless the driver is very well terminated, some of the backward crosstalk will be reflected from the driver. Reflected backward crosstalk is usually larger in amplitude and much longer in duration than forward crosstalk. Unless a differential driver has nearly equal source impedance to both the high and low states, common mode backward crosstalk is not equally reflected from both lines. This results in unequal reflected crosstalk traveling toward the receiver as both common mode and differential crosstalk. Line terminations to control interference may be more important for high speed interconnections than terminations to control signal ringing. Source termination of line drivers is useful both for reducing the output drive levels so less interference is generated, as well as controlling reflections of crosstalk that travels toward the source. Furthermore, reasonable source termination means that ringing can be controlled with load terminations that are less well matched to the line impedance and it will reduce currents in clamping diodes.

8.4.3 Unbalanced line drivers

Many line driver circuits have an inherent tendency to have a different dynamic impedance in the different states. For example, TTL and CMOS drivers tend to have lower impedance in the low state than in the high state, and the low state is not well matched to a typical line impedance. ECL outputs may have higher impedance in the low state because there is significantly lower output current loading in the low state, especially with an underdamped termination to −2.0 V. If drivers are not specified for impedance matching in both states, the typical design solution is to add series resistance to the driver output so that the added resistance provides a better (although compromised) match to the line impedance for both states. The impact of series resistance includes the additional components required, and frequency capability (bandwidth) is limited.

Crosstalk and reflections of crosstalk are often a more serious noise problem in high speed networks than ringing of signals. Fast rise times generate more crosstalk in less coupled length, and faster clock rates allow less time for signals to settle. Signal ringing may combine with crosstalk reflections so that the worst noise occurs much later than the arrival of the incident wave of the signal. Although the noise will eventually settle to acceptable levels for synchronous data, high speed networks do not have the luxury of waiting for transients to settle.

Most higher speed networks will need to control crosstalk and reflections of both signals and crosstalk to achieve reliable operation and incident wave switching at high data rates. All clock signals that can change the state of memory devices must be carefully controlled so that there cannot be noise or signal distortion that could cause faulty operation. It is reasonable to expect that crosstalk will cause problems in high speed wired interconnection networks unless the system has been carefully designed to operate properly even with the maximum crosstalk noise that can occur. A portion of the system noise margin available must be allocated for crosstalk noise (including crosstalk reflections), and the interconnection network should be designed to reduce the maximum crosstalk to within the noise margin allocation.

8.4.4 Mutual coupling

Some analysis defines the capacitance and inductive coupling separately, where $K_C = C_m / C_0$ and $K_L = L_m / L_0$ are the capacitance and inductance coupling, respectively.[6] In this case, K_B is approximately $K_S(K_C + K_L)/4$. A more general form which is more valid for larger values of C_m and assumes that $C_m / C_0 \approx L_m / L_0$ is the following:

$$K_B = \frac{K_S * C_m}{2C_0 + C_m} \tag{8.8}$$

It may be noted that this form reduces to $K_B \approx K_S * C_m/2C_0 = K_S(K_C + K_L)/4$ when $C_m << C_0$, and never exceeds unity, even when $C_m >> C_0$.

In some cases the mutual coupling C_m may be large compared to C_0. One example occurs if multiple printed circuit signal layers are stacked above a single ground plane to reduce the total number of layers, as illustrated in Fig. 8.5. Only the signal conductors that are routed adjacent to a continuous ground return plane (or dc power plane decoupled to the ground plane) will have a defined impedance and a common reference path for the return current. Any signals attempting to travel along the conductors farther away from the ground plane will couple the signal into the nearest signals. The broadside coupling between signals on the adjacent layers farthest from the ground plane represents an extreme case that must be avoided to reduce crosstalk. It is important to control crosstalk so that all signals travel close to a continuous return path.

Another case of serious coupling between signals may occur if careful attention is not paid to maintaining a continuous ground plane adjacent to the entire path of a signal. For example, a ground plane may be cut out to allow a signal track to cross multiple signals flow-

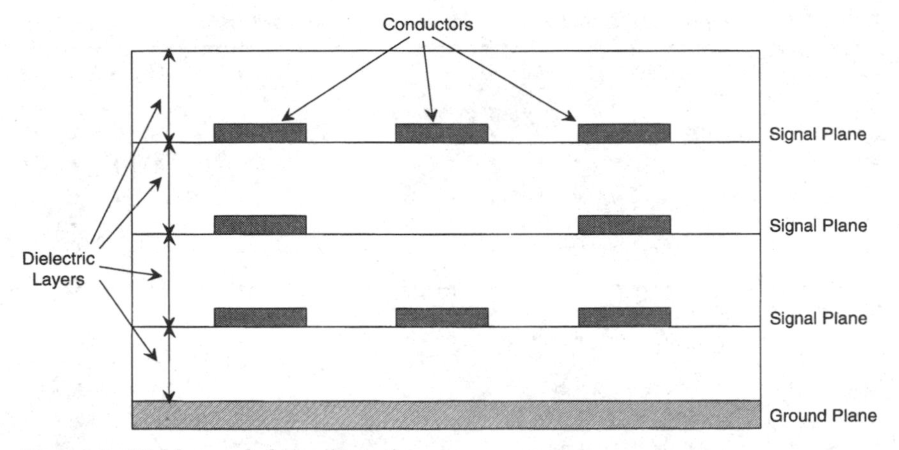

Figure 8.5 Highly coupled circuit routing.

ing toward a connector field. Since ground return current cannot flow through the open part of the ground plane, the return current will attempt to flow through the nearest signals.

An important rule is never to allow ground (or power plane) reference planes to be cut out for a group of vias or pins, and never allow signals to be run in the ground reference plane. A similar condition would occur in a connector pin field if the ground pins are only on one side, or even just on the ends. The ground plane must continue through pin fields to allow return currents to flow around individual clearance holes for signal pins. If signal lines are run through pin fields, the ground return conductor must also run continuously between the same pin fields. The general principle is that a signal return must run adjacent to every signal line. Ground pins must be located throughout connector pin fields to provide a path for signal currents to flow as the signal transitions travel through connectors. The designer should always remember that for every electron that flows in one direction in the signal line, another electron must flow in the opposite direction. If there is no adjacent return path through the ground reference, the current may stop flowing in the signal line (a signal reflection), or the return current may flow through adjacent signal lines as crosstalk.

Typical single-ended signal interfaces require that both dc power and the returns for high speed signals share the common ground pins and ground planes. The designer is cautioned that "high speed signals" means fast transition times (rise times), not necessarily high data rates or clock frequency. It is critical to keep signal return lines close to signal lines throughout the signal transmission line. This includes package power and/or ground pins, continuous power and/or

TABLE 8.1 Examples of Ground Pin Allocations in Connector Pin Fields

Single row 50% grounds	G	•	•	G	•	•	G	•	•	G	•	•	G	•	•	G
Two row staggered 20% grounds	G	•	•	•	•	•	G	•	•	•	•	•	G	•	•	G
	G	•	•	G	•	•	•	•	•	G	•	•	•	•	•	G
Three row offset 30% grounds	G	•	•	G	•	•	G	•	•	G	•	•	G	•	G	
	•	•	•	•	•	•	•	•	•	•	•	•	•	•	•	
	G	•	G	•	•	G	•	•	G	•	•	G	•	•	G	
Four row offset 20% grounds	G	•	•	G	•	•	G	•	•	G	•	•	G	•	G	
	•	•	•	•	•	•	•	•	•	•	•	•	•	•	•	
	•	•	•	•	•	•	•	•	•	•	•	•	•	•	•	
	G	•	G	•	•	G	•	•	G	•	•	G	•	•	G	
Five row staggered 25% grounds	G	•	•	G	•	•	G	•	•	G	•	•	G	•	G	
	•	•	•	•	•	•	•	•	•	•	•	•	•	•	•	
	•	G	•	•	G	•	•	G	•	•	G	•	•	G	•	
	•	•	•	•	•	•	•	•	•	•	•	•	•	•	•	
	G	•	G	•	•	G	•	•	G	•	•	G	•	•	G	

ground reference planes on printed circuit boards, and adjacent power and/or ground pins in connector fields. Typical single-ended signals require about 20 to 50 percent additional power and/or ground reference pins in connectors. Table 8.1 illustrates some examples that provide adjacent power and/or ground signal return reference pin arrangements in connector pin fields. Ground return pins are indicated by a G, although they may be dc power pins that are properly decoupled to ground with high-frequency capacitors. The examples in the table indicate the approximate ratio of ground pins to available signal pins, which depends on various conditions, including how well the repetitive pattern fits at the connector ends. The patterns in these examples are based on ground pins at the corners and ground pins in the outer rows. This is intended to contain the signal fields inside the connector field to reduce the interference to and from other signals or effects of the connector shell. The ground pins may be considered a partial coverage shield, or orienting the pin field somewhat like stripline between ground planes, although the effect for less than three rows is not significant.

8.4.5 Dual stripline

A typical method for reducing broadside coupling in printed circuit layout is to restrict the routing of conductor lines on adjacent layers to orthogonal directions, so that signals only cross at 90° angles to

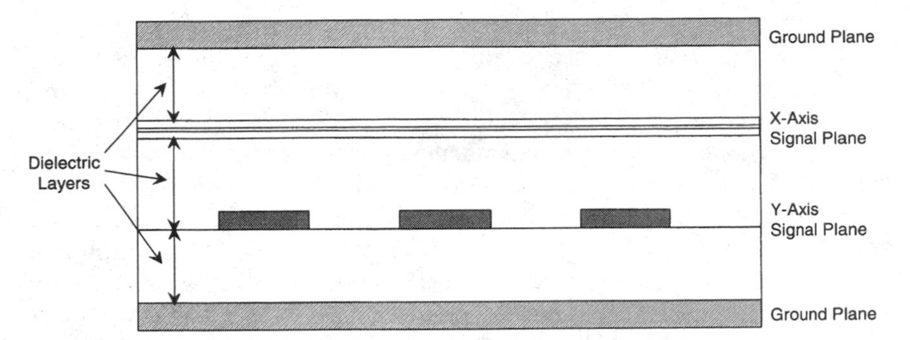

Figure 8.6 Dual stripline illustration.

minimize the coupled length. Even fringe field edge coupling between signals on the same layer may be severe if the conductors are not significantly closer to the ground plane than the open space between the conductors. Although it is preferred to place each layer of conductors between ground planes in the stripline configuration, one compromise is to locate two signal layers between ground planes and restrict the routing to orthogonal directions. This is called *dual stripline* or *offset stripline* and is illustrated in cross section in Fig. 8.6. The horizontal lines represent conductors routed parallel to the paper (left to right, or x axis). The crosshatched rectangles represent the orthogonal conductors routed into the paper (y axis in a plan view). Each signal layer has a near ground plane and a more distant ground plane. The two orthogonal layers allow a high degree of interconnection capacity, with interconnections between the two layers, called *vias.* Vias that interconnect only between adjacent layers and do not extend through the circuit board are called *buried vias.*

Exceptions to the orthogonal routing restriction may be allowed in areas where it can be assured that no parallel signals are routed in the adjacent layer. One of the signal ground planes may be a dc voltage plane, if it is decoupled to the other ground plane near every location where signals change to different reference layers. Multiple ground planes must be shorted together near every location where signals change to different ground layers as the signal return reference.

A minor nuisance for the designer of dual stripline is estimation of the transmission-line impedance because the reference planes are at two different heights, which is not the configuration for the centered stripline as defined in Eq. (7.6). Although equations exist for stripline not centered between the reference planes, they are more complicated and are not yet supported by an industry standard.[6,7] However, Eq. (7.6) may be used to calculate a lower-bound estimate with H the height to the *nearest* reference plane. Similarly, an upper bound can be estimated with H to the farthest reference plane. It is suggested that substituting the *average* of the two values of dielectric height into the centered stripline equation will be an adequate preliminary estimate for line impedance for most engineering purposes, considering the uncertainty of coupling to adjacent signals and manufacturing tolerances.

A high-frequency device that is designed for a high backward crosstalk coefficient is called a *directional coupler.* If the coupler is designed so that the forward crosstalk coefficient is negligible, only signals traveling in one direction will couple to the adjacent port. A directional coupler may be useful in separating signals or reflections that travel in one direction from those that travel in the other direction. A directional coupler may be a useful instrument for investigating reflections on high-frequency lines. Although observing the signal waveforms can only detect the combined result of incident and reflected waves, they can be separated by a directional coupler.

8.4.6 Mutual coupling between active signals

Mutual coupling affects signals on both of the lines that are near each other, although the crosstalk is different in the forward and backward directions. The following discussion considers the one line that has a traveling signal transition as the primary line. The other line that has mutual coupling to the primary line is called the *adjacent* line. Sometimes the "other line" is called the *victim* line. Cases are considered where the adjacent line may also have a signal traveling on it. Of course, when there is also a signal traveling on the other line, the roles could be reversed and the adjacent line could have been called the primary. A signal traveling on the adjacent line also affects the primary line, including both crosstalk and a change in impedance.

If the adjacent line has a signal transition that occurs at the same time as the transition on the primary line, the primary line impedance is either increased or decreased. Which case occurs depends on whether the transition on the adjacent line is in phase or opposite phase as the transition on the primary line. The impedance for in-phase mutual coupling is called *even mode* impedance; opposite phase mutual coupling is called *odd mode* impedance. Mutual coupling to a signal that is in phase reduces the loading effect of the capacitance to ground, so the impedance on the primary line is increased to about $Z_{\text{even}} = (1 + K_B)Z_0$. When the mutual coupling is to a signal that is opposite phase, the primary line impedance is reduced to about $Z_{\text{odd}} = (1 - K_B)Z_0$. The impedance of the primary line when the adjacent line is quiet is the geometric mean of the even and odd mode impedance, so the quiet mode impedance Z_{quiet} is about the following:

$$Z_{quiet} = \sqrt{Z_{even} * Z_{odd}} = Z_0 \sqrt{1 - K_B^{\ 2}} \tag{8.9}$$

The presence of mutual coupling can be considered to have an effect on impedance similar to lumped capacitance, which will ordinarily cause some reflections on the driven signal line. It is usually not practical to compensate for mutual coupling effects, partly because the effect depends on the conditions for the other signals. It should also be noted that unlike lumped capacitance loading, mutual coupling in a uniform dielectric does not have a significant effect on propagation delay because it also decreases inductance. However, waveshape distortion may increase the total delay for the signal to achieve the level necessary to cross the threshold of the receiver. An effective method to reduce mutual coupling to control both crosstalk and impedance mismatches for printed circuits is to keep the signals spaced farther apart than the conductor-to-ground spacing (maintaining a spacing-to-height ratio of 3:1 is recommended). Shielding and twisted pairs with balanced differential signals are effective methods for limiting crosstalk in cable bundles.

8.5 Differential Pairs

A differential line pair operates in odd mode because the signals traveling on the two lines are of opposite polarity. This effect on impedance is consistent and can be estimated from crosstalk calculations. Part of the signal may be considered to travel as individual signals in the dielectric between each of the two signal conductors and a ground reference. Mutual coupling between the two conductors means that part of the signal travels in the dielectric between the two conductors. When both signal conductors are in close proximity to a shield or ground plane, the dominant (lower) impedance is each line to ground. Printed circuit conductors in a microstrip or stripline configuration have broadside coupling to the ground plane(s), with only edge coupling between conductors. Therefore, the mutual impedance between lines is higher than the impedance to ground for usual line spacing. The differential impedance is twice the reduced impedance of each line, or approximately $Z_{diff} = 2(1 - K_B)Z_0$, where Z_0 is the intrinsic impedance of a single line to ground. The physical layout is illustrated in Fig. 8.7*a*, with the equivalent differential impedance indicated in Fig. 8.7*b*. It should be noted that the typical Z_0 of each line to ground is much lower than Z_M, so most of the transient differential current actually flows through each Z_0 and through the ground plane.

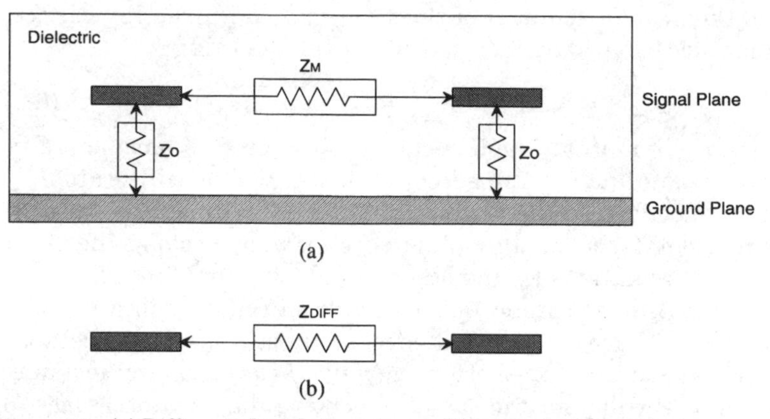

Figure 8.7 (*a*) Differential mode impedance. (*b*) Equivalent circuit.

8.5.1 Circulating ground currents

Although very little static current is carried in the ground for balanced differential lines, it is important to maintain a continuous ground reference along the differential line for high speed signals. The ground carries circulating currents that flow as the rise time of the transient travels along the line and maintains the constant line impedance necessary to prevent reflections and signal distortion. Although the equivalent circuit shown in Fig. 8.7 shows only an impedance between lines, the transient currents circulate in the ground plane. This results because the actual dynamic impedance to ground is much lower than the mutual impedance. Increasing currents flow under the leading edge of a signal transition, then decrease after the trailing edge as the loop current begins to flow in the signal conductors. The transverse circulating ground current under the leading edge flows from under the positive-going signal toward the negative-going signal. This displacement current flows in the dielectric between each conductor and the ground reference and diminishes as the steady-state signal level is established.

The ground reference is often the shield for long cables. The ground reference also carries whatever ground currents flow between units. It is important to control the ground offset between ends of directly wired interfaces, because most high speed receivers have a very limited common mode offset voltage tolerance. Although isolators such as transformers or optical couplers may significantly increase the offset voltage tolerance, even these have a finite voltage limit and may seriously restrict performance.

Even well-isolated interfaces (including low-frequency signals and power distribution) should provide a discharge path for leakage currents or static charges. An interconnecting cable should not be the only ground interconnection, especially if the only ground reference could be lost when the cable is disconnected. Typical interface circuits may be damaged by static charges. All signal interconnections that can be disconnected should be provided with an auxiliary ground discharge path that remains connected during the engaging or disengaging of the signal connectors. One method for achieving this is to use recessed signal pins, with longer contacts to engage the ground before the signals and disengage the ground last when disconnecting.

It should also be considered that other currents may share the common ground system. The entire ground system must be designed to control the interference to sensitive signals. The ground system must be sufficiently robust to carry any power-line currents that may flow through it (including fault currents), as well as transients such as static discharge, electromagnetic radiation, solar flares, and lightning surges.

8.5.2 Common mode of differential pairs

The opposite polarity signals of a differential pair mean that ideal signals average to a constant for all points an equal distance from both conductors. The accuracy of this cancellation depends upon how exactly equal the signals are in timing, waveshape, and amplitude. This signal cancellation at points of equal distance means that a "virtual ground plane" (of zero thickness) exists at the midpoint between differential lines. This is illustrated in Fig. 8.8 for two flat conductors over a ground plane.

This is the same virtual ground plane concept described previously for impedance calculations, owing to the symmetry between the single wire over ground and the two-wire electric fields. However, it should be emphasized that most of the transient displacement current flows in the ground reference plane, not through the virtual ground illustrated in Fig. 8.8. The virtual ground plane remains at the same

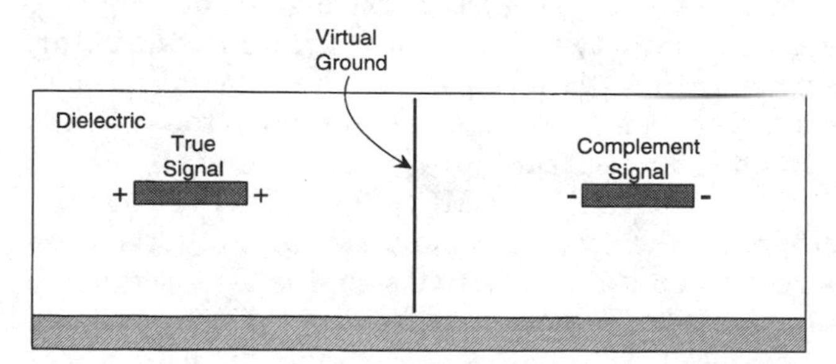

Figure 8.8 Virtual ground between differential pairs.

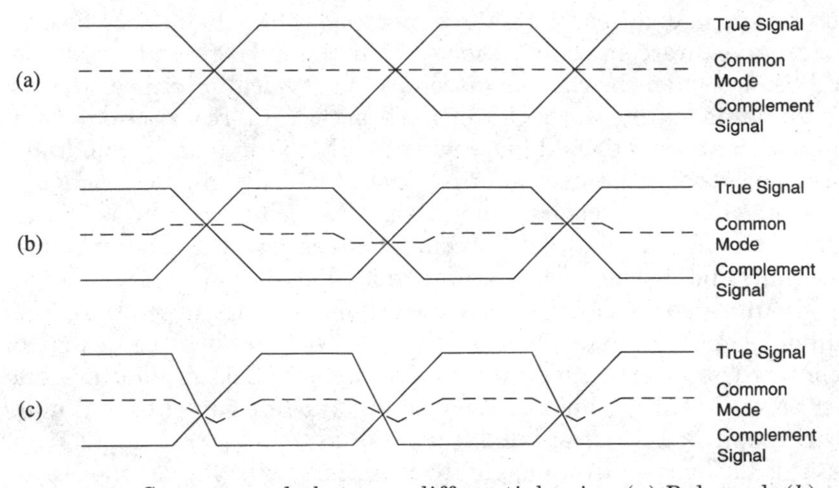

Figure 8.9 Common mode between differential pairs. (*a*) Balanced. (*b*) Complementary skew. (*c*) Unbalanced rise time.

potential only when the signals on the two conductors are exactly opposite polarity (including waveshape during transitions). The average of the two signal levels is called the *common mode* voltage level. For example, if the two levels are ECL signals of −0.7 and −1.8 V, the average common mode level is −1.3 V. For CMOS levels of +5 V and ground, the average common mode level is +2.5 V.

For exactly balanced differential signals, the common mode level remains constant as the signals change state. Figure 8.9 illustrates the exactly balanced case with constant common mode and two unbalanced cases. Case *b* has the same waveshape, but one signal is slightly shifted in time, which is called *complementary skew.* Case *c* has a different rising edge transition rise time than the falling edge. When the signals are not exactly matched, a common mode signal results from every transition. Common mode signals are a common cause of crosstalk and EMI radiation in differential interfaces.

The cancellation of signals does not occur for single-ended signals, even if they are paired and twisted with a ground conductor. Furthermore, the dynamic impedance Z_0 is not the same as if there were a virtual ground between the conductors, because the common mode voltage level changes each time the signal changes state, as illustrated in Fig. 8.10. For the 5-V CMOS example, the common mode level is +2.5 V only when the active signal is at +5 V; the common mode level is zero when the only active signal is also at ground. Therefore, the common mode is about 50 percent of the active signal. The signals do not cancel at points an equal distance from both conductors. Furthermore, even the ground conductor carries current and

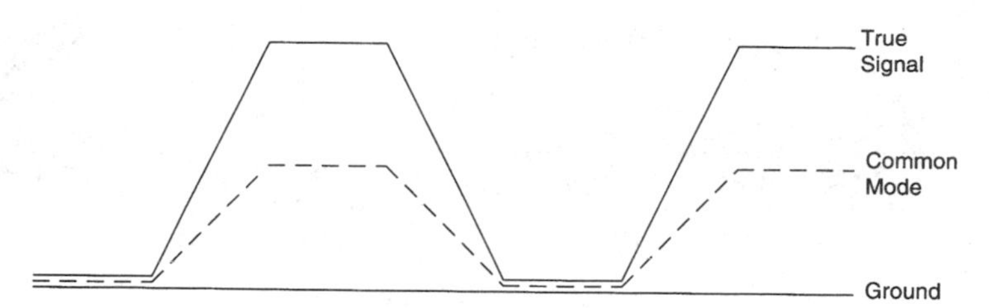

Figure 8.10 Common mode between single-ended signal and ground.

cannot be considered a perfect ground, owing to finite resistance and length. The ground itself may even result in standing waves at resonant frequencies.

Differential signals cancel exactly only at points exactly an equal distance from the two-signal conductors, along the "virtual ground" that lies between the conductors. The fields do not exactly cancel to one side of the signal pair. However, as the distance from the conductors becomes large compared to the separation between conductors, the signals nearly cancel for most directions. Therefore, the electric field surrounding a differential pair decreases rapidly at greater distances away from the two conductors even for most directions not exactly an equal distance from the two conductors. The exception is for points that lie in (or near) the same plane as two flat conductors, where the electric field is influenced by only one conductor. In this case, the far conductor essentially lies in the "shadow" of the near conductor, so there is little influence from the far conductor. This means that differential printed circuits (including "flex" interconnection cables) do not provide any significant reduction in crosstalk to adjacent conductors on the same layer.

A common approach for controlling crosstalk into adjacent circuits on the same printed circuit layer is to maintain sufficient spacing between conductors. Ground planes on both sides of conductors (called stripline for printed circuits) greatly reduce crosstalk compared to a ground plane on one side only. Particularly critical signals can be isolated with grounded guard lines between conductors. Grounded guard lines is often an effective technique on flex or ribbon cable interconnections, where the conductors are at a predetermined spacing, usually to match connector pins. A foil shielded ribbon cable is illustrated in Fig. 8.11. This shows a ground wire between each of the differential pairs, including on both ends. Although it may be inconvenient, the grounded lines of a long cable should be connected to the ground plane or a low-impedance shield at sufficiently short intervals to avoid standing waves or resonance which reduce the shielding effectiveness. Routing the cable adjacent to other ground

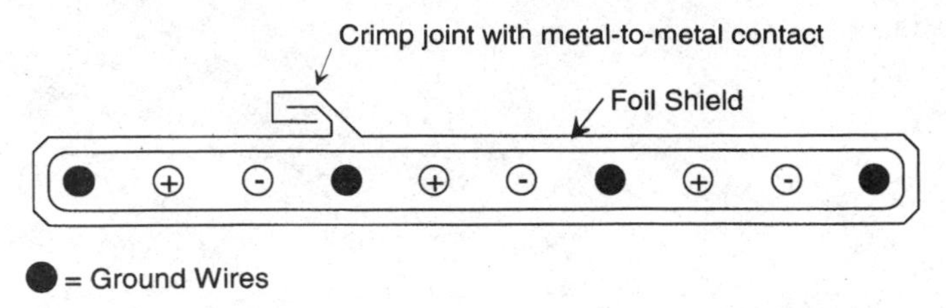

Figure 8.11 Ribbon cable with foil shield.

planes or grounded structural members may also reduce the high-frequency resonance of the shield of a long cable.

It is desirable that the shield be in contact with the ground wires (especially the end wires) to carry any shield currents. The foil joint should be made to ensure that there is full contact of the foil metal-to-metal along the seam, not simply overlap. Contact to a shielded connector at the ends of the cable is especially critical to terminate any shield currents into the enclosure to minimize voltages generated by shield currents.

The shield (typically foil) surrounding a flat ribbon cable may be formed almost around each of the conductors to provide an effective shield between conductors. An example is illustrated in Fig. 8.12. This construction can be used to avoid the need for some of the guard wires between each pair. This construction is also appropriate for transmission of single-ended analog or digital signals that would otherwise require individually shielded wires. If the narrow gap between shields between conductors is much less than the conductor-to-ground spacing, only very high frequency components could pass through the aperture. However, the signal conductor-to-ground spacing determines the upper limit to frequencies that will travel along the line. Therefore, all frequency components (dc to the upper limit) that can travel along the line are blocked from coupling to adjacent lines nearly as effectively as if the conductor were completely surrounded by a shield.

A household example of a nonsolid shield is the window of a microwave oven. The holes in the metal mesh in the window pass visible light but are small enough to block the microwave frequency.

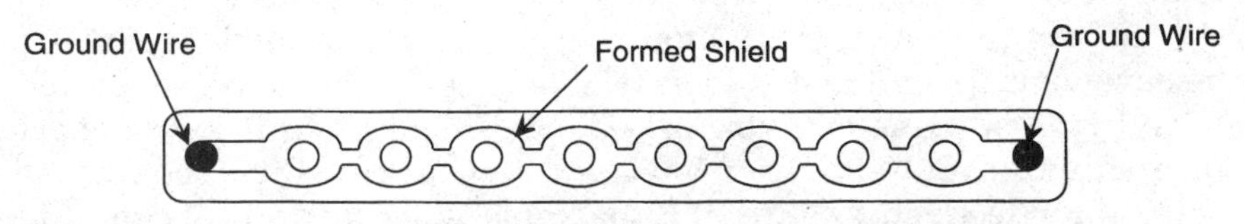

Figure 8.12 Ribbon cable with formed shield between conductors.

Similarly, the large air traffic control radar antennas at many airports are not solid (to reduce weight and wind resistance), but the mesh size is small compared to the radar wavelength.

8.5.3 Shield effectiveness and transfer impedance

Although thin foils may be used as signal shields, it should be understood that the impedance of the foil alone is not low enough to be an effective shield for high levels of interference for long cables, or to carry ground currents that may flow between units. Furthermore, it is important that a low-impedance connection be made from the shield to the connector at the end of the cable so the shield integrity is maintained through the connector and to the mating shielded enclosure. Therefore, ground wires must be included in the cable to carry currents, and the shield must maintain reasonable contact with the ground wires, which are sometimes called *drain wires.*

The effectiveness of a shield may be measured by *transfer impedance,* which is the ratio of voltage induced on the signal conductors to current flowing through the shield (*including the connectors*).[8,9] Any voltage generated on the shield is induced or transferred onto the signal conductors and appears as noise on the signals. A transfer impedance envelope for a braid shield on a round cable bundle is illustrated in Fig. 8.13. At low frequency, the transfer impedance is simply the dc resistance of the shield, but impedance is more complex at higher frequencies owing to inductance and standing waves which depend on

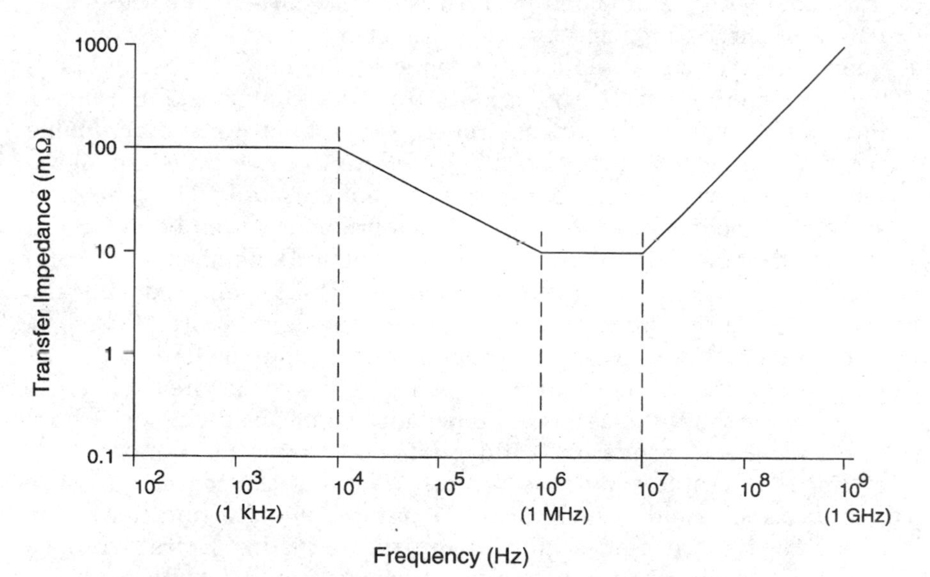

Figure 8.13 Transfer impedance envelope.

the shield construction, location of adjacent grounds, and length. Good cable shields have low impedance (higher shield effectiveness) in the 1- to 100-MHz range where much of the digital noise is generated. However, it is nearly inevitable that the envelope of the shield transfer impedance rises after the cable length becomes significant compared to the wavelength of the frequency. The illustration indicates the envelope of the upper limits of an actual impedance function of frequency. An actual transfer impedance function has characteristic peaks and nulls due to interaction with nearby structures, resonance effects, internal ground lines, and standing waves on the shield.

It should be realized that the shield itself acts as a transmission line when the cable length is long compared to the wavelength of the frequencies involved. This means that it is very difficult to provide effective shielding on long cables for the high-frequency components of fast rise time digital signals. It is usually more effective to avoid EMI than it is to shield EMI once a problem is discovered. EMI may be any combination of problems, including excessive emissions, susceptibility to external energy (including ESD and lightning), internal crosstalk and sneak paths, and power disturbances. The use of balanced differential signals and maintaining close spacing of signals and power distribution to ground planes, shields, and grounded structural members are usually effective to reduce EMI. Printed circuit (and flex cables) of stripline between ground planes is more effective than microstrip. Shielded twisted pairs are more effective for reduction of EMI than flex or ribbon cables. Keep the longest dimension of any openings in a shielded enclosure much shorter than the wavelength of the frequencies involved. Reducing the rise time (and frequency) is a classical method for reducing the high-frequency bandwidth of digital signals, but this is a luxury that most high speed interconnections cannot afford. Source termination (including double termination for long lines) does help reduce the rise time and signal amplitude as well as ringing of signals and crosstalk. The transmission-line impedance of the shield at high frequencies can be reduced if the distance to a large ground plane or structural member is low compared to the diameter of the cable shield. This means reducing the H/D ratio for the round wire over ground impedance in Eq. (7.4).

The effectiveness of shields depends on shield construction, and critically on the shield connections to the connector at the cable ends. Solid metal conduit has lower impedance than braid shields, whose effectiveness depends on braid pitch and coverage. Temperature cycling of the cable and connector may relieve initial contact pressure and degrade shield effectiveness. Therefore, shield connections may need to be tightened after initial temperature cycling. Spiral-wrapped shields (metallized tape or wires) with poor contact between turns will have more inductance and resonate at lower frequencies than a braid shield that provides nearly continuous contact along the entire length of the shield.

A perfect shield would allow unlimited external interference to induce currents in the shield at no voltage and therefore no induced voltage on the internal signals. Of course, a shield is intended to work both ways, to reduce external interference from inducing noise on the internal signals as well as to reduce the external emissions that result from the internal signals. The shield is imperfect to the extent that when electric fields on one side of the shield induce currents in the shield, voltages on the shield cause electric fields on the other side. Therefore, lower induced voltage on the shield means lower electric fields on the other side. The high impedance of foil shields means that they are far less effective than braid shields or solid conduit at high frequency. Very low noise applications may use more than one shield layer to improve shielding effectiveness.[10]

A concept that may be used to understand a cable shield is to consider the shield to be an antenna, although effectiveness as an antenna is undesirable in this case. As for transfer impedance, the lower the antenna effectiveness, the better the shielding. The effectiveness as an antenna depends on the length compared to the wavelength of the frequency of the signals. At very short lengths, the shield is a poor antenna, but at appropriate lengths relative to the frequency, the shield becomes more effective as an antenna but less effective as a shield. As a receiving antenna, an imperfect shield converts some of the external electromagnetic fields to electrical interference on the internal signals. As a transmitting antenna, an imperfect shield converts some of the internal signals to external electromagnetic radiation. Routing the shielded cable close to other ground planes and grounded structures, as well as using cable straps grounded to the structure, should reduce the high-frequency shield impedance and the tendency of the shield to radiate at high frequency. High speed digital processors and interface connections generate high levels of electrical noise over a very wide frequency spectrum. Preventing wideband digital noise in long cables from interfering with any nearby sensitive receivers, sensors, or other processing equipment is often a challenging shielding application.

8.5.4 Twisted pairs

The differential lines in many longer cables are run as twisted pairs to reduce the resulting electric field at all locations around the twisted pairs. When the twist interval is short compared to the distance that the signal travels during the rise time transition, approximately equal lengths of the true and complement are exposed to any point away from the twisted pairs. Therefore, the electric fields approximately cancel, so the field strength drops rapidly at distances farther away (called the far field) from the twisted pairs. However, the twisting is not always effective for adjacent conductors in the near field. For example, adjacent pairs twisted at exactly the same pitch may have matching signal polarities aligned at each twist, so that the mutual coupling is not canceled. Most of the electric field lines between the differential signals are concentrated between the twists. At any significant distance away from the cable, the external field is nearly canceled by alternating turns of the signal wires. However, this is not true where the cable ends, as the last turn is not matched by another turn. Furthermore, the wires typically are routed without twisting for a short distance to facilitate the connections at the end of the cable.

The propagation delay for twisted pairs is longer than coaxial cable because the twisting increases the length of the twisted wires compared to the straight-line distance between ends. The total propagation delay (electrical length) of twisted wires increases for shorter twist intervals. If a pair of wires is cut to a certain length before twisting, the physical length of the pair shortens as the wires are twisted. Of course, the same effect occurs in winding the strands of rope, but there is little concern for the total length of the strands in rope (except to start with enough extra length to finish with the desired length). One of the concerns for the electrical length of a twisted cable is that the timing accuracy of multiple signals will depend on the accuracy of the twist interval. Furthermore, if a larger round cable is built up with spiral-wound layers, the propagation delay depends on the spiral pitch interval in each layer. High speed interfaces typically require timing accuracy which is tight compared to the delay of longer cables. Careful attention must be paid to the design and assembly of complex cables to maintain accurate timing relationships between signals.

8.5.5 Estimating crosstalk

Calculating the values for crosstalk is usually quite difficult to perform manually. It is uncommon to be able to estimate the ratio of mutual coupling to intrinsic capacitance of lines. Published data is useful, but it is typically sensitive to the assumptions used to simplify the analysis. For example, the analysis may assume that all lines are perfectly terminated at both ends to eliminate reflections and that there is only one primary line as the source of only one driven signal. A "rule of thumb" that may be used to scale published crosstalk data is that crosstalk may be twice the single source line level if there may be sources of crosstalk from two sides, and crosstalk may double because of reflections. Therefore, a crosstalk budget should allow for at least 4 times the published levels; 5 times is recommended to allow some margin for uncertainty.

Rule of thumb for estimating multiple line crosstalk without source termination: at least 4 times the published data for single lines terminated at both ends.

The analyst is cautioned against placing too much confidence in either published or "rule of thumb" crosstalk data. A continuous, low-impedance ground plane is assumed for most analysis. Care must be taken to assure that this is appropriate, especially if signals change reference planes, travel through connectors, or transition to cables. Furthermore, most analysis is based on a single transient, which begins with static conditions. High speed networks and interconnections typically have high speed, repetitive signals that are not likely to settle to static levels between transitions. Real systems also usually have multiple signals that may switch simultaneously and share the same ground system. Multiple, repetitive signals may combine to produce more serious results than predicted for a single transient. The effects may be similar to standing waves or resonance which may respond to excitation frequencies.

8.6 Grounding Systems

Ground or shield resonance may be nearly inevitable when the interconnection is long compared to the clock frequency or data rates. It is usually not practical to include damping resistance in the ground or shield or to shorten the propagation delay to less than the clock frequency or data rate. A common approach to reducing the effect of high-frequency circulating currents is to implement a sufficiently large and low-impedance ground or shield to reduce the voltage transients that result from the currents that do flow. The interface drivers and receivers must be selected to operate properly in the presence of the resulting voltage transients, including resonance or standing waves on the ground system itself.

Particular attention must be paid to the design of the grounding philosophy for complex systems, especially when different technologies are expected to perform properly in close proximity, sharing resources, and with potentially conflicting requirements. The integration of more digital processing into complex sensor, data processing, and communication systems often results in even more difficulty in meeting the requirement that the wide-bandwidth noise produced by digital processors does not degrade the operation of sensitive circuits.[11]

The concept of "single-point ground" was developed primarily for low-frequency power interconnections, where a wire can control where currents can flow. Resistance (not inductance) is usually the primary concern for distribution of high currents. However, high-frequency signals are not limited to wires but may travel in any shorter, lower-impedance dielectric path of convenience, sometimes called "sneak paths." High-frequency ground systems are usually designed to provide a very low impedance path for currents to flow within the unit without interfering with other sensitive units. These "mass ground" systems allow current to flow in any shortest path of low impedance but limit the voltage that might result.

Typical high speed digital and rf systems integrate the ground and shielding systems with the heat transfer and chassis structure. This provides a very low impedance path for internal ground currents and terminates the shield of external cables. DC power distribution to th ground system should be connected by a single-point ground if possible to avoid large ground return currents flowing through the ground system and structure.

8.6.1 Separation of ground areas

Signal ground planes may be connected to the chassis with a modified single-point ground philosophy. It is usually important to connect the chassis to the signal ground near the area where the external interface signals enter and/or exit the interface circuits. A modified single-point ground philosophy may be used to improve isolation between high-level noise generation sources and sensitive circuitry. As an example, TTL and CMOS logic with fast rise times generates significant high-frequency noise, which tends to exceed the noise margin of ECL logic and interferes with sensors and communication receivers that operate at very low levels.

Mutually incompatible technologies can be physically separated to reduce interference by the same principles as reducing crosstalk, that is, to separate the signals by significantly more than the signal space to ground. The signals and transients are effectively kept within their respective areas by a close ground plane that provides a shorter, lower impedance path than into the other area. If it is necessary to route some potentially interfering signals through a sensitive area, the signals should be shielded, typically by routing on a different printed circuit layer, and separated by a ground layer as a shield. The grounds may be connected between the areas in a modified single-point ground, preferably where there are signal interfaces. If the signal interfaces are external, the ground is connected to the chassis; if there are interface signals between the areas, the grounds should be connected between them at the location of the interface signals.

It is generally preferable to cluster the interfaces to a small area if possible. However, if equally sensitive interfaces must be placed at different locations, it is generally recommended to provide ground connections at each interface, with the ground impedance between the connection areas as low as practical. Low-impedance connections need to be wide (minimum length-to-width ratio) to reduce high-frequency impedance. Typical interfaces between signal areas include logic translators and analog-to-digital converters. Interface circuits that minimize ground currents also reduce ground noise.

8.6.2 Mass grounds vs. isolation

There may be concern as to how massively the grounds between these areas should be connected versus how well they should be isolated to avoid interference. A more massive ground interconnection is usually preferred to reduce the ground noise that ground currents may generate. One method that may be used to provide flexibility is to make the ground interconnections between critical areas as removable links. If there is concern that performance may be adversely affected by the ground connections, they may be removed and performance compared.

Mass ground systems are designed to control ground voltage by providing a low-impedance path for any currents that flow. Single-point grounds are designed to limit where low-frequency currents may flow by restricting the location of ground connections; the ground voltages that result are determined by the impedance of the ground connections. Even the commercial ac power distribution (60 Hz in North America) typically uses a mass ground philosophy with an earth ground connection at each load point to minimize the chance of stray voltages.

Interfaces to sensitive units should be designed to reduce the opportunity for interference, typically with high-impedance interfaces that draw very low currents that flow between units. Differential interfaces with line-to-line terminations (not to the receiver ground) is a technique to avoid termination current from flowing between units. Special shielding and isolation techniques may be needed to isolate sensitive inputs from ground currents.

When the ground or shield length exceeds about 5 times its width, inductance becomes significant and the ground impedance increases for higher frequencies. Ground planes in flex cables or ground straps between units should minimize the length-to-width ratio to minimize inductance. DC resistance is usually of less importance unless the ground is also carrying large power supply currents. The reason that width is important is the skin effect. Mutual repulsion of like changes in current means that high-frequency current transients spread out from the source of the transient and are crowded to the outer edges of flat planes. The farther apart the like currents can spread, the lower the self-inductance and current density, and the lower the voltage transients that result from current transients. It is not essential that the wide ground interconnection be solid; multiple wires in parallel may be nearly as effective if the open space between them is small enough compared to the quarter wavelength for the highest frequency of concern.

The same effect applies at interface connectors and at the smaller scale of device package leads. Ground and power pins should be spaced throughout the connector pin field, not clustered together. A single package ground or power lead with a high length-to-width ratio cannot support high-frequency current changes without significant voltage transients. Package lead inductance may be reduced by the same method needed at other connectors: multiple ground leads are spaced so that all signal leads are reasonably close to ground lines. This allows the high-frequency current transients to spread out over leads spaced farther apart. Multiple package pins are usually required for both ground and power leads, especially for rail-to-rail CMOS line drivers. Power and ground package lead inductance is typically a significant limitation to how rapidly a high speed driver can switch transmission-line load currents.[12]

8.6.3 Computer analysis

Computer programs that use interconnection geometry, dielectric properties, signal rise times, and the input-output characteristics of drivers and receivers are usually a good source of estimates for line impedance and crosstalk noise. However, it should be understood that all analysis programs (including any programs that analyze transmission line or circuit characteristics) are only as good as the assumptions and models and algorithms used, and how well these represent the conditions, complexity, uncertainties, and variability of the system in actual operation. Although the computer program may provide results to several digits of numerical precision, the careful analyst should realize that the computer results are only estimates. There are no laws of physics or Maxwell's equations that promise that electrical signals in real systems have to behave just as a computer printout says they should.

8.6.4 Typical estimates: Some 10 percent rules of thumb

The following typical results are intended as "rules of thumb" guidelines that may be used to understand the basic concepts, or as order of magnitude preliminary estimates when no other analysis or measurements are available.

Twisted pairs couple about 10 percent of the differential signal as common mode to adjacent lines. Likewise, twisted pairs receive about 10 percent of an adjacent signal transition as a differential signal. Well-terminated, balanced differential lines convert about 10 percent of common mode coupling into differential noise. Braid shields should attenuate signals to less than 10 percent at low frequency. However, high-frequency (above about 10 MHz data rates) performance of twisted pairs and shields is sensitive to the shield design configuration and cable length, and is extremely vulnerable to degradation of the termination integrity.

Microstrip (thin conductors near a single ground plane) should have less than 10 percent coupling to adjacent lines if the clear space between conductors is 3 times the height of conductors above the ground. Stripline configurations (thin conductors between ground planes) may reduce the spacing by a factor of 2; a space of 1.5 times the height to each ground plane for less than 10 percent coupling. However, a typical layout without concern for coupling may have a spacing to height above ground ratio of about unity, and driven signals may run on both sides of a "victim" line. For this case, backward crosstalk coupling of up to 50 percent of the driven signal step is possible. The two ground planes of the stripline configuration provide only a little advantage at close line-to-line spacing because the mutual coupling becomes more significant.

The distance traveled during a 1-ns rise time is about 10 to 15 cm (4 to 6 in). Therefore, if the coupled length can be reduced to less than a few centimeters, the crosstalk can be reduced significantly. For example, even if the spacing in a connector or dense pin field is about the same as the distance to the nearest ground, the crosstalk can be limited to about 10 percent if the length of the close spacing is limited to less than 2 cm and the rise time is at least 1 ns. Of course, it is important that there is a continuous ground plane near all signal conductors and that ground pins be located throughout connector pin fields so that all signal pins are near grounds (never cluster power and ground pins together!). A continuous ground reference should be carried through connectors and cables to maintain control of impedance and crosstalk.

8.7 Ground and DC Power Planes in Printed Circuits

DC power pins in connector fields or solid dc power planes in printed circuits may be used as an impedance reference, but they must be decoupled to the ground plane with good high-frequency capacitors, especially at locations where signals may transition to or from a ground reference and a dc power plane reference. About a 0.1-μF multilayer ceramic (MLC) high-frequency capacitor (or equivalent) with lead length kept to an absolute minimum is typically recommended for rise times of at least 1 ns. A smaller value such as 0.01 μF for lower effective series inductance (ESL) and low effective series resistance (ESR) may be appropriate for faster rise time signal transitions. Larger high-frequency capacitors (about 1.0 μF) should be located within a few centimeters of each of the smaller capacitors. Decoupling capacitors should be placed so that a capacitor is connected to the reference planes within about 1 cm (including total capacitor lead lengths) of the location of any signal transition to or from dc power and ground references. This is illustrated in Fig. 8.14, which shows a decoupling capacitor near a signal transition.

The top layer of the multilayer printed circuit board (MLPCB) is pads for through-hole vias and component leads. The second layer is a ground plane; layers 3 and 4 are *X* and *Y* signal layers in a dual stripline configuration for orthogonal routing. The fifth layer is a dc power plane also used as the signal reference, with the sixth and seventh layers other *X* and *Y* signal layers as dual stripline. An actual

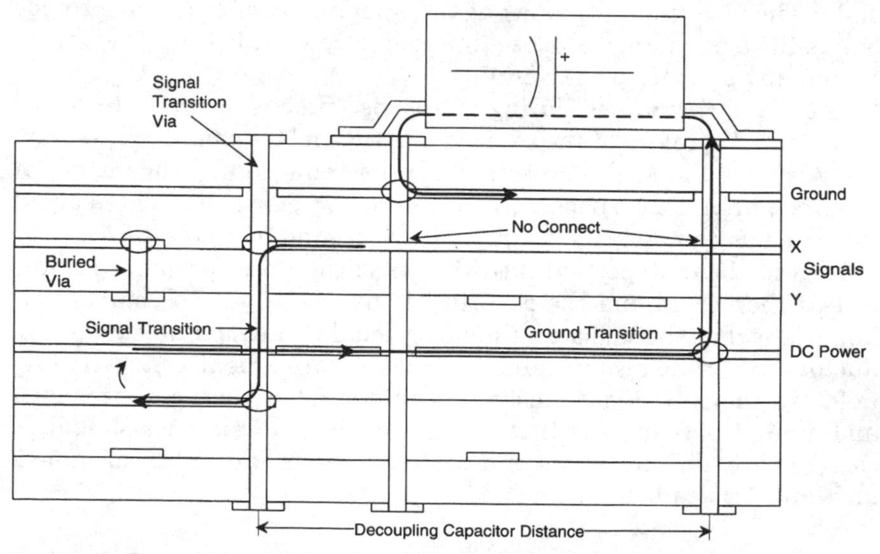

Figure 8.14 Decoupling capacitors for signal transitions.

MLPCB may have several more sets of dual stripline between reference planes than this illustration. The sequence of ground and power planes separating *X* and *Y* dual stripline signals as illustrated tends to require about three layers for each two signal layers. Connections to the vias are indicated by ellipses in this illustration.

The signal flow shown by the heavy arrow is routed from right to left in the *X* direction in layer 3, which has a signal return reference to the ground plane of layer 2. The signal transitions through the through-hole signal transition via to layer 6, which has a signal return reference to the dc power plane of layer 5. The capacitor is connected between layers 2 and 5 and is located near the signal transition. The decoupling capacitor allows the return current which was flowing in ground plane layer 2 to transition to dc power plane layer 5.

8.7.1 The return current path for signal transitions

The signal flow shown is quite direct and would be controlled by the signal routing rules. However, the return path is less direct and obvious, even in this illustration. Furthermore, most signal routers or even the designer reviewing critical signal routing may ignore the signal return path in the reference planes. It is just too easy to assume that ground is everywhere, so the ground reference is just "there" wherever it is needed.

The signal current is considered positive, entering the figure from the right on *X* signal layer 3. The return current for this signal then exits the figure from the right on ground layer 2. The initial dynamic current flows as displacement current in the dielectric between signal layer 3 and ground layer 2. However, when the signal transitions to signal layer 6 below the dc power plane, dynamic current no longer continues to travel to the left in the ground layer. The return current that was flowing in the ground plane must transition to the dc power plane. Therefore, the signal current is reduced until the new return path is established.

The return current sequence as is shown in this illustration now finds a nearest path up through the ground via to the − terminal of the capacitor, through the decoupling capacitor to the + terminal, then down through the dc power via to the dc power plane, then from right to left through the dc power plane to the area where the new signal runs below the dc power plane.

The signal transition cannot begin to travel in the dielectric to the left below the dc power plane until the transient return current begins to flow in the path through the decoupling capacitor. Both the signal current path and the ground return current path are shown in Fig. 8.14. The curved arrow between the signal and the dc power plane indicates the current in the dielectric as the transient wave travels to the left in the figure. At a sufficiently fast rise time, the difference in the path length would be seen as a high impedance for the additional length of the ground return path. It should also be noted that any inductance in the leads or internal connections of the capacitor will delay the change in transient current through the capacitor. Therefore, the decoupling capacitor must be a low-inductance, high-frequency capacitor with leads as short as possible. The total decoupling capacitance should be large compared to the total line and lumped capacitance that may be charged during signal transitions. If the goal is for the power supply to change less than 1 percent during signal switching, the local high-frequency decoupling capacitance should be greater than 100 times the total load capacitance.[13] A 1 percent change in the capacitor voltage also means a 1 percent difference in the signal voltage with respect to ground. The decoupling capacitor must be located close enough to all signal transitions so that the additional path length in the reference return path is not significant compared to the signal rise time. If this is not practical for very high speed or critical signals, the signals should be routed along a single, continuous reference plane.

A buried via is shown to the left of the figure, between X signal layer 3 and adjacent Y layer 4 of the dual stripline signal pairs. A signal transition between these two adjacent layers has the same reference plane transition issue, but to a lesser extent because a portion of the return current flows in each of the two adjacent reference planes both before and after the buried via transition. Decoupling capacitors near signal transitions are still required because the greater proportion of the total return current changes to the reference plane nearest the signal layer.

8.7.2 An alternate MLPCB stack-up for dual stripline

An alternate to the layer stack-up sequence that was illustrated in Fig. 8.14 is to use the same voltage for the reference plane both above and below each pair of dual stripline signal layers. A typical layer stack-up would alternate dual stripline signal layers between two ground layers, with dual stripline layers between two dc power layers. This will tend to require about four layers for each two signal layers. However, since this sequence is an even number, the odd-to-even layer sequence tends to remain the same. This permits buried vias on each pair of dual stripline signal layers because buried vias are normally allowed only between the original two-sided circuit layers (called "C-stage"), before they are stacked together to form the multiple layers. Buried vias are preferred for critical high speed signals because they have less lumped capacitance than a through-hole via.

An additional benefit is that placing ground and dc power back-to-back tends to provide a large-area, high-frequency bypass capacitance between power and ground. The advantage for high speed signal transitions is that the decoupling between reference planes for each pair of dual stripline signal layers is simply a via to short the same voltage together. This can be located closer to the signal transitions, eliminating the physical size and the internal and lead inductance of the capacitor, although power supply capacitance is still needed.

An example of a dual stripline layer stack-up with common reference planes is illustrated as follows:

Outer layer 1: Pads

Layer 2: Ground

Layers 3 and 4: Dual stripline signals (with buried vias)

Layer 5: Ground (same as layer 2)

Layer 6: DC power

Layers 7 and 8: Dual stripline signals (with buried vias)

Layer 9: DC power (same as layer 6)

Layer 10: Repeat layer 2 sequence (ground or power) (or pads if the outer layer)

8.8 References

1. Hugh H. Skilling, *Electric Transmission Lines,* McGraw-Hill, New York, 1951, chap. 9.10, p. 208.
2. M. E. Van Valkenburg, *Network Analysis,* Prentice-Hall, Englewood Cliffs, N.J., 1955, chap. 13-4, p. 324.
3. Samuel Seely, *Electron Tube Circuits,* McGraw-Hill, New York, 1958, chap. 4-14, p. 138.
4. Joseph M. Petit and Malcolm M. McWhorter, *Electronic Amplifier Circuits, Theory and Design,* McGraw-Hill, New York, 1961, chap. 4.9, p. 109.
5. Richard A. Matick, *Transmission Lines for Digital and Communication Networks,* McGraw-Hill, New York, 1969, chap. 7.
6. James E. Buchanan, *Signal and Power Integrity in Digital Systems: TTL, CMOS, and BiCMOS,* McGraw-Hill, New York, 1996, chap. 6.
7. Robert E. Canright, "A Simple Formula for Dual Stripline Characteristic Impedance," *IEEE SOUTHEASTCON 1990 Proceedings,* vol. 3, pp. 903–905.
8. S. A. Schelkunoff, "The Electromagnetic Theory of Coaxial Transmission Lines and Cylindrical Shields," *Bell System Technical Journal,* vol. 13, 1934.
9. Albert R. Martin and Steven E. Emert, *Shielding Effectiveness of Long Cables,* 1979 IEEE International Symposium on Electromagnetic Compatibility, New York, IEEE Publication 79CM1383-9 EMC.
10. Albert R. Martin, "An Introduction to Surface Transfer Impedance," *EMC Technology,* July 1982.
11. Daryl Gerke and Bill Kimmel, "The Designer's Guide to Electromagnetic Compatibility," *Supplement to EDN,* Jan. 20, 1994.
12. Richard A. Quinnell, "Ignore Packaging Effects at Your Peril," *EDN,* June 9, 1994.
13. C. H. Chen (ed.), *Computer Engineering Handbook,* McGraw-Hill, New York, 1992, p. 7.15.

Chapter

9

Signal Reflections, Bandwidth, and Losses

This chapter describes the reflection factor and its relationship to the voltage standing wave ratio (VSWR), the signal rise time (t_{rise}), and signal levels. The important relationship of rise time to the harmonic content of digital signals is described. High-frequency line losses and the relationship of termination resistance to waveshape on long lines are discussed and illustrated.

9.1 The Reflection Factor

The ratio of the reflected wave to the incident wave is called the *reflection factor* or *reflection coefficient*. The reflection factor can be applied to either current or voltage but is commonly understood to be the voltage reflection factor. The current reflection factor is simply the negative of the voltage reflection factor. The reflection factor is the fraction of a wave that is reflected at a discontinuity. Complex transmission lines with multiple discontinuities may have multiple reflections, including reflections of the reflections, called *ringing*.

Any crosstalk that is coupled onto a transmission line may also travel on the line and be subject to reflections. Common mode crosstalk may reflect unequally if the impedance change is not exactly equal for both lines, causing common mode crosstalk to be converted to differential crosstalk. Of course, the magnitude of the reflection factor varies from zero (no reflection) to a maximum of unity, when all of the energy is reflected. The fraction of the *energy* reflected is the square of the voltage reflection factor. The reflection factor may be either positive or negative, indicating whether the reflected voltage is the same phase or the opposite phase of the incident wave. A change to higher impedance in the direction of travel of the incident wave results in a positive reflection factor; the reflected wave is in phase, so the voltage is increased. Conversely, a change to lower impedance in the direction of travel results in a negative reflection factor; the reflected wave is the opposite phase, so the voltage is reduced.

A typical digital load is a higher impedance than the transmission line, so the reflection factor at the unterminated load end is positive and the signal voltage level is increased. However, the typical source is a lower impedance, so the reflection factor at the source end is negative and the signal voltage level is reduced. This sequence of voltage increase, then voltage decrease is the beginning of ringing that occurs on typical unterminated lines. Since practical lines have some losses, the reflection factors are less than unity and the ringing sequence is dampened and will eventually settle to a final value. Some common logic drivers and receivers have nonlinear output characteristics and overshoot clamp diodes on inputs, so the reflection factor is not constant.

The common symbol for the reflection factor is ρ (Greek letter rho), although some papers use Γ.[1] The reflection factor symbol may include a notation to indicate the location of the discontinuity. For example, ρ_S indicates the source reflection factor; ρ_L indicates the load reflection factor. The impedance of the transmission line on which the incident wave travels is designated by the symbol Z_0. The new impedance can be either the end of a line with an impedance of R or a new transmission line with a different Z_0. The reflection factor at a change in line impedance may be designated ρ_Z. Resistance at the source and load end may include the notation R_S and R_L, respectively.

The equation for the reflection factor is

$$\rho = \frac{R - Z_0}{R + Z_0} \tag{9.1}$$

The reflection factor can vary from a limit of +1 for R approaching infinity for an open line, to zero when R is equal to Z_0, to the limit of −1 for R approaching zero for a shorted line. Practical lines cannot have perfectly open or shorted ends, so the magnitude of the reflection factor is always less than unity. The reflection factor approaches zero as R approaches Z_0 from either above or below Z_0. When R is nearly Z_0, the denominator is nearly $2Z_0$, so the sensitivity to small differences is about 50 percent of the mismatch. For example, a ± 10 percent difference from Z_0 results in a reflection factor of only about ± 5 percent. The general form (difference over sum) is similar to the form for the hyperbolic tangent $(e^x - e^{-x})/(e^x + e^{-x})$, so the shape of the curve is similar. The value of the function = 0 when $x = 0$, or $R/Z_0 = 1$.

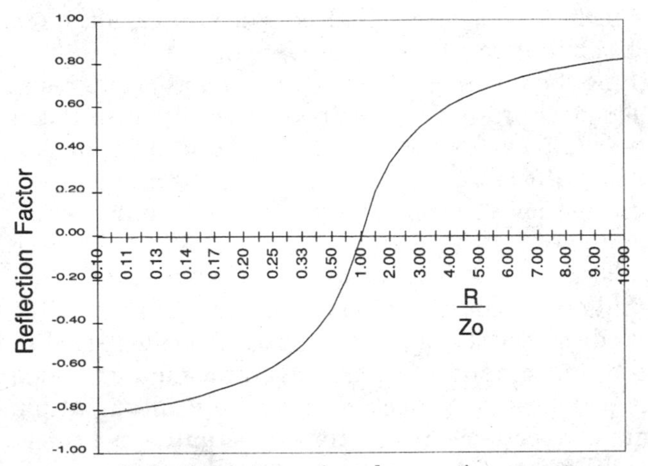

Figure 9.1 Reflection factor vs. impedance ratio.

The reflection factor as a function of the R/Z_0 ratio is shown in Fig. 9.1. Note that the scale for $R < Z_0$ is linear in the inverse of the ratio. To determine the reflection factor at the load, substitute $R = R_L$; for the reflection factor at the source, $R = R_S$; and for the reflection factor at a change from Z_0 to Z_1, $R = Z_1$.

The equation for the reflection factor is based on resistive impedance, but actual loads can be complex (include capacitance or inductance reactive components), and the reflection factor may be complex, including purely reactive. If the line impedance has a reactive component, the matching impedance for zero reflection should be equal to the line impedance, so that $Z_{load} - Z_0 = 0$. For example, for the common case of a capacitance component to the effective line impedance (usually a result of series resistance losses in the line and capacitance loading), the ideal termination for no reflection would include a capacitance equivalent to the line capacitance. This can be understood intuitively by the analogy to an ideal line. If two sections of an ideal line have the same (complex) impedance, there is no reflection at the junctions of the two sections. Likewise, there is no reflection if the line is terminated with a complex network matching the line impedance.

The termination for maximum power transfer would be the complex conjugate of the line impedance, so the reactive terms cancel at the frequency of interest. The maximum power transfer termination for a typical line with capacitance loading would include an inductance to provide the complex conjugate to the line capacitance, so the opposite reactive components cancel (or series resonate) at some desired frequency. Complex termination is not practical in most cases of digital data transmission which must pass a wide frequency range, although a fixed-frequency clock might be a possible exception.

Reasonably matching the line impedance with an appropriate resistance is sufficient for most ordinary high speed digital transmission lines. A complex termination impedance called *ac termination* is sometimes described as providing proper line termination and eliminating the power dissipation of an ordinary resistor termination.[2,3] Although ac termination does eliminate power dissipation for very low frequency signals, it cannot eliminate the power dissipation on the line as the signal transition travels on the line. The capacitor value must be selected so that the time constant is long compared to the rise time, but the time constant should be short compared to the data interval for the highest frequency. This is often incompatible with high speed interconnections. If the time constant is not short enough so that it is fully charged to the static level before each new transition, the termination and signal levels will be vulnerable to the duty cycle and data patterns. It should be understood that the dynamic response to a signal transition does not necessarily result in a static level at the load. When a signal transition is launched onto a line, the signal transition is less than the open-circuit voltage due to source resistance. An ac termination with the resistor matching the line impedance will initially terminate the signal level at the initial level on the line. However, the ac termination approaches an open circuit as the capacitor charges, so the signal level will eventually charge toward the open circuit level, if enough settling time is allowed. Therefore, an ac terminated line eventually responds as an unterminated line that is open at the load end. However, if the capacitor does not fully charge before each new transition, the signal levels at the load and the load on the driver depend on the previous signal level(s). The initially less than steady-state levels at an ac termination may be partly compensated by a resistor greater than the line Z_0. The higher the resistor, the more the transient response of the line behaves as a source terminated line with an open load. The use of ac termination may require compromises between source impedance, the ac termination resistor and capacitor, and the most critical data rates. Step response and pulse distortion for various other improper terminations are illustrated in *Transmission Lines for Digital and Communication Networks.*[4]

AC termination might be considered for signals continuously operated at a fixed frequency and duty cycle (usually 50 percent). In this case, the time constant may be large compared to the period, so that the capacitor remains charged at a constant level at the middle level between the two logic levels. However, this still leaves the transient issue of stabilization time at startup, and the termination may not default to a defined state when the signal is unpowered unless otherwise biased. AC termination with a long time constant does reduce total power dissipation because there is lower voltage across the termination resistor. The power dissipation for AC termination is equivalent to the average power dissipation per resistor for rail-to-rail signals with an equal resistance split termination to the power rails. However, a split termination is not recommended for CMOS inputs that could be disconnected because the default signal level is in the input threshold region between the two logic levels.[2]

The reflection factor for a load that is capacitance only is an exponential that changes from -1 (initially a low impedance) to $+1$ as the capacitance charges to the new signal level if the line is long enough that no reflections are returned from the source during the charging interval. The resulting voltage waveform for a step input is an exponential rise time with a time constant of $Z_0 * C_L$. It is of interest to note the relationship to an ordinary lumped circuit with a source impedance equivalent to the line impedance, with an RC time constant. Therefore, a capacitance load on an otherwise unterminated load will have a rise time and bandwidth limitation consistent with the $Z_0 * C_L$ time constant. If the load is terminated, the resulting impedance at the end of the line is Z_0 in parallel with the termination resistor. For the case where the termination resistor is equal to Z_0, the time constant is $Z_0 * C_L/2$.

It should be noted that although the time constant for lumped loading is less with a terminated line, the reduced reflection also reduces the voltage amplitude. Therefore, the faster time constant does not always mean a corresponding reduction in the transition time to the receiver threshold. Of course, if the input signal to the load is not an ideal step input but is bandwidth limited and has a finite rise time, the resulting waveshape is limited by the combination of both the input signal arriving from the line and the time constant of the load capacitance.

9.2 Impedance Ratio: VSWR

The reflection factor is commonly as described above for the ratio of the reflected voltage to the incident voltage. A change to higher impedance means an increase in voltage and a decrease in current. Conversely, a change to lower impedance means a reduction in voltage and an increase in current. The proportion of the signal reflected depends on the ratio of the change in impedance. The total signal voltage after a reflection is complete is the sum of the incident wave voltage and the reflected wave voltage, or $(1 + \rho)$ times the incident wave. The total signal voltage can be expressed as follows:

$$V(\text{sum}) = V(\text{incident})\frac{2R}{R + Z_0} \tag{9.2}$$

When R is very large compared to Z_0, the total signal voltage is nearly doubled; when $R = Z_0$, the total voltage is equal to the incident signal voltage.

Waves traveling toward a driver will reflect from the driver in the same manner as reflections from a load. The typical difference is that the load at the receiving end is usually well matched or higher than the line impedance; however, the impedance of the driver is usually lower than the line impedance and often not well specified. Even when series impedance is added to a driver to provide better line matching, the resistance is kept below the line impedance to increase the signal amplitude. Therefore, the typical reflection factors for transmission-line systems not exactly matched are for the source reflection factor to be negative and the load reflection factor to be positive. This results in an underdamped line response, with larger initial and final signal values than an exactly matched line. The underdamped response results in some initial overshoot, which provides slightly faster response and some compensation for lumped capacitance loading and skin-effect losses.

If the equation for the reflection factor is normalized by dividing numerator and denominator by Z_0, the result is the following form:

$$\rho = \frac{R/Z_0 - 1}{R/Z_0 + 1} \tag{9.3}$$

The impedance ratio R/Z_0 is also known as the voltage standing wave ratio (VSWR) because the impedance ratio is also the amplitude ratio which results when high-frequency incident and reflected waves combine on a long line. Because the waves are traveling in opposite directions, the relative phase changes with distance from the location of the reflection. At locations where the two waves are in phase, they combine for an amplitude maximum. For example, when the load reflection factor is positive (the load is higher-impedance than Z_0), the reflection is in phase at the load. Conversely, at locations where the two waves are of opposite phase, the reflected wave partly cancels the incident wave for an amplitude minimum. For positive reflection, the waves will be at opposite phase at odd multiples of the quarter wavelength from the location of the reflection. Although the two waves are traveling in opposite directions, the location of the maximums and minimums is stationary, which is the origin for the term "standing" waves. The amount of cancellation and reinforcement depends on the reflection factor because it is the ratio of the amplitude of the reflected wave to the incident wave.

9.2.1 Maximum and minimum modulation

VSWR is also the ratio of the maximum to the minimum of the amplitudes of the standing or combined waves. When the load impedance matches Z_0, there is no reflection to modulate the amplitude of the incident wave, so the VSWR is unity. As was shown above, the sum of the amplitude of the incident wave and the reflected wave is the incident wave times $2R/(R + Z_0)$. Therefore, this is the maximum amplitude for a positive reflection coefficient when R is greater than Z_0. Similarly, the minimum is the amplitude of the incident wave minus the reflected wave, or $(1 - \rho)$ times the incident wave, which reduces to $2Z_0/(R + Z_0)$. The minimum and maximum are of similar form, with R in the numerator for the maximum and Z_0 in the numerator for the minimum. Therefore, the ratio of the maximum to the minimum is simply R/Z_0 for R greater than Z_0, or Z_0/R for R less than Z_0. This is simply the impedance ratio, expressed as the ratio of the larger to the smaller impedance.

VSWR is often considered a more convenient way of expressing the impedance ratio directly than the reflection coefficient. VSWR is more popular for applications where the line length is long relative to the wavelength of the frequency. This incident and reflected waves can combine on long lines for both maximum and minimum modulation at different locations along the line. The also means that the input impedance seen by a driver on a long line may be different from Z_0 if the far end is not well terminated. For example, the input impedance of a quarter wavelength line (ignoring losses), terminated at the far end by a load resistor of R_L, is $(Z_0)^2/R_L$. Therefore, the minimum impedance is inversely proportional to the load resistor.

For digital applications where the line length is shorter than the wavelength of the clock or data frequency and all transients settle to essentially static levels before another transition occurs, standing waves do not occur. For this case, the reflected wave has decayed to essentially zero before another incident wave is launched, so incident and reflected waves cannot combine. However, at higher clock and data rates on longer lines, the possible effects of standing waves should be considered.

The magnitude of the reflection factor may be calculated from VSWR as follows:

$$\pm \rho = \frac{\text{VSWR} - 1}{\text{VSWR} + 1} \tag{9.4}$$

TABLE 9.1 Impedance Ratio, VSWR, Reflection Factor, and Reflection Factor Squared

Impedance ratio	VSWR	Reflection factor, %	(Reflection factor)^2, %
1000:1	1000	+ 99.80	+ 99.60
100:1	100	98.02	96.08
10:1	10	81.82	66.94
5:1	5.0	66.67	44.44
4:1	4.0	60.00	36.00
3:1	3.0	50.00	25.00
2.5:1	2.5	42.86	18.37
2.0:1	2.0	33.33	11.11
1.5:1	1.5	20.00	4.00
1.4:1	1.4	16.67	2.78
1.3:1	1.3	13.04	1.70
1.25:1	1.25	11.11	1.23
1.20:1	1.20	9.09	0.83
1.15:1	1.15	6.98	0.49
1.10:1	1.10	4.76	0.23
1.05:1	1.05	2.43	0.06
Exactly 1:1	1.00	0.00	0.00
0.952 = 1:1.05	1.05	−2.43	0.06
0.909 = 1:1.1	1.1	−4.76	0.23
0.833 = 1:1.2	1.2	−9.09	0.83
0.667 = 1:1.5	1.5	−20.00	4.00
0.500 = 1:2	2.0	−33.33	11.11
0.200 = 1:5	5.0	−66.67	44.44
0.100 = 1:10	10	−81.82	66.94
0.010 = 1:100	100	−98.02	96.08
0.001 = 1:1000	1000	−99.80	99.60

This equation is of the same form as Eq. (9.2), but VSWR is defined as only greater than unity, so the appropriate sign must be used to define positive or negative reflection factors. The voltage reflection factor is positive when the wave travels toward a higher impedance; it is negative when traveling to a lower impedance.

Various values for resistance ratio VSWR and reflection factor are shown in Table 9.1. VSWR is always just the impedance ratio, regardless of whether the ratio is greater than 1 or less than 1. The sign of the reflection factor depends on the impedance ratio in the direction of the traveling wave. The table includes values of the reflection factor squared. The square of the reflection factor is significant because it determines the first undershoot on lines terminated with resistance at one end only. The square of the reflection factor also determines the first overshoot on lines terminated with the same VSWR relative to Z_0 at both ends (lower impedance at the source, higher impedance at the load).

9.3 The Short-Line Scaling Factor

When the line length is short compared to the rise time of the incident wave, the reflection begins to return to the source before the initial rise time is complete. When the reflection begins to overlap the initial rise time, the subsequent portion of the source signal adjusts to partially match the load instead of the transmission line Z_0.

A line is considered a short line when the reflection partly overlaps the initial rise time, and a long line when the two-way propagation delay is longer than the initial rise time. The reflection is completely separated from the initial rise time for a long line. The line length at which the two-way delay just matches the rise time is called the *critical length*. Because the critical length is proportional to the rise time, the reflection amplitude increases more rapidly at faster rise times. Networks with faster rise times will tend to have more lines with reflections and ringing, even if the worst ringing does not increase for long lines. Nothing "critical" or dangerous happens at the critical length; it is just the line length at which full reflection amplitude is reached. Lines longer than the critical length do not have any greater amplitude of reflection, although it takes longer for the reflection to arrive back at the source and for any ringing to settle.

Although it is not analytically exact, it is convenient to scale the reflection factor by the ratio of the round-trip propagation delay time to the initial rise time. The short-line scaling factor is the following for a 0 to 100 percent linear rise time:

$$K_S = \frac{2T_0'}{t_{\text{rise}}} \quad \text{if less than 1}$$

$$K_S = 1 \quad \text{otherwise} \tag{9.5}$$

$T_0' = K_L * T_0$ is the total one-way loaded propagation delay, including the delay effects of lumped capacitance loading.

9.3.1 Rise time definitions

There are various ways to define rise time. Many computer analysis programs define a 0 to 100 percent linear rise time, so rise time is 100 percent of the signal excursion. It has been traditional to define the 10 to 90 percent of a capacitance charging waveform as the rise time, so this rise time is 80 percent of the signal excursion. Defining less than 100 percent of the excursion is appropriate for waveforms that only approach the final value, because the rise time to 100 percent is indefinite. About 2.3 time constants (*RC* for a capacitor charging circuit) is required to achieve 90 percent of the final value, so a 10 to 90 percent rise time is about 2.2 time constants. The exponential time constants are shown in Fig. 9.2. The time delay from the 0 to 50 percent level is about 0.7 time constants, as is the time between the 80 and 90 percent levels. The time to reach the 80 percent level is about 1.6 time constants, so the time from 80 to 90 percent is about 0.7 time constants. Just that last 10 percent takes as long as the time for the first 50 percent of the excursion.

It is recommended that the more conservative definition of rise time at the 20 to 80 percent interval be used for high speed analysis. This definition more accurately represents the central, more linear

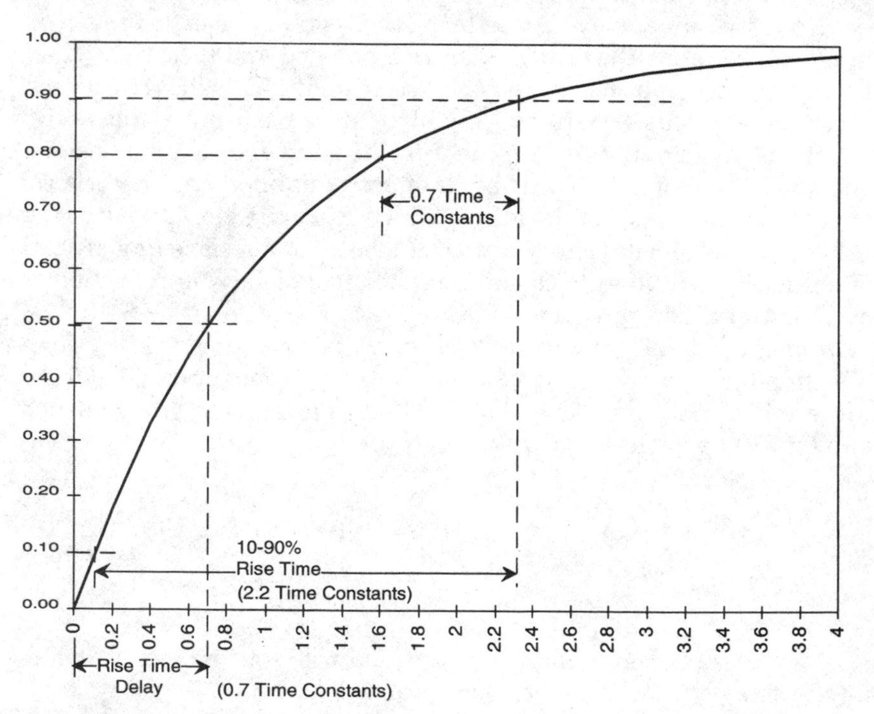

Figure 9.2 Exponential time constants.

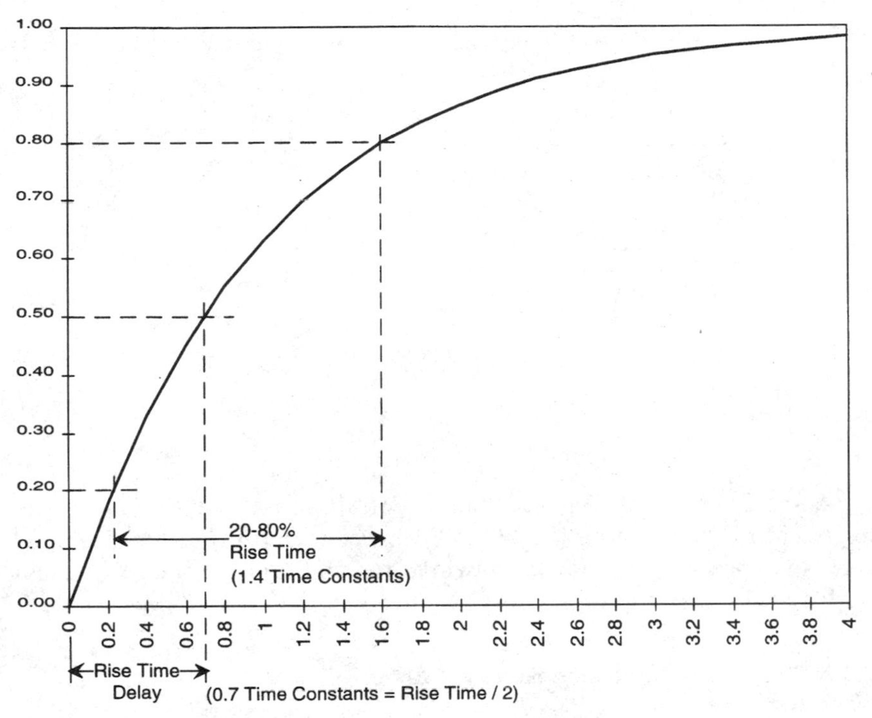

Figure 9.3 Exponential rise time definitions.

portion of a typical high speed waveshape. This definition is 60 percent of the excursion and about 1.4 time constants for an *RC* charging waveform. These rise times on the same exponential waveshape are shown in Fig. 9.3.

The 20 to 80 percent rise time definition is also more convenient for relating rise time to delay and frequency. Because the time to 50 percent of the full excursion is the typical time to the threshold of a receiving circuit, it is reasonable to consider about 0.7 of a time constant as the rise time delay. Sincc about 1.4 time constants are required for the 20 to 80 percent rise time, the rise time delay to the 50 percent threshold level is about one-half of the rise time. This is a convenient relationship to remember and is the same relationship as for the totally linear rise time typically used in computer simulations.

> A rule of thumb for relating 0 to 50 percent rise time delay to the 20 to 80 percent rise time is that the rise time delay is one-half of the rise time for an exponentially shaped waveform.

Figure 9.4 illustrates the 0 to 100 percent rise time usually used for linear rise time analysis, with the delay to the 50 percent level exact-

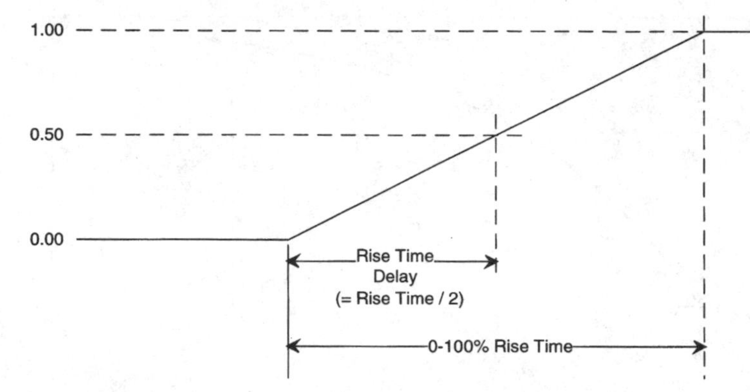

Figure 9.4 Linear rise time definitions.

ly one-half of the rise time. Although not all actual waveforms are an exponential waveshape, the 20 to 80 percent definition for rise time remains convenient and appropriate for the measurement of most high speed waveshapes.

9.3.2 The short-line scaling factor for 20 to 80 percent rise time

The definition of the short-line scaling factor in Eq. (9.5) is based on a 0 to 100 percent linear rise time. However, most practical waveforms of high speed transitions have significant rounding. Actual transmission lines have skin-effect losses that attenuate the trailing portion of the waveshape. Defining the 20 to 80 percent levels of the central 60 percent of the signal transition is a more linear representation of the signal slew rate, or slope. If the central 60 percent rise time convention is used in the short-line scaling factor based on a 100 percent linear transition, it will significantly underestimate the signal overlap. Therefore, the short-line scaling factor should be scaled according to the appropriate rise time convention. The scaled definition for K_S using the 20 to 80 percent rise time is the following:

$$K_S = \frac{1.2T_0'}{t_{\text{rise (20-80)}}} \qquad \text{for } K_S < 1 \tag{9.6}$$

The rounding of the trailing portion of typical high speed signal transitions usually results in less signal undershoot than estimated with the scaled K_S. The three propagation delay times before undershoot occurs usually both slow the rise time and round the trailing edge because of skin-effect attenuation.

9.3.3 Rise time and bandwidth

Another convenient relationship of the 20 to 80 percent rise time convention is that 60 percent of the excursion is just under two-thirds of the total excursion. Therefore, this rise time is a conservative (slightly faster) estimate of the third harmonic frequency of a repetitive triangular wave with the same rise time definition. The traditional relationship of bandwidth is the following: $t_{rise} = 0.35/F_b$, or equivalently $F_b = 0.35/t_{rise}$. F_b is the 3-dB bandwidth of an *RC* circuit (a single-pole, low-pass network), and t_{rise} is the 10 to 90 percent rise time of the same *RC* circuit when driven with a perfect step (zero rise time input, which requires infinite bandwidth). At the 3-dB (half power loss) frequency, the signal voltage response has lost about 30 percent of the voltage amplitude.

Few high speed applications have the luxury of either providing perfect step inputs or losing 30 percent of the primary signal amplitude, so the 0.35 ratio between bandwidth and rise time is optimistic. A designer troubleshooting a high speed circuit with an oscilloscope is faced with a similar dilemma: What oscilloscope bandwidth is needed to accurately display suspected problems? The designer may be taking a risk that small 1-ns spikes will not be accurately displayed with a 350-MHz oscilloscope. A more practical interpretation of the bandwidth rule is that if the signal were a zero rise time pulse 1 ns wide, the 350-MHz scope may display it as a 1-ns rise time pulse of only 70 percent of the true amplitude. Small 1-ns spikes or overshoot of finite rise time will be distorted (smoothed and attenuated), so they are likely to be overlooked and may not trigger the oscilloscope sweep.

A more conservative relationship between rise time and frequency bandwidth is to use the 20 to 80 percent definition for rise time and include the third harmonic in the frequency bandwidth. The use of the faster rise time definition tends to compensate for the actual finite input rise time, and including the third harmonic reduces the distortion of the actual waveshape. If we consider a repetitive triangular wave, the full cycle time (or period) of the fundamental frequency is 2 times the linear rise time of each transition. This is illustrated in Fig. 9.5 for a continuous pure triangular wave. Alternately, the rise time of the triangular wave is one-half of the period. Since the 20 to 80 percent rise time is 60 percent of the full transition rise time, the 20 to 80 percent rise time of the triangular wave is 30 percent of the period of the triangular wave. Therefore, the 20 to 80 percent rise time is a little less than the period of the third harmonic.

The third harmonic of the triangular waveshape contains about 10 percent of the fundamental frequency amplitude because the amplitude of the nth harmonic of a triangular wave decreases as n^2.

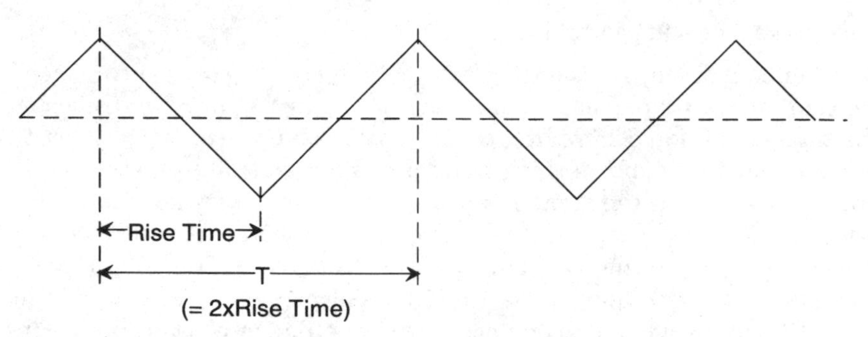

Figure 9.5 Triangular waveform.

Therefore, the third harmonic amplitude is reduced to one-ninth of the primary. Including the third harmonic in the bandwidth results in a reasonable representation of the triangular waveshape. The more conservative third harmonic bandwidth required to reasonably represent signals with a 20 to 80 percent rise time is the following:

$$3\,F_b = \frac{1}{t_{\text{rise (20–80)}}} \tag{9.7}$$

Including the third harmonic provides a convenient inverse relationship between bandwidth and rise time. This allows a convenient translation between the frequency and time domains, because some analysis is more direct in the frequency domain. Many high-frequency engineers are more comfortable with frequency, although the typical digital engineer may be more familiar with the rise time of signal transitions. The simple inverse relationship of rise time and the third harmonic bandwidth is convenient to remember for the relationship of frequency and rise time.

> A simple rule of thumb for relating bandwidth to the 20 to 80 percent rise time is that the bandwidth is simply the inverse of the 20 to 80 percent rise time.

A common misconception is that the clock frequency or data rate determines the frequency requirements of digital data processing systems, but the rise time is more important. For example, signals with 1-ns (10^{-9}) 20 to 80 percent rise times require about 1 GHz of bandwidth and generate significant energy at and above 1 GHz (10^9), even if the clock rate is only a few MHz. Very high speed circuits with subnanosecond rise times require at least a few GHz of bandwidth. It should be obvious that high speed circuits with rise times at or below 1 ns require design techniques that used to be typical only of those frequencies called "radio frequency" or just "rf." This frequency range covers the radio and television broadcast spectrum and approaches the radar frequency range.

Another concern for this frequency range is electrostatic discharge (ESD). Static charge may build up on machinery but is commonly generated by humans, who can easily build up a static charge of many thousand volts if no discharge path is provided. Although a static discharge is a single event, the typical rise time is a few nanoseconds. Therefore, the discharge has frequency components up to nearly a gigahertz. High speed circuits respond to this bandwidth, so the discharge energy may propagate throughout the circuits and ground connections, causing damage and/or upsets far from the location of entry. It also means that because the circuits are designed to be responsive to these frequencies, they are very vulnerable to ESD damage and/or upset.

9.4 Harmonic Content and Bandwidth

Triangular or trapezoidal waveforms contain odd harmonics of the primary (or fundamental) frequency, which is the repetition rate of the waveform. Triangular harmonic amplitudes decrease as $1/n^2$, which is 40 dB per decade on a logarithmic scale. The first harmonic present is the third, so it is down to about one-ninth, or about 11 percent of the primary. The ninth harmonic is down to 1/81 of the primary; if there were a tenth harmonic, it would be down to 1 percent of the primary. This continues, with each 10 times increase in frequency reducing the amplitude by a ratio of 100:1.

A trapezoidal waveform is illustrated in Fig. 9.6. If the trapezoidal waveform is lower-frequency than the triangular wave (has a longer repetition period of T) but has the same rise time, then there are lower-frequency components down to the primary at a frequency of $1/T$. However, the harmonics of the repetitive trapezoidal wave only

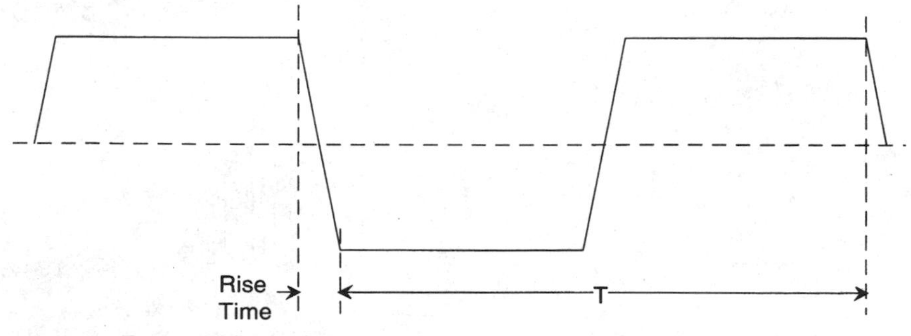

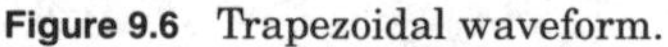
Figure 9.6 Trapezoidal waveform.

decrease as $1/n$, so they are reduced by a ratio of only 10 for each decade of frequency, until the frequency for the rise time is reached at $1/\pi t_{rise}$. After the rise time frequency, the harmonic amplitudes drop by a ratio of 100 per decade of frequency, the same as for the pure triangular wave.

Because the envelope of harmonic amplitude does not drop as rapidly until after the frequency for the rise time is reached, the bandwidth limit is determined primarily by the rise time. The bandwidth limit is both the upper frequency at which the interference to other units is reduced, as well as the frequency bandwidth needed to transmit the signal without significant waveshape distortion. Because typical circuits that generate fast rise times also have comparable input bandwidth response, the bandwidth limit also indicates that the circuit is vulnerable to interference in this frequency range.

The envelope of the harmonic amplitude for the triangular wave is illustrated in Fig. 9.7. Note that the spectral lines are drawn to illustrate that the magnitude of only some of the lines is as high as the envelope. The lines are drawn for illustration only and are not drawn to scale for actual amplitude or spacing. The logarithmic scale compresses the spectral lines at the higher frequencies.

The envelope of the harmonic amplitude for the trapezoidal wave is illustrated in Fig. 9.8. The trapezoidal waveform has a lower primary frequency and therefore more energy at the lower frequencies.

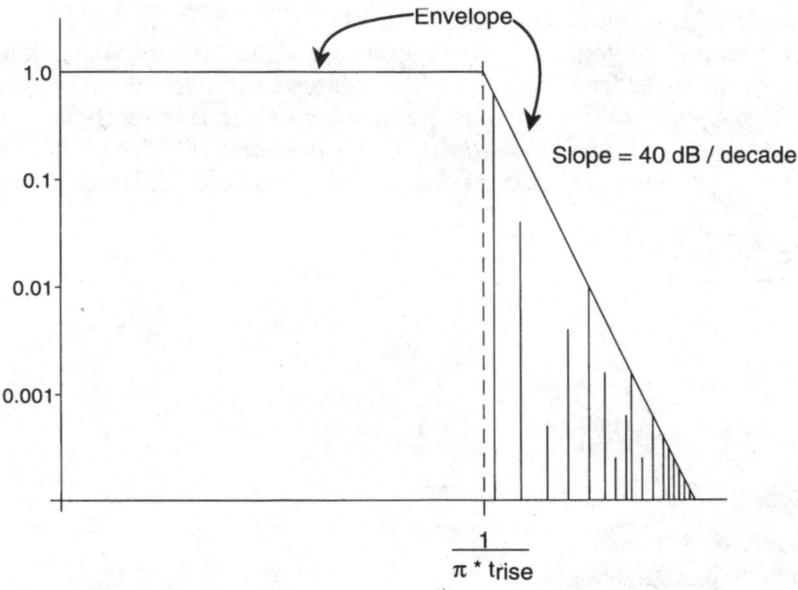

Figure 9.7 Triangular waveform harmonic content.

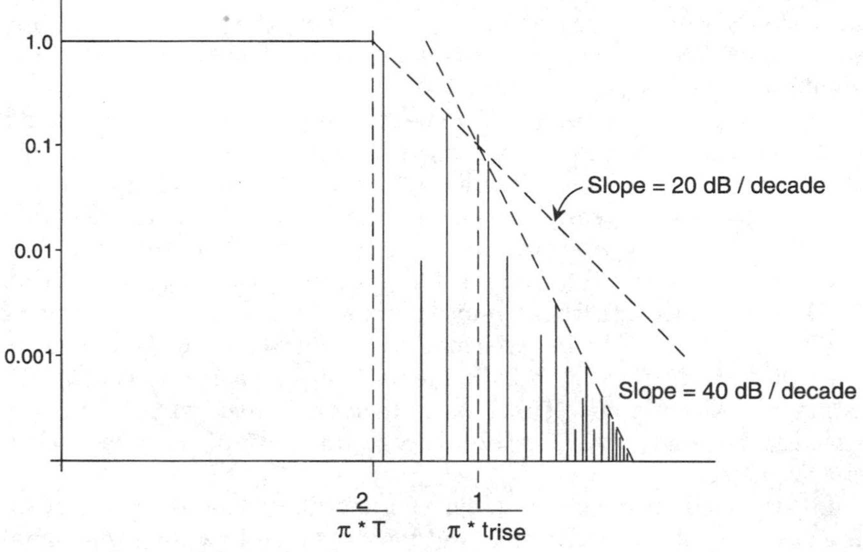

Figure 9.8 Trapezoidal waveform harmonic content.

However, the rate of reduction of amplitude for the higher harmonics is the same as the triangular wave. Actual signals containing data information would not be an indefinitely repeating waveform, so actual data waveforms also contain significant *average* energy at lower frequencies.

The clock frequency, or data rate, is a 20-dB relationship to bandwidth; at 10 times the clock frequency, the harmonic amplitude is reduced by a factor of only 10. The rise time is a 40-dB relationship to bandwidth; at 10 times the inverse of rise time, the harmonic amplitude is reduced by a factor of 100. Therefore, rise time is a more important indicator of the potential for interference from high speed signals. Although controlling rise time may be beneficial, this is not practical for typical high speed systems. Most high speed interconnection systems will need to isolate fast rise time signals with electrical design techniques including ground planes and shielding.

The slope of the harmonic content for the high frequencies is the same for the trapezoidal (and mixed-frequency data signals) as the triangular waveform, although the average energy level is a little lower because the repetition rate is lower. The energy in each transition is determined by the rise time, regardless of whether the transition only occurs once (as in an ESD event), occasionally as for a low data rate signal, or at the high repetition rate of a high-frequency clock. Proper transmission of every signal transition of the required data is not a 20-dB issue or a 40-dB issue; it is *all* of the issue.

Although frequency and rise time affect the average rate at which average energy declines at higher frequency, the transmission-line interconnection system must pass every signal transition so that the signals can be properly received.

To use the analogy to an ESD event, the potential for damage for each event is essentially the same whether the ESD event occurs only once a year or if it occurs once a day. There is no reduction in stress because the once a year discharge energy cannot be "averaged" to a lower level per day. A system may be overstressed by the first ESD event, independent of how long it has operated previously without an ESD event. Although there may be some "memory" effects at very short time scales (possibly seconds to recover from an ESD event), every day starts anew with little memory of the past if no permanent damage has been done. There is no memory of benign prior history that can be used to reduce the average intensity of the effect when the event occurs.

Likewise, the response to a signal transition depends primarily on the rise time of that transition and very little on how long the signal has been at a static level prior to the transition. However, the repetition rate (or frequency) of the transitions does determine how much time is available for each response to settle. If there is not sufficient time for the signals to settle nearly to steady-state static levels before each new signal transition, then the remaining effects of the previous transition(s) interfere with the new signal transition. However, if sufficient time is allowed for the effects of the previous transition to settle, the response to each new signal transition depends primarily on the rise time of the transition.

It may seem that the obvious solution to high-frequency problems is to slow down the output rise time and reduce the input bandwidth. It is true that wherever it is practical, reducing output rise time and input bandwidth can be very effective. A little series resistance (about $Z_0/2$) in a logic output significantly reduces transient currents, ground bounce, and line ringing, with little effect on speed performance.[5] Similarly, a little resistance in series with input clamp diodes significantly reduces overshoot current and eliminates extended line recovery times. When the termination components are integrated into the circuits, this means that the device performance (including the terminations) can be specified. However, few devices are adequately specified for transmission-line performance with significant overshoot, and few control minimum rise time. Even using older part numbers offers little frequency bandwidth control, as replacement parts with older part numbers on the package may actually contain equivalent high speed newer parts on the inside. The newer parts may be fully specification-compliant, since minimum rise time and input pulse width rejection are seldom specified. The continuous rate at which semiconductor technology advances means that parts will continue to perform faster.

The increase in device performance means that increasingly more processing can be performed in smaller spaces and at higher speeds. This spells opportunity to the system designer who needs to pack more and more processing performance into cost-effective systems. However, digital designers will find that they have to deal with frequency issues that could be ignored in days past with slower parts and clock rates. Logic design must be properly implemented with electrical design that maintains signal integrity. Interconnections and high speed interfaces will need to consider transmission-line effects at shorter distances, dealing with issues that were previously considered the domain of "radio frequency" (rf) and microwave.[6] System integrators may need to understand how to interface with radio, radar, and microwave receivers, and how high speed digital signal transitions interfere with sensitive receivers. As digital technology advances from separated preprocessors and postprocessors, into transmitters and receivers, and integrated into antennas, complex systems must be designed for integrated compatibility with reliable performance. Highly integrated complex systems require careful attention to compatibility at the initial architecture and systems design stage because redesign and rework is often time-consuming and difficult if not nearly impossible.

9.5 Design for Test and BIST

Another "good word" will be included here on the need for testability to be included in the performance requirements for complex high speed systems. The days of long ago when rise times were measured in microseconds and clock rates in kilohertz are obviously not adequate for complex high speed systems. Likewise, the days are gone when performance verification could be achieved with physical test points on every important signal. An extensive set of test points would allow an experienced engineer with such test equipment as oscilloscopes, logic analyzers, and spectrum analyzers connected to selected test points to stare into CRT screens to try to comprehend how well the signals functioned. If the system was not performing properly, the engineer would need to determine the cause(s) of the problem and determine whether parts replacement or redesign and rework were needed to improve performance. Waveshape observation is still an important tool for verification that critical signals such as clock distribution have adequate design margin, but most high speed systems are so complex that performance verification defies the comprehension of a single human.

Performance verification has several parts: The field usage issue is "does it still perform the way it used to?" The manufacturing issue is "does this unit perform the same way the others do (or like the 'golden' design standard)?" Design issues may be both much more subtle and more complex and are often more costly and time-consuming than the design and fabrication of the first engineering evaluation unit(s). Design verification issues include: "Are all of the bits correct all of the time for all input data patterns and sequences, for all modes?" "How do we know what 'correct' is?" "If the system results are not what someone thinks or says are expected, is the system wrong, or are the expectations wrong, or are the requirements wrong?"

If design verification results do conclude that the system performance is not proper, then the concern may be how to determine the cause(s) of the problem and how to redesign and rework the system to change the operating performance. This can be costly and time-consuming for complex high speed systems, particularly if redesign cycles for custom integrated circuits are required.

Many high speed functions are particularly difficult to evaluate because the addition of physical monitors may significantly affect the high speed performance. Conventional production test machines often do not contain enough memory to test complex functions at full speed, especially if the new technology under test is intended to operate faster than the capability of available testers with older technology. Built-in self-test (BIST) can be very cost effective for complex high speed systems, especially for design evaluation. However, it is critically important that the design for testability requirements be included in the original design requirements and part of the system architecture. Design for testability can be very efficient when integrated into the system architecture but may be very difficult to add on after being ignored in the original design. Built-in self-test typically includes a few additional pins for test access, full speed deterministic pattern generators, and compression registers to provide a reduced representation of the test result.[7,8,9] Although the logic required for BIST may have been significant for older generations of integrated circuits, the impact of BIST logic may be insignificant for the higher levels of integration.

9.6 Skin-Effect Losses

Skin-effect losses attenuate the higher-frequency components of a signal transition more severely than the lower-frequency components, so there is distortion of the waveshape. The effect can be considered as somewhat similar to *RC* charging, but the typical waveshape is different. The exact effect depends on many parameters, and it is often bet-

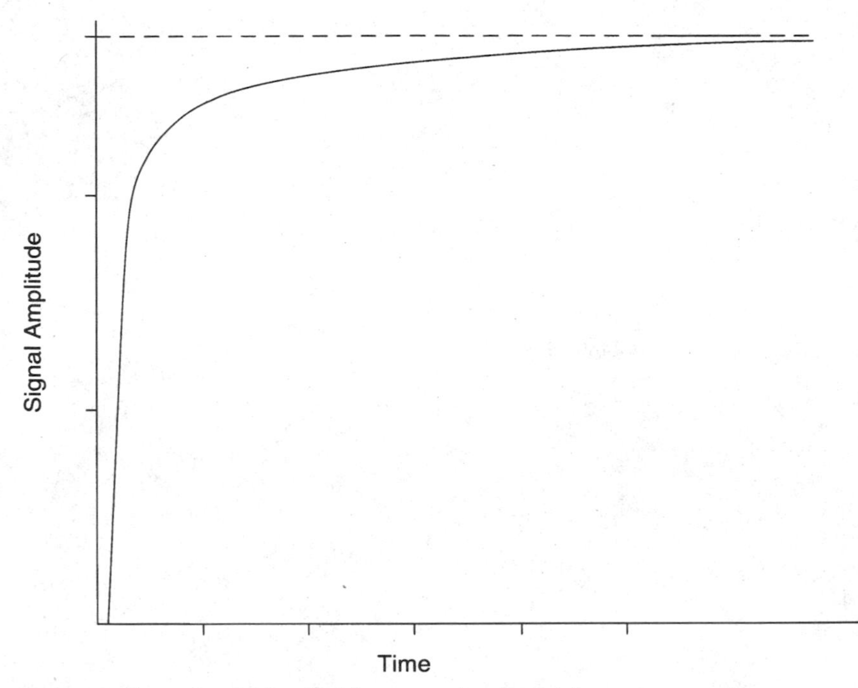

Figure 9.9 Example of skin-effect losses on signal waveshape.

ter measured than calculated. Matick provides a mathematical basis for calculations and several examples of waveshapes.[4] Barna provides a similar analysis but reports that the results may be in error by as much as a factor of 5 either way.[10]

Figure 9.9 is an illustration of a typical response to a fast rise time transition at the end of a line dominated by skin-effect losses. A comparison to the exponential shown in Figs. 9.2 and 9.3 indicates that for a given initial rise time, the exponential charges more quickly toward the final value. The skin-effect deterioration of the transition results in very slow line charging as the signal level nears the steady-state static levels. For example, for the exponential waveshape, the ratio of the time to reach 90 percent of the final value to the time to reach 50 percent is about 3. However, for skin-effect distortion, the ratio of the time to 90 percent to the time to reach 50 percent may be greater by an order of magnitude, or a factor of 30. Skin-effect losses tend to result in even slower charging to reach the last few percentage points toward a steady-state static signal level.

Figure 9.10 illustrates a square-wave signal at the open end of a line with series termination at the source. This figure illustrates actual signals recorded of a differential twisted pair line. This line is short

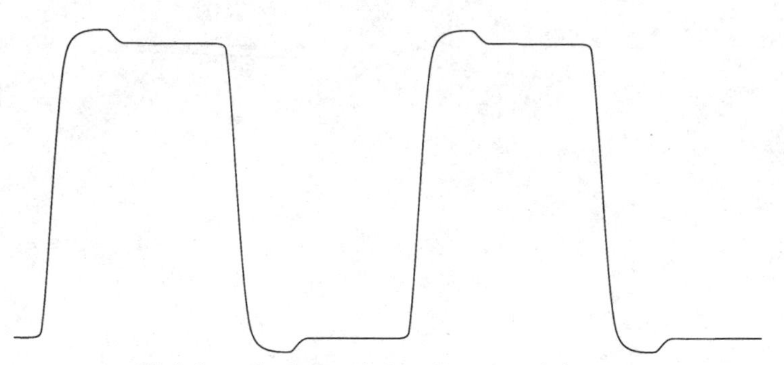

Figure 9.10 Slightly underdamped short-line waveform.

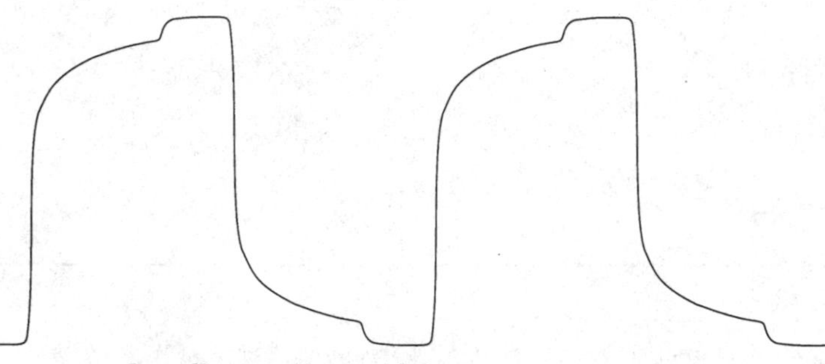

Figure 9.11 Long-line waveform with losses.

enough that the source termination initially results in a small overshoot at the load, but any subsequent ringing is not observable.

The losses on a longer line are illustrated in Fig. 9.11. This waveform is similar to the shorter line illustrated in Fig. 9.10, but the line is much longer and the square wave frequency is lower, so the actual time scales are different. The losses on the longer line are greater, so that the signal level at the load does not approach the final signal levels until the arrival of the second incident wave.

Figure 9.12 illustrates the waveforms at the source of a long line that is well terminated at the source and open at the load. The upper signal is the ideal step waveform that would result from the initial and reflected transients on an ideal lossless line. The terminated source would launch an initial signal level exactly one-half of the steady-state static signal levels; the open load would reflect this transition back to the source with no losses or distortion. The sum of the initial signal level plus the reflection would exactly match the steady-state static signal level, so equilibrium is achieved and the signal level remains there until the next signal transition.

The lower signal in Fig. 9.12 illustrates that the reflected signal is significantly distorted after the round trip on the long line. The initial

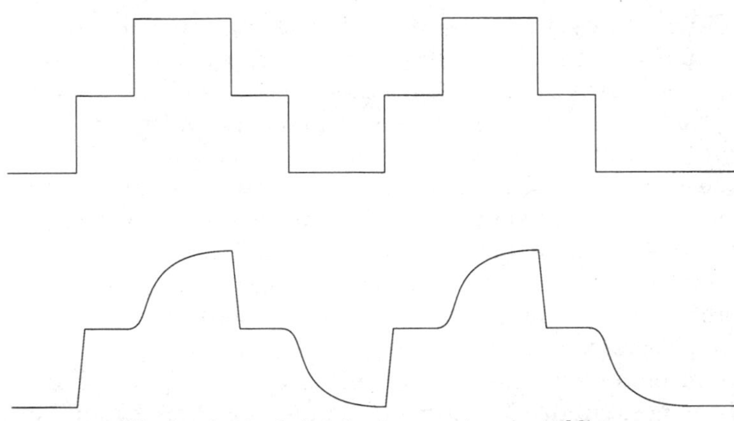

Figure 9.12 Ideal and actual signals at source terminated line.

fast rise time signal level is about half of the final value. The initial signal excursion and the signal during the round-trip propagation delay time for the actual signal are similar to the ideal signal. However, the returning reflection has a slower rise time and severely rounded waveshape that gradually approaches the steady level.

The steady-state static level is never quite reached before the next transition of the square wave in this illustration, so the final level is partly determined by the data rate and line losses. When the data is not a fixed square wave, the final level will be slightly different for different data or clock frequencies or different data patterns. Lower frequencies or data patterns that remain in the same state longer will reach higher levels than for the narrower pulses.

The waveshape deterioration is not as obvious at the load because only the one-way losses have affected the signal. The separation of the initial and reflected waves at the source makes the deterioration more obvious. The amplitude of the initial signal excursion and the losses on the line do affect the final waveshape at the load. The source continues to provide energy into the line until the sum of the initial signal level and the reflection equals the steady-state static signal level. Lower source termination impedance and lower line losses will reach the steady-state static signal level sooner. Although this may result in some overshoot, signals on a long line with losses will usually be attenuated so that undershoot after the signal travels three times on the line is insignificant.

The waveforms in Fig. 9.12 also illustrate the line impedance seen by the source when the far end is an unterminated open circuit. The high and low differential levels are the steady-state static levels when essentially no current is flowing in the line. The round-trip propagation delay on the line is the time of the “flat” portion of the waveform, where the differential signal is zero. As the relationship of the line length to the pulse width becomes longer, the proportion of the time at the flat zero signal is greater. The static level at the source disappears when the round-trip time exactly equals the pulse width. At this line length and pulse width, the ideal signal is always at the flat zero signal level because the reflection returns to the source at the same time the source transmits the next transition of the opposite polarity. Of course, the actual signal shown in the figure has losses, so that the reflection does not exactly cancel the next transition transmitted. Therefore, the actual signal is not exactly zero, even when the reflection coincides with the next transition transmitted.

When the outputs of the differential driver are equal, the load on the driver is a short circuit. The driver is supplying short-circuit current to the line during the time that the output to the line is a zero flat differential signal level. The longer the unterminated line, the longer the driver supplies short-circuit current to the line. When the length of the line results in a signal that is always a zero flat differential signal level, the one-way propagation delay is one-fourth of the period of a continuous square wave. This line length is referred to as a quarter wavelength line. Because the reflections are opposite phase with the transmitted transitions, the impedance that is seen by the source is the inverse of the actual impedance at the end of the line. This is exactly true only on an ideal line with no losses or distortion so that the reflection exactly matches the transmitted transitions. Actual lines with losses will have a minimum impedance when the one-way propagation delay of an unterminated open line is one-fourth of the period of a continuous input signal.

Although the maximum current loading occurs at the quarter wavelength, it is important to realize that this short-circuit current flows every time the input signal transition is driven on the line. This is one of the reasons that it becomes much more difficult for higher-voltage drivers to drive lines at higher frequencies, that higher-voltage drivers generate larger current transients, and that lower-impedance drivers generate larger current transients. Series termination at the source tends to reduce the problems by reducing the magnitude of these transient currents. Although not practical in many cases, parallel termination at the load reduces the transient currents at the expense of more constant current at any frequency, compared to an open line. Lower-voltage excursions will have a dramatic effect on reducing current transients and power dissipation at the driver. Therefore, high-frequency interconnections on long lines are more compatible with lower-voltage excursions. However, this requires closer design attention to terminations and avoiding interference and disturbances on the line. Reduced noise margins make it more difficult to reliably recover the signal at the receiver and provide a challenge to the design and distribution of the termination voltage.[11,12]

9.6.1 Data patterns on long lines

The previous figures illustrated square-wave signals on ideal and actual lines. Practical applications usually involve arbitrary data patterns, so that line losses have different effects on signal levels, depending on the data patterns. Pattern sensitivity often represents a limit on the rate at which a long line can be operated with reliable detection of the data at the receiving end of the line. The following figures are illustrations derived by computer overlay of an arbitrary waveform for a limited set of data patterns. The results indicate the time jitter due to pattern variations and receiver threshold range. The illustrations do not address other issues such as line and receiver noise, or timing jitter or drift between data and the receiver clock. The overlay of multiple bit times illustrates various signals that may be observed by a long-persistence oscilloscope trace triggered by the receiver clock.

Figure 9.13 illustrates two data patterns as ideal line inputs in the upper portion of the figure. Both patterns begin at the LOW state; the upper one remains at the HIGH state until it returns LOW at *b* as indicated. The line output is shown aligned with the line input for clarity of illustration, ignoring the actual line delay that would exist.

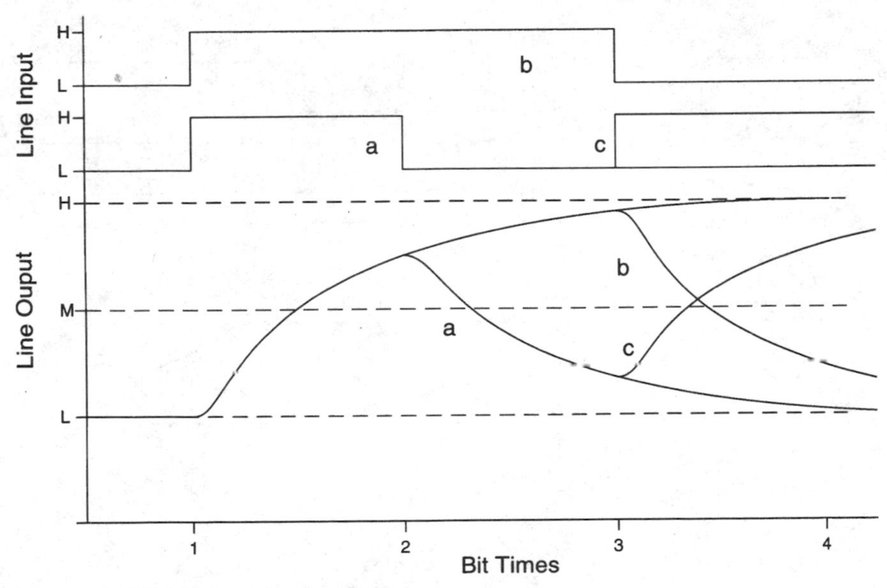

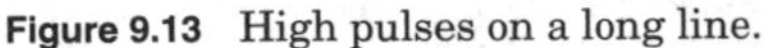

Figure 9.13 High pulses on a long line.

The line output signal is shown as a slow rise time signal that begins at the LOW steady-state level and approaches the HIGH steady-state static levels without overshoot or ringing. The signal level is fairly close to the steady-state static level when the *b* transition occurs, so the signal transition at *b* would not have significantly different characteristics than a signal transition starting from exactly steady-state static levels. The *b* transition begins a little lower than steady-state static levels, so it will cross the middle level indicated by the *M* a little sooner than if it started from exactly steady-state static levels. If this line were operated at a data rate that had no pulse widths less than the upper line input signal, there would be only a little pattern sensitivity jitter at the receiver.

Another data pattern is illustrated as the lower line input signal. This signal also starts at a LOW level but returns to the LOW level at *a* as indicated. The pulse width at *a* is only one-half the pulse width at *b* so the line output signal is not as close to the steady-state static level. The line output signal has crossed the midpoint well before the *a* transition, so a receiver threshold at exactly *M* should properly detect the signal before the *a* transition. However, the *a* transition begins well below the steady-state static HIGH level. Therefore, the *a* transition crosses the *M* level more quickly than the *b* transition. This reduces the timing window of opportunity for reliably detecting the HIGH pulse prior to the *a* transition. The lower line input pattern also shows an optional return to the HIGH level at a *c* transition.

Figure 9.14 illustrates the complement of the pattern of Fig. 9.13

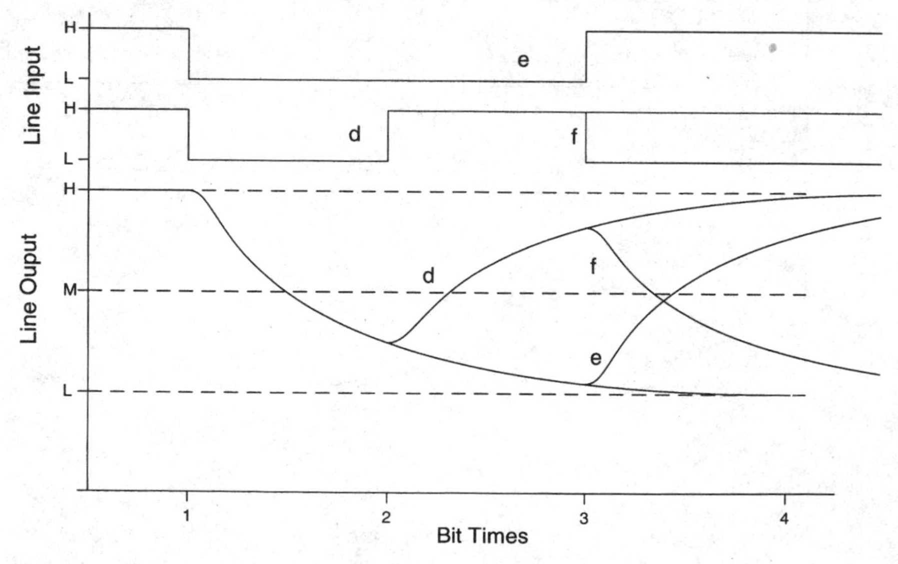

Figure 9.14 Low pulses on a long line.

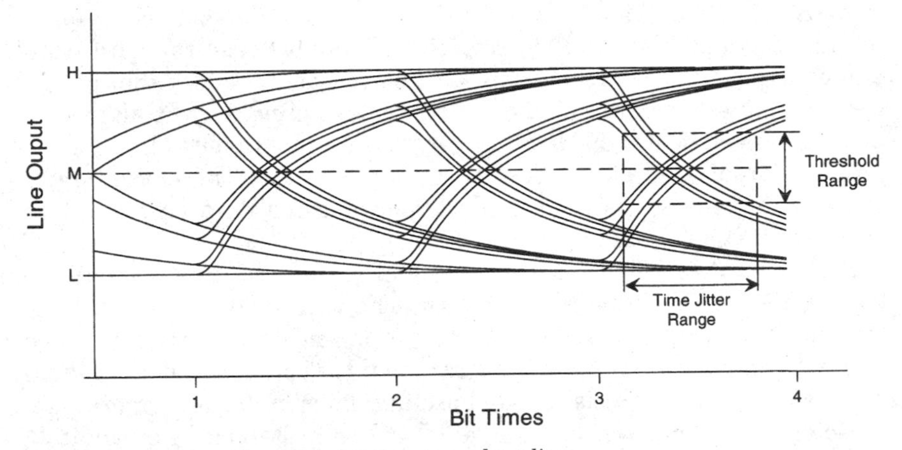

Figure 9.15 Overlay of multiple bit times on a long line.

with transitions indicated as e for the long pulse, and d and f for the short pulse. The d transition crosses the M level in the positive direction at the same position as the a transition crosses in the negative direction. This pattern does not contain any different information but is shown to illustrate that these two patterns will be combined by an overlay to generate the combined pattern.

The overlay patterns of Fig. 9.15 were generated in two stages. First, the complementary patterns of Figs. 9.13 and 9.14 were overlaid to combine the transitions for both the positive and negative directions. Next, the pattern with the combined transition directions was shifted forward and backward enough bit times to overlay a reasonable variety of transition times in the figure. Although many of the possibilities have been omitted for clarity of illustration, the transitions shown indicate a range of signal levels and the times at which they cross the M level.

Patterns as illustrated in Fig. 9.15 are sometimes called *eye patterns* because the open area in each bit time resembles an open eye. If there is not sufficient open area that the receiver can reliably detect the data, the eye pattern is said to be closed. Eye patterns also indicate when the data should be sampled by the receiver to provide the best margins for signal level uncertainty and timing jitter. As shown in this figure, the maximum signal amplitude occurs just as the signal transitions begin. However, the middle of the time interval between the last and first crossing of the M level is a little earlier, owing to the difference in the rise time of the signals. Most high speed digital transmission lines operated near the maximum data rate for the line losses operate best when sampled just slightly before the new data transitions begin. It may be noted that "center sampling" is a practice for some slow speed interconnections. The data is sampled halfway between the data transitions, typically by using the opposite polarity of the clock that generated the data transitions. Although center sampling may be aligned to provide proper operation, it is usually more vulnerable to line losses, propagation delays, and clock duty cycle tolerances. Therefore, typical center sampled lines are operated at lower rates than possible with sampling just before the new transition times.

Figure 9.15 also illustrates a threshold range and time jitter range box for the last bit time in the figure. The threshold range is a little large for emphasis in the illustration, but practical receivers have some tolerance on the input switching threshold. Typical receivers should also have input hysteresis to avoid oscillation or multiple triggering on noise on slow input rise time signals. If the hysteresis is adequately specified to guarantee that the receiver will not switch before the input has crossed the *M* level, the left edge of the time jitter range box can be moved to the right. Hysteresis essentially increases the delay in detecting the threshold crossings, which allows the data to be sampled a little later with respect to the beginning of the new data transitions.

9.6.2 Underdamped data patterns on long lines

An underdamped line termination crosses threshold faster and may result in some overshoot beyond the line losses. Figure 9.16 illus-

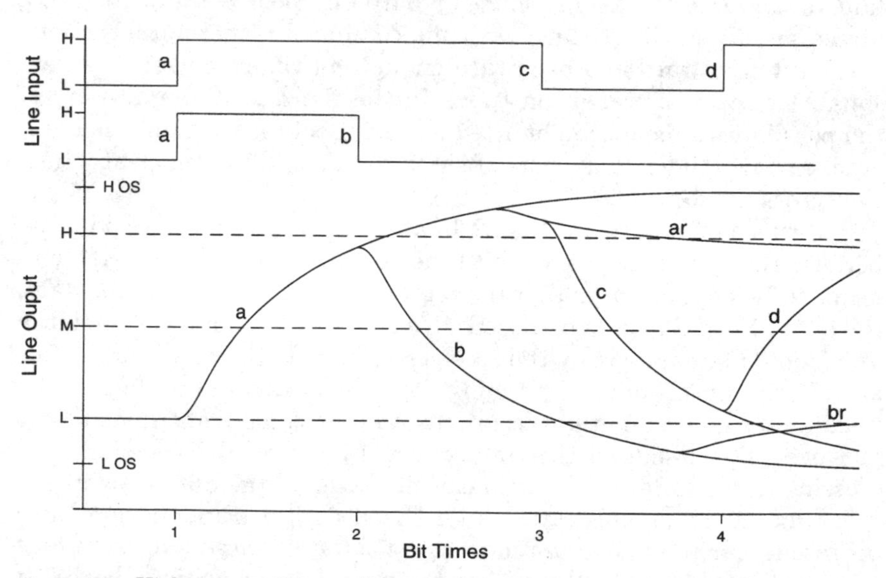

Figure 9.16 High pulses with overshoot on a long line.

trates two patterns of line input signals and the corresponding line outputs. The signal waveshape is similar to that illustrated in the previous figures but includes other worst patterns for this case. The overshoot was selected at 25 percent beyond the steady-state static levels for convenience in illustration. The overshoot levels are indicated by HOS for the HIGH level and LOS for the LOW level. Although ringing would be negligible on most long lines, this illustration includes the theoretical undershoot of 1/16, or about 6 percent. The figure shows the continuation of the *a* and *b* transitions as if there were no reflection. The corresponding reflections are labeled *ar* and *br* to indicate a reflection that begins just before the second bit time after the original transition.

The line output waveforms with overshoot were overlaid with the complement pattern (not shown) and then shifted multiple bit times to result in the overlay patterns illustrated in Fig. 9.17. The higher signal levels due to the overshoot result in a more open eye pattern, with less time jitter range than the no overshoot case illustrated previously.

Figure 9.18 illustrates a similar case of underdamped termination with overshoot, but the overshoot lasts longer before the reflection returns the signal level back toward the steady-state static levels. The worst case patterns are a little different to include a maximum overshoot on transition *a* before the reflection.

The line output waveforms for the longer line with overshoot were overlaid with the complement pattern and then shifted multiple bit times (as for the previous figures) to result in the multiple bit times overlay of Fig. 9.19. Very close inspection would indicate that the time

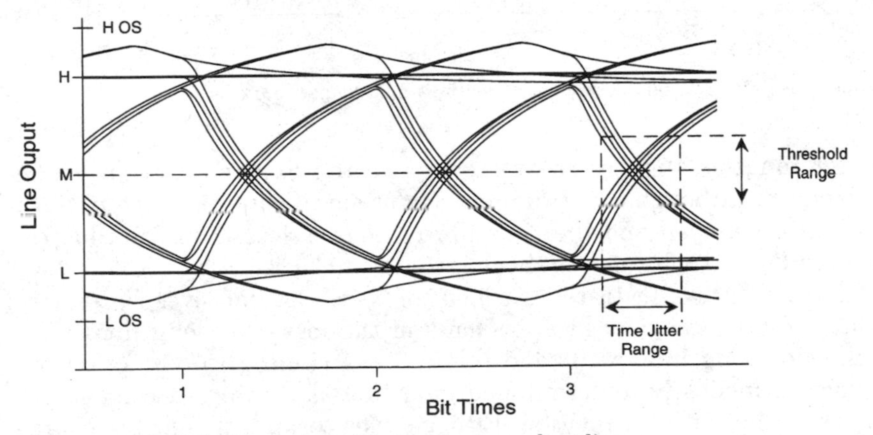

Figure 9.17 Multiple bit overlay with overshoot on a long line.

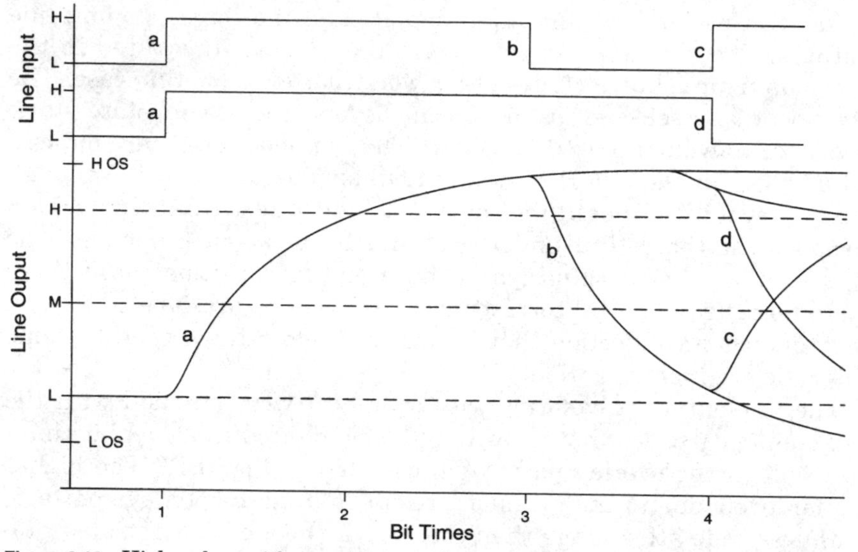

Figure 9.18 High pulses with overshoot on a longer line.

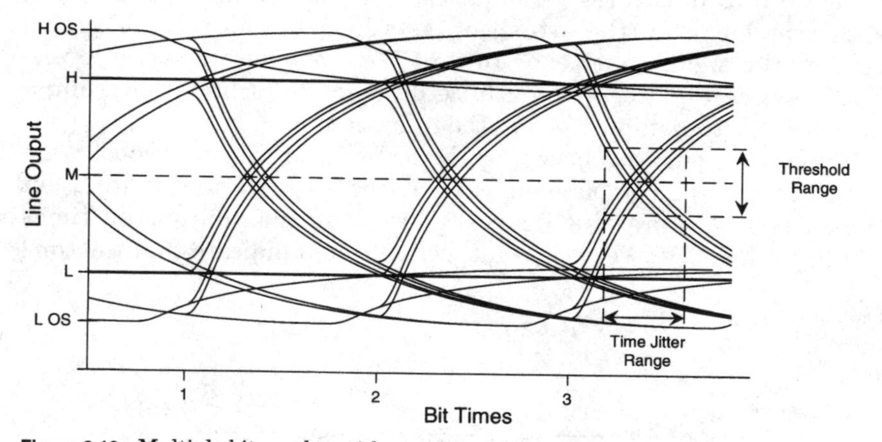

Figure 9.19 Multiple bit overlay with overshoot on a longer line.

jitter range is slightly greater owing to the longer duration of the overshoot, although the difference is not significant. The eye pattern is still more open and the time jitter range is less than the illustration in Fig. 9.15 for no overshoot.

These figures illustrate that modest overshoot improves signal rise times and can compensate for some of the losses on long lines. The losses on long lines will tend to suppress ringing, so the principal effect on modestly underdamped long lines is an increase in amplitude for the initial transition. Either series termination at the source or parallel termination at the load can be shown to be suitable for long lines with a load (or clustered loads) only at the end. However, it is expected that most long lines will use some termination at both ends to control interference, ringing, and crosstalk reflections. Terminations at both ends also relax the termination tolerance at each end.

Although it is generally considered that the terminations will be resistors, some applications may integrate Schottky diode clamps to control overshoot and ringing. Termination resistors and diodes may be integrated directly into circuits to avoid the need for placement of external components. Of course, the integrated circuit designer must consider the possible signal range, including overshoot that may occur on transmission lines. Generally, this means that terminations must be adequately isolated from substrate or other bias voltages so that other diodes do not conduct, even for signals beyond power rails or when power is lost. Typical termination resistors should be oxide isolated to avoid bias or substrate currents, should accommodate and properly recover from a wide range of input levels, and should tolerate ESD rating requirements.

9.7 References

1. Robert E. Canright, Jr., and Arden R. Helland, "Reflections of High Speed Signals Analyzed as a Delay in Timing for Clocked Logic," *24th ACM/IEEE Design Automation Conference,* Miami Beach, Fla., June 1987, paper 7.4.
2. James E. Buchanan, *Signal and Power Integrity in Digital Systems: TTL, CMOS, and BiCMOS,* McGraw-Hill, New York, 1996, p. 159.
3. Daryl Gerke and Bill Kimmel, "The Designer's Guide to Electromagnetic Compatibility," *Supplement to EDN,* Jan. 20, 1994, p. S55.
4. Richard A. Matick, *Transmission Lines for Digital and Communication Networks,* McGraw-Hill, New York, 1969, chap. 5.
5. David Wyland, "Resistor Output Logic Gives High Speed with Low Noise," *Quality Semiconductor Databook,* Quality Semiconductor, Inc., Santa Clara, Calif., 1991, p. 6-61.
6. James E. Buchanan, *Signal and Power Integrity in Digital Systems: TTL, CMOS, and BiCMOS,* McGraw-Hill, New York, 1996, chap. 1.
7. Paul H. Bardell, William H. McAnney, and Jacob Savir, *Built-in Test for VLSI: Pseudorandom Techniques,* Wiley, New York, 1987.
8. C. H. Chen (ed.), *Computer Engineering Handbook,* McGraw-Hill, New York, 1992, pp. 2.10, 6.9.
9. Dan Strassberg, "BIST: Pie in the Sky No Longer," *EDN,* Sept. 2, 1996, p. 77.
10. Arpad Barna, *High Speed Pulse and Digital Techniques,* Wiley, New York, 1980, chap. 5.5.
11. Greg Edlund, "Noise Budgets Help Maintain Signal Integrity in Low-Voltage Systems," *EDN,* July 18, 1996, p. 111.
12. Samuel H. Duncan and Robert V. White, "Designing 2.1V Futurebus+ Termination System Requires System—Engineering Approach," *EDN,* May 26, 1994, p. 117.

Chapter

10

Transmission Lines: Reflection Diagrams

This chapter deals with the issues of transmission lines and drivers and receivers that are not completely linear. The issue of reverse recovery for minority carrier semiconductor devices is described. This chapter deals extensively (but not exclusively) with the control of transmission-line overshoot and ringing by clamp diodes to the power rails, which is typical of advanced CMOS technology. The application of series Schottky diodes to permit operation of a party line bus with partial power down is illustrated. Reflection diagrams are used to illustrate the high speed transmission-line response to clamp and series diodes under various conditions. Diode recovery snap off undershoot and standing waves are described and illustrated. The use of resistance in series with both the driver outputs and clamp diode inputs is shown to provide nearly ideal high speed transmission-line response with no dc power dissipation or loss of steady-state static signal levels. The small series resistance can be internal to the integrated circuits to eliminate the need for external components.

10.1 Linear Analysis: The Principle of Superposition

The equations for signal levels, including overshoot and undershoot, and exponential decay envelope of signal ringing are based on linear equations that reasonably describe electrical performance over an appropriate range of operation. These equations depend on the characteristics of the circuit elements being *stationary* and *linear*. *Stationary* means that the characteristics of the circuit elements do not change with time (sometimes called *time invariant*). *Linear* means that the response of the circuit element is always in the same proportion to the stimulus, so that the response to multiple inputs is the sum of the responses to each input.

The ability to sum the response to multiple inputs is called the *principle of superposition.* It means that a specific change in the input stimulus results in the same change in the output response, regardless of the previous output level or the frequency content of the stimulus. A linear transfer function plots as a straight line on a linear scale; it is of the common mathematical form $a + bx$. In this equation, a is a constant that defines the response when the independent variable $x = 0$. b is the slope of the line, which defines the change in the output response for a unit change in x. The constants themselves (either a or b or both) may be the sum of individual components. Therefore, any change in the value of a increases the value of the function by the amount that a increases, regardless of the previous value. Likewise, any increase in b increases the sensitivity of the output response to a unit change in x by the same amount that b increases, regardless of the previous value of b or of the value of the output response.

An "ideal" power supply is an example of a function whose output voltage is considered a constant, independent of output load current. The constant output voltage may be changed by a reference adjustment for a regulated power supply, or by changing the number of primary cells stacked in series for a battery. An ideal resistor is an example of a function whose voltage drop is proportional to current flowing through the resistor. Of course, there are practical limits to the range of currents that allow the resistance to remain constant. For example, temperature may change the resistance value, and excessive current may damage the resistor, permanently altering the resistance value.

An ideal voltage source in series with a resistance is an example of a function whose output voltage is a constant at no load (or open circuit) but whose output voltage drops in proportion to the load current. An ideal battery will also accept current into the battery, with a voltage rise at the terminals proportional to the charging current. Of course, there are practical limits to the range of currents that can be provided or accepted by a battery of a certain size. However, battery voltage is not necessarily stationary because the voltage depends on the state of the battery charge level. This means that the battery voltage at a specific time depends on conditions prior to that time. A response that depends on prior conditions is referred to as a *memory effect.* Of course, the memory effect is used in memory systems: The logic state sensed when a memory cell is selected depends on the state previously stored in that cell.

10.2 Semiconductor Devices

Semiconductor diodes are an example of a device with both nonlinear and nonstationary characteristics. Diodes may be considered to have a "threshold" voltage. At moderate voltages below threshold (but within the breakdown limit), the current is very low and nearly independent of voltage. At voltages near the threshold, the current increases rapidly as forward voltage increases. For ordinary semiconductor *p-n* junctions, the current increase is very rapid owing to minority carriers at the junction. Therefore, it is usually reasonable to model semiconductor junction diodes as a battery voltage equal to the threshold voltage when the diode is in forward conduction, and an open circuit at any voltage below the threshold.[1] The diode forward conduction is not a perfectly constant voltage as current increases, so it is more accurate to include a small series resistance to the threshold voltage for forward conduction.

The forward voltage of a semiconductor junction diode depends on device parameters and temperature. Typical silicon junction diodes that interact with transmission-line signals have an equivalent threshold voltage of about 700 to 800 mV at room temperature. The voltage increases about 20 mV for each 10°C *decrease* in temperature. An approximate range for silicon junction diode forward voltage at typical forward currents is about 900 to 600 mV over the temperature range of −55 to +125°C.

10.2.1 Minority carrier devices

Semiconductor junction diodes, as well as bipolar (junction) transistors conduct forward current with minority carriers.[2] These semiconductors conduct heavily once turned "on," so they perform well as switches, However, bipolar semiconductors cannot sustain reverse voltage (i.e., turn off) until the minority carriers are swept out of the junction area by reverse current. This means that the dynamic behavior of semiconductor junction diodes and bipolar transistors depends on forward conduction conditions. The speed at which these devices can be turned off is limited by the previous forward conduction current. Faster turn-off is possible if forward conduction is avoided or if forward current is strictly limited. When reverse current in a semiconductor *p-n* junction finally sweeps the minority carriers out of the junction area, the diode becomes essentially open circuit and the reverse current ends very abruptly.

Diodes that have a very rapid drop in reverse current are called *step recovery* or *snap-off* diodes (Ref. 2, pp. 200ff.). The application to high speed transmission lines is that any semiconductor junction diodes present at transmission-line interfaces may have reverse recovery effects if signal levels such as overshoot result in heavy forward currents. An abrupt change in current flowing through a transmission line causes a corresponding change in voltage. The amplitude of the voltage change is the product of the change in current and the transmission-line impedance.

10.2.2 Majority carrier devices

Metal semiconductor diodes, also called Schottky barriers or Schottky diodes, conduct forward current mainly by majority carriers.[3] Majority carriers (electrons for an *n*-channel device, holes for a *p*-channel device) are also the conduction mechanism for MOSFETs (metal oxide semiconductor field-effect transistors). This means that Schottky diodes and MOS transistors do not have the reverse recovery problems associated with minority carriers in junction semiconductors. This means that Schottky diodes and MOS transistors have high speed switching capability. Schottky diodes have a lower threshold than silicon *p-n* junctions, which also makes them more suitable for transmission-line signal clamps and in circuits for low-voltage applications. However, majority carrier devices do not turn on as rapidly as junction devices, and their forward current characteristics are more resistive. The forward voltage, or "on" resistance, depends on temperature and device parameters, especially size.

Schottky diodes and bipolar transistors have been combined to provide the advantages of the fast turn on and low "on" resistance characteristics of bipolar transistors with the fast reverse recovery characteristics of Schottky diodes. A Schottky diode clamp from the collector to the base of an *npn* transistor prevents heavy saturation, so both the bipolar transistor and the Schottky diode can recover quickly from forward conduction. Of course, the recovery speed is limited by other conditions, including capacitance and the common "Miller effect" which provides negative feedback from the collector to base (Ref. 2, p. 428). The same effect causes negative feedback from the drain to gate of MOS inverter transistors. Schottky diodes are integrated with bipolar transistors for the logic families popularly called Schottky TTL. Schottky diodes are also used for output and input transmission-line clamps to control overshoot, and in series with party line bus drivers to block reverse power supply charging currents and ground offset currents.

Most transmission-line drivers are reasonably modeled as a voltage source with a series resistance. The resistance of the source may be the total of the inherent internal resistance of the switching transistors and any resistance added in series with the line driver to limit overshoot. Linear analysis is adequate to estimate the performance of these line drivers if the signal levels remain within the range at which the source resistance remains essentially constant.

Some typical departures from linear operation of drivers and receivers involve signal voltages that are beyond the range at which the output transistors in a driver conduct load current, or signal voltages that exceed power rails and may turn on clamp diodes. Many receivers have input clamp diodes to protect them from ESD (electrostatic discharge), but the diodes may not respond well to signals that overshoot beyond the power rails. These clamp diodes will also limit input signals to nearly ground if the receiver power supply is lost and faults to ground. Clamp diodes may have transient response characteristics that are different from static response, so the response is not stationary. Although the diode is essentially open circuit for static reverse voltage, this is not achieved until any charge stored during forward conduction is eliminated by reverse current. Therefore, diode characteristics may include a "memory" effect that means operating characteristics depend on previous conditions. Simple linear analysis is generally not adequate to model diode behavior through the threshold region or during reverse recovery.

10.2.3 Transmission-line drivers

Bipolar transistors in ECL and TTL bipolar driver outputs may not conduct current if the output signal level is in a region that results in a cutoff condition. For example, a negative excursion below a normal LOW output of a Schottky TTL may cause the Schottky clamp to divert the base current from the LOW output inverter transistor. A positive excursion above the normal HIGH output of TTL may turn OFF the output emitter follower pull-up transistor. TTL compatible CMOS with an NMOS source follower pull-up will respond similarly to a positive excursion above the normal HIGH output. It is possible for the ECL output emitter follower to turn OFF for transient positive excursions, but the pull-down current keeps the transistor ON for static conditions.

Rail-to-rail CMOS output drivers with inverter transistors to both power rails are usually quite linear over the normal range of currents, including reverse currents that result from reflected overshoot. Although rail-to-rail CMOS has diodes to the power rails inherent in the CMOS inverter transistors, these usually have no effect on outputs because the transistors conduct reverse currents at voltages less than the diode threshold. Therefore, most CMOS drivers are reasonably linear for signal levels up to a diode drop beyond the power rails. However, the diodes in CMOS transistors will clamp unpowered output signals to ground if the V_{CC} supply is lost and defaults to ground.

Most actual response functions are not exactly linear over an unlimited range, but many are reasonably linear over a useful range of interest. ECL and rail-to-rail CMOS are usually quite linear over most operating conditions. Transmission lines themselves are not exactly linear. For example, skin-effect losses affect frequency components differently and alter the shape of the waveform. Low-loss transmission lines are close enough that a linear assumption is sufficient to model basic operation, although not waveshape on long lines. Not all drivers have a similar ON resistance for the two logic states, but each transition can be analyzed with an appropriate source impedance. CMOS drivers eventually become current limited (high impedance) at heavy loading. CMOS drivers designed for 50-Ω transmission lines should remain linear for up to 100 mA of load current for 5-V signal transitions, or 70 mA for 3.3-V operation.

ECL (emitter coupled logic) drivers with bipolar emitter follower outputs and pull-down load-termination resistors usually have very low internal output impedance that remains quite linear if the pull-down is sufficiently low resistance to a voltage more negative than the LOW output state. Capacitance loading or positive-going undershoot on negative transitions may result in transients that cut off the emitter follower.

Schottky TTL drivers and TTL-compatible CMOS drivers for 5-V operation typically have very different impedance characteristics for the two logic states. The HIGH state impedance is usually much larger than the LOW state impedance. The HIGH state usually has almost no drive capability (very high impedance) for signal levels above about 3.5 V. This is about equivalent to defining the no-current (or open-circuit) voltage as about 1.5 V less than the power supply voltage. When the HIGH state output signal level is less than this, the driver sources current at a modest impedance level, typically 40 to 60 Ω. However, essentially no current is drawn for any voltage between the no-current value below the power supply and about the same value above the power supply. Therefore, for HIGH state output voltages near the power supply voltage, the output impedance is very high, and the driver is essentially OFF. If the output signal reaches any level in this region, the signal level will tend to remain at that level, unless there is some other load on the line, or until the driver switches to the LOW state.

This means that the output impedance depends on both the output signal level and the logic state of the driver. Linear analysis is based on parameters such as impedance remaining constant, independent of logic state and signal levels. This highly nonlinear characteristic means that linear analysis of ringing would not model the typical line response, at least for general conditions. The mismatch of the LOW and HIGH impedance (even for normal static conditions) also means that signal reflections or crosstalk traveling toward an unbalanced source will result in a different response depending on the state of the driver and will result in common mode to differential mode conversion on differential lines.

10.3 Nonlinear Analysis

There are some approaches and assumptions that can be used to estimate the performance of nonlinear circuits under certain conditions. Two of the basic methods for dealing with nonlinear analysis are numerical methods and graphical methods.[4] Computer analysis, or computer aided design (CAD) tools typically use numerical methods. Mechanical and thermal analysis often uses forms of finite element analysis in which the problem is divided into very small parts, so that each of these small elements can be analyzed in a more convenient manner. Electronic circuit analysis is often performed by computer simulation using models of nonlinear device characteristics.[5] One approach to deal with the different impedance for the two logic states of a digital line driver is to perform separate analysis for the different logic transitions, starting conditions, and/or signal levels. This is an approach to nonlinear analysis called *piecewise linear analysis.*

The usually lower impedance of the LOW state for some logic families (such as TTL) is used to analyze the HIGH-to-LOW transition. The impedance of the HIGH state for the lower signal levels is used to estimate the initial signal transition for the LOW-to-HIGH transition. The load impedance is used to estimate the reflection from the load (if the load is nearly open the reflection doubles the initial signal transition). If the initial signal plus the reflection results in a signal level within about ± 1.5 V of the power supply, the driver impedance is considered to be nearly open, so there is no further change in signal level or ringing if the reflection returned from an open line. This is because an equilibrium condition results when there is no current at the load and no current at the source. It should be noted that traveling waves do not necessarily double whenever they arrive at an open circuit. A reflection results from an unbalance between the voltage and current relationship on the line and the voltage and current which the end of the line will accept.

Because the initial impedance of the HIGH state is typically a reasonable match for most line impedance of typical transmission lines, there is typically little ringing on the LOW-to-HIGH transition for TTL drivers. However, the range of potential HIGH state signal levels depends on the tolerance ranges on the previous LOW signal level, the transmission-line impedance, the driver output impedance, any load impedance, and the power supply voltage. This range of HIGH state signal levels, in turn, affects the potential ringing on the HIGH-TO-LOW transition. For the typical case of LOW state output impedance much lower than the transmission-line impedance, the worst ringing on the HIGH-TO-LOW transition occurs when the HIGH state signal level is the highest, which causes the greatest initial signal excursion. This results in greater ringing excursions and more reflection cycles to settle toward the static LOW level.

A traditional method for stabilizing HIGH signal levels and dampening the ringing on the HIGH-TO-LOW transition is to add series resistance on the driver output. Increasing the output impedance of the LOW state reduces the initial HIGH-TO-LOW signal excursion and quickly dampens any ringing toward the static level. Although the LOW-to-HIGH transition is now typically overdamped, this keeps the HIGH signal from reaching the high-impedance signal levels near the power supply voltage. Although the output impedance of the two logic states remains unbalanced in absolute terms, the ratio of the HIGH state impedance to the LOW state impedance is reduced. This means that the signal levels stabilize more quickly for both logic states and that common mode to differential mode conversion on differential lines is significantly reduced. Additional benefits of driver series resistance are reduced transient currents and signal excursions, so that both ground bounce at the driver package and crosstalk along the transmission line are reduced.

Special precautions should be noted in the application of series resistance at the driver output. One consideration is that the additional resistance reduces the amount of current that the driver can sink and maintain a signal level near the no-load LOW level. This can be significant in such applications as single-ended (one signal line, referenced to ground) transmission lines with standard TTL, high input current Schottky TTL, or if there are several loads, or a pull-up or termination resistor at the load end. TTL input thresholds are typically nearer to ground than the midpoint of the power supply, so it is important to achieve an adequate signal margin in the LOW state. When the load current is significant, the series resistance value must be selected to limit the voltage drop across the resistor in the LOW state.

The input currents for typical CMOS and ECL loads are usually so low that the loss of dc signal level is not significant with series termination. Loading limitations are usually determined by initial signal level, load placement, capacitance loads, and maximum frequency requirements when the dc loading is very low. Of course, it is important that the pull-down resistor for ECL outputs must be located *before* the series resistor so the dc current in the pull-down does not flow through the series resistor.

Both series termination and a parallel load resistor may be appropriate for differential lines. The series termination is in both lines, so the change in signal level is the same on both lines. The lines may be terminated individually, or the load resistance may be line-to-line to eliminate ground return currents. Termination at both ends is called *double termination* and can control both signal ringing and crosstalk reflections. Double termination also reduces ringing more effectively with less accuracy required on matching the resistor values to the line impedance. Therefore, double termination of long lines is more tolerant of manufacturing uncertainty and may allow the termination resistors to be integrated into the driver and receiver circuits.

10.3.1 Ferrite shield beads

One of the alternatives to series resistance for waveshape control of single-ended signals of modest length is ferrite shield beads in series with the driver output. These parts are made with a lossy ferrite material that surrounds a wire, so that the dc impedance is low, but the high-frequency components of a fast rise time signal are attenuated to control ringing. Ferrite shield beads in series with the driver are often effective when the line is much shorter than the quarter wavelength of the highest frequency to be transmitted, so that signal levels settle to stable levels before the signal repeats. It is difficult to analyze the effect of ferrite shield beads in a transmission line, since the impedance is both resistive and reactive, and varies with frequency.

A typical guideline for selecting an appropriate shield bead characteristic is that the series impedance should be low enough at the highest signal repetition frequency to be transmitted, so that the voltage drop of a resistor of the same impedance would result in acceptable static signal levels. Ferrite shield beads typically show an impedance that increases with frequency owing to the inductive reactance at lower frequency, but achieve a maximum impedance at high frequency, where it becomes more resistive. The impedance at high frequency (typically more than 10 times the highest frequency to be transmitted) should be about 50 percent of the line impedance but generally should not significantly exceed the line impedance. Clock distribution within a digital unit is a typical application using series termination.[6]

10.4 Reflection Diagrams

Graphical methods are a convenient and useful method for solving common nonlinear problems in electronic circuit analysis. One of the advantages is that it provides a visual result that usually provides the system analyst with an insight into process, as well as waveforms of the results. Understanding of the process involved is often useful for modifying the system to improve performance for a particular application.

The transmission-line analysis technique called reflection diagrams is a graphical method for determining the voltage and current relationships during the iterations of incident and reflected waves on a transmission line. Reflection diagrams are sometimes called *Bergeron plots,* named for the French hydraulic engineer who developed the method to study the propagation of water hammer effect in pipes (another analogy between water and electricity!).[7]

The reflection diagram analysis is based on the principles of superimposing load lines on output operating characteristics. For digital output drivers, there are two output operating characteristics, one for a logic LOW, the other for the logic HIGH. Any of the output characteristics or the load line may be nonlinear. Engineering textbooks that deal with such nonlinear circuit elements as incandescent lamps, ballast tubes, vacuum tubes, diodes, and transistors often illustrate the principles of operation with graphical analysis of load lines superimposed on operating characteristics.[2,4,8,9]

An example of a load line superimposed on the operating characteristics of a metal oxide semiconductor field-effect transistor (MOSFET) is illustrated in Fig. 10.1. The MOSFET characteristics are shown as curves of source-to-drain current versus voltage for a few examples of gate-to-source voltage. The traditional convention seems to be that current is chosen for the vertical scale, with voltage on the horizontal, so the slope of the load line is negative. In general, the MOSFET

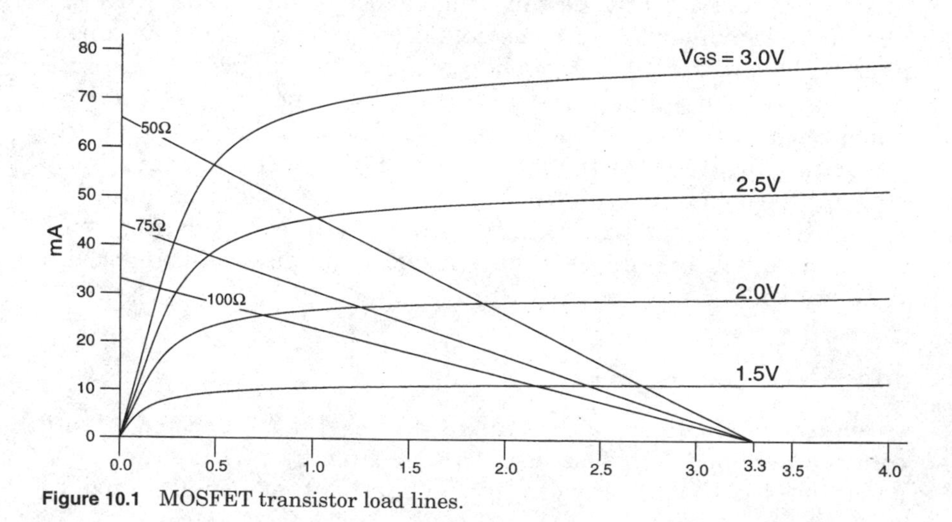

Figure 10.1 MOSFET transistor load lines.

transistor is capable of conducting significant current when high gate voltage is applied, but essentially zero current when the gate voltage is very low. These two extremes would define the two operating characteristics for the two logic states of a digital line driver transistor. Complementary MOSFET (CMOS) drivers are implemented with a complementary pair of transistors so that one of the two transistors is ON in either of the two logic states.

The voltage and current scales of Fig. 10.1 illustrate values that may be appropriate for a transmission-line driver operated at 3.3 V. Linear load lines are drawn for 50, 75, and 100 Ω. The intersection of the load lines with the operating characteristic for the gate voltage of 3.0 V indicates that the transistor is capable of driving a 50-Ω load. However, the 50-Ω load is nearing current limiting, as indicated by the curvature of the operating characteristic near the intersection with the 50-Ω load line. When the transistor is ON (the gate voltage is high), the transistor conducts sufficient current that most of the voltage drop is across the resistor. Conversely, when the transistor is OFF (the gate voltage is low), essentially all of the voltage drop is across the transistor. If the static load is very low current, the transistor voltage switches between static voltage levels of nearly zero volts to the power supply voltage, with nearly no current flowing at either level. However, significant current may flow in a transmission line until static levels are reached. The selection of current for the vertical axis and voltage for the horizontal axis has been traditional for electronic devices. However, it is more convenient to reverse the axes for reflection diagrams so that the voltage waveforms can be directly extracted in a companion diagram.

The opposite slope for source and load lines is a result of associating a voltage rise to the source and a voltage drop for the load, and the voltages sum to zero. This is illustrated in Fig. 10.2 for the loop current of a voltage source with an internal impedance and a load impedance $R_L = Z_0$, which represents the dynamic impedance of a transmission line. The voltage rise is from the voltage source, with a

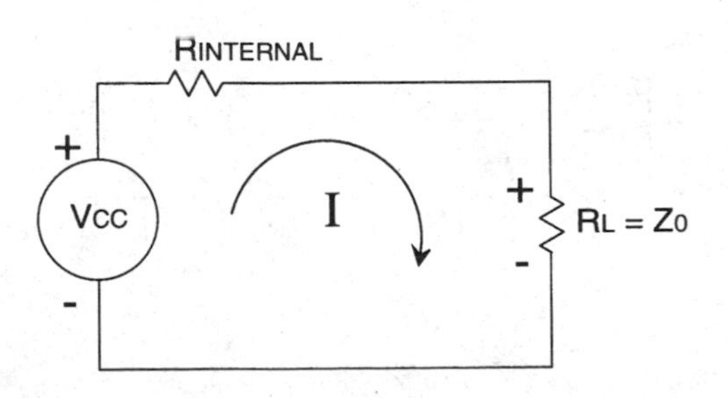

Figure 10.2 Loop current illustration.

drop in the internal resistor that depends on the load current. When the load current is very low, the output voltage approaches the open-circuit, or no-load, voltage.

Figure 10.2 is drawn for a single internal voltage source and positive current flowing from the positive terminal of the source. If there is some external source of current that flows into the source, the direction of net current flow through the internal impedance may be reversed. Nonlinear sources may result in an internal resistance that may be more like a diode, so it is a very high impedance to reverse current. The internal voltage drop from the load current results in the operating characteristic of the source, which can be observed at the connection to the load impedance. Of course, if the internal source impedance were a constant resistance, the operating characteristic would be a straight line. Graphical analysis would simply indicate that the point at which the two lines intersect determines the voltage and current which result from that load. Although graphical analysis also works for linear circuits, the equations for linear operation may be used to solve for the results in a mathematical form.

10.4.1 Load lines on operating characteristics

Figure 10.3 is an example of output operating characteristics for LOW and HIGH output levels, an input load curve with clamping

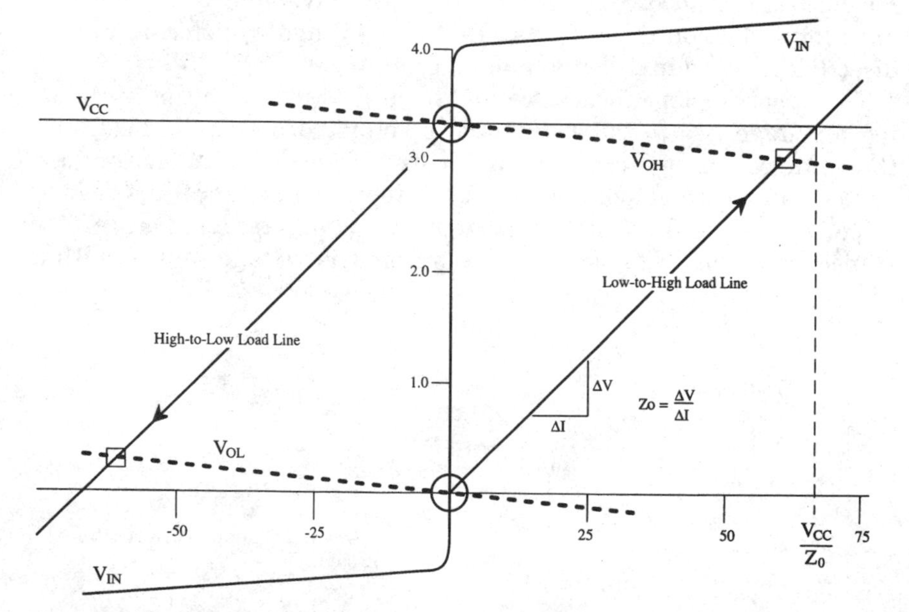

Figure 10.3 Input and output operating characteristics with Z_0 load lines.

diodes beyond both of the power rails. The figure indicates dynamic load lines for the transmission-line switching load during both LOW-to-HIGH and HIGH-to-LOW transitions. It may be noted that the convention chosen for reflection diagrams is that voltage is shown on the vertical axis, with current on the horizontal axis.[7,10] This is chosen for convenience in developing the voltage waveform on an accompanying figure. The positive current convention chosen is that current out of the driver is positive to the right in the figure. Therefore, the V_{OH} output slope is downward toward the right from the zero-current steady-state point. Similarly, the V_{OL} output slope is in the same direction, but normal operating load current is negative to the left of the zero-current axis. The slope of load lines from the driver is positive, with the zero-current V_{IN} load line coincident with the zero-current axis.

The example shown in this figure is intended to represent a CMOS output driver with reasonably linear, low output impedance output characteristics over the range of positive and negative load currents. Voltage is shown on the vertical scale for about a 3.3-V power supply. The static input load current is intended to represent a CMOS logic input, so the input current is essentially zero for all signal levels between the power rails. The two steady-state static operating points where the V_{IN} load line coincides with the two-driver output characteristics are indicated with large circles. Clamp diodes conduct to limit input voltage beyond the rails if the load is powered at the same level as the driver.[11] The current scale is chosen by personal preference so that the dynamic transmission-line impedance plots as a 45° slope for convenience, but any slope may be used. The values shown in this example are scaled for $Z_0 = 50\ \Omega$, so the transient currents drawn for a 3.3-V power supply exceed 50 mA.

The intersection of the static V_{IN} line with each of the two output characteristics are generally assumed to be the static operating points that exist before any transition begins. However, it should be noted that this is only an assumption and may not be valid for high speed lines that do not fully settle to static levels before each signal transition. A further caution is that although these curves are intended to be realistic for many practical applications, few device manufacturers adequately specify the limits of the full range of dynamic operating characteristics to permit worst case analysis.[10] Therefore, the system analyst is cautioned that reflection diagram analysis is appropriate for illustrating typical characteristics under some conditions but is not necessarily an assurance of actual operation.

The transient operating points for each transition are indicated by squares at the intersection of the dynamic load line from the static operating point to the opposite operating characteristic. This represents the current and voltage conditions at the driver when the signal transition is first launched into a transmission line after static levels have been previously established. Obviously the transient operating points are far from the static operating points, so there must be a significant reflection at the load to satisfy the conservation of energy principle at the load end of the line.

The analyst using reflection diagrams is cautioned against applying general principles that may not be valid for some nonlinear applications. For example, it is generally true that when a traveling wave arrives at a higher-impedance discontinuity, the voltage is increased. However, this does not necessarily apply to reflections returning to nonlinear drivers. If a reflected wave returning to a driver results in cutoff, the voltage does not continue to rise without limit. Once the driver is in cutoff, it is high-impedance and cannot provide further energy into the line. In the case of an open load, the reflection established zero current as an equilibrium condition at the load. When the reflection returns to a cutoff driver, zero current will remain as the equilibrium condition at the driver, so there is no further change in current on the line.

10.4.2 The conservation of energy

A more general principle for a traveling wave arriving at a discontinuity is that the principle of conservation of energy applies so that any change in voltage and current is driven closer to equilibrium conditions. Further traveling waves cease when equilibrium conditions are achieved. To the extent that conditions approach equilibrium, further traveling waves become insignificant. To the extent that the traveling waves become insignificant, they also disappear because of small but finite losses in the line. It should be noted that equilibrium is not necessarily at a single point for nonlinear devices. A nonlinear driver is compatible with zero current over a range of output voltage for which the driver is cut off. An open load or even a load clamped beyond the power rails is also compatible with zero current over a range of input voltage. Therefore, an equilibrium condition may exist over a range of voltages at which both driver and load are compatible with zero current.

The energy conservation principle for an open line may be considered as follows:

- A transition of V_1 is driven on a line for the round-trip time of $2T$, where T is the one-way propagation delay. The driver becomes open circuit when the reflection from the open end returns at the end of the round-trip time.

- One half of the energy initially launched into the line is stored on the line as capacitance energy = $C(V)^2/2$, the other half as inductance energy = $L(I)^2/2$. It can be shown that these two are equal when $L = C(Z_0)^2$ and $I = V/Z_0$ are substituted into $L(I)^2/2$.
- When the traveling wave is reflected at the open end, the inductance energy is converted into additional capacitance energy, doubling the voltage on the capacitance. The driver continues to supply energy into the line until the reflection returns to the source, so the total energy driven into the line during the round-trip time is the product of power and time, or $(2T) * (V_1)^2/Z_0$.
- The final energy stored on the line is = $C(V_2)^2/2$. If we substitute $T = \sqrt{LC}$ and $Z_0 = \sqrt{L/C}$ and set the final energy equal to the total energy driven into the line during the round-trip time, the equation is

$$\frac{C(V_2)^2}{2} = \frac{(V_1)^2(2\sqrt{LC})}{Z_0}$$

Simplifying this results in $(V_2)^2 = 4(V_1)^2$, so the final voltage $V_2 = 2V_1$ satisfies the conservation of energy on the line. Therefore, there are no further reflections and equilibrium conditions at the load and the source are also satisfied.

10.4.3 Reflections for a driver with a series diode

The signal levels that result from an unterminated line that goes into cutoff at the HIGH level can be illustrated with a reflection diagram. Consider an example of the driver circuit illustrated in Fig. 10.4. This

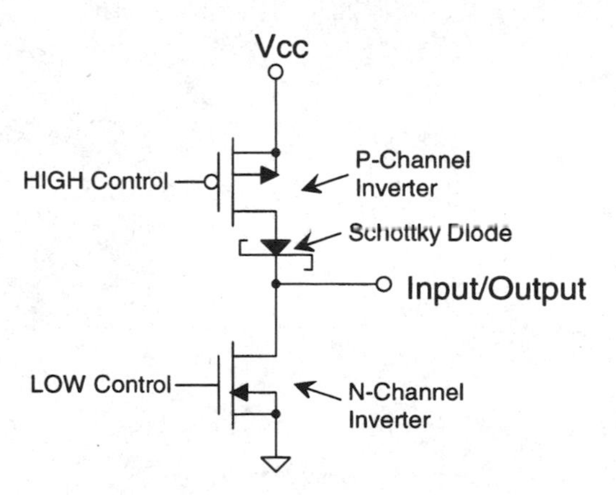

Figure 10.4 CMOS output driver with series diode.

is a CMOS inverter driving (nearly) rail-to-rail signal levels, but with a high-current Schottky diode in series with the *P*-channel inverter. The Schottky diode prevents ringing on the LOW-to-HIGH transition and blocks V_{CC} charging of party lines with multiple power supplies, or for "live insertion" into active systems.

Without the diode, a powered driver in the HIGH state remains a low impedance to overshoot, converting it to undershoot. Furthermore, the internal body diode of the *P*-channel inverter (as indicated by the arrow in the transistor in the figure) would result in charging current from any output higher than V_{CC} of the driver, even with power off. Low impedance charging into a power bus would seriously overload active signals. Forward biasing of internal diodes is also a risk of damage or latch-up due to currents injected into the semiconductor substrate.

10.4.4 Rising edge reflection diagrams

Illustrated in Fig. 10.5 are the V_{OL}, V_{OH}, and V_{IN} operating characteristics and a LOW-to-HIGH transition load line. The V_{OL} line is a linear low-impedance characteristic of the *N*-channel inverter, which conducts heavily for either forward or reverse currents. The V_{OH} line is the *P*-channel inverter in series with the diode, so the output is about a diode drop below V_{CC} for positive current out of the driver. Positive current is to the right in Fig. 10.4. However, the series diode blocks negative current, so the V_{OH} line is coincident with the zero-current vertical axis above V_{CC}. The input characteristic is no current for inputs above ground, with no clamp diodes to V_{CC}. Therefore, the V_{IN} and V_{OH} characteristics are coincident above V_{CC}. Therefore, any

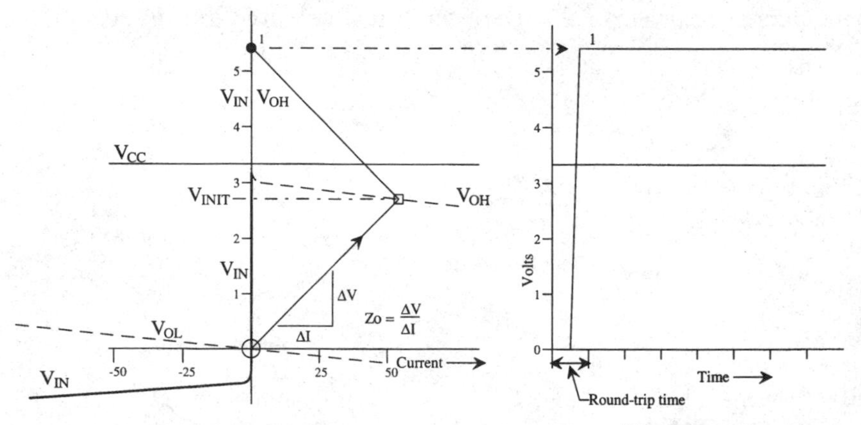

Figure 10.5 Rising edge reflection diagram for driver with series diode.

HIGH level at or above V_{CC} represents steady-state static levels with zero current. The input characteristic is illustrated with a diode clamp to ground in Fig. 10.5, but this does not affect the LOW-to-HIGH transition.

The LOW-to-HIGH transition intercepts the V_{OH} line at the level that results in the initial voltage, V_{INIT}, on the line. This level arrives at the load and is reflected to double the voltage at zero current. This voltage level is indicated by the dot with the number 1 to indicate the first incident wave signal level at the load. The slope of the reflection load line is the negative of the load line from the driver. The triangle formed by the initial steady-state static level indicated by the large circle, the intercept of the V_{OH} line, and the intercept of the zero-current line marked by the dot is an isosceles triangle, so the voltage level at the dot is exactly double the initial level at the intercept of the V_{OH} line. This satisfies the conservation of energy, and steady-state conditions are achieved at zero current on the line. The waveform illustrated on the right side of the figure indicates that the signal voltage at the open load rises to the static level above V_{CC} and remains there. The diode in series with the driver has prevented any conversion of the overshoot to ringing, but the signal level overshoots to well above V_{CC} and remains at that level.

The V_{OH} output characteristic illustrated in Fig. 10.5 is nonlinear in the region between the intercept of the Z_0 load line and the reflection to the zero-current intercept above V_{CC} at the dot. The output impedance in the small region around the Z_0 intercept is very low, but the output impedance in the small region around the zero-current intercept at the reflection is very high. Figure 10.6 illustrates that this

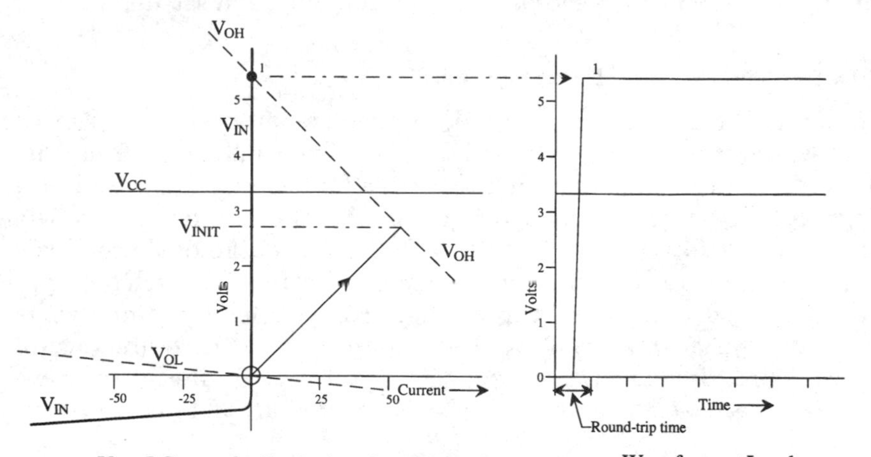

Figure 10.6 Linear equivalent rising edge reflection diagram.

nonlinear combination is equivalent to a linear V_{OH} output characteristic through these same two points. The resulting equivalent output impedance between these two points is equal to Z_0, so the driver acts as if it is exactly source terminated. One half of the final signal travels to the load, where the voltage level doubles. The steady-state static equilibrium level results when the reflection returns to the driver.

The waveform at the load is a single transition from the previous steady-state static level to the new equilibrium level. However, the final level is ambiguous because a range of voltages satisfies the zero-current equilibrium condition. Any significant change in starting conditions, Z_0, or V_{OH} may result in a different final level. Although the V vs. I curves illustrated in these figures represent the CMOS driver circuit of Fig. 10.4, similar nonlinear V_{OH} characteristics may be typical of other logic families. The familiar TTL families (including Schottky TTL, BiCMOS, and TTL-compatible CMOS outputs) typically do not conduct negative current because the outputs are bipolar emitter followers or NMOS source followers. Output drivers may include Schottky diodes in series to prevent reverse current, even when V_{CC} is lost.[12]

The minimum output HIGH level at which TTL can maintain a zero current equilibrium is about the same as the 3.3-V CMOS circuit illustrated in Fig. 10.4, although the TTL circuits are typically powered by 5 V. However, overshoot on the LOW-to-HIGH transition typically results in static HIGH levels well above the minimum level at which significant current could be driven. Even if the inputs have clamp diodes to limit excursions above 5 V, the high input impedance allows the HIGH level to remain at a range of levels, depending on previous signal history, actual loading conditions, and settling time.

10.4.5 Falling edge reflection diagrams

Illustrated in Fig. 10.7 is the falling edge transition for the V vs. I curves and the steady-state static HIGH level established from Fig. 10.4. The initial transition at the source falls nearly to ground at a high current level of almost 100 mA in this illustration. The reflection at the load is limited by the diode clamp, so the diode conducts significant forward current until repeated reflection cycles finally reach zero current at the third reflection cycle, or five line delays after the initial transition for this example. Then there are several more cycles of damped ringing levels as the signal alternates above and below the steady-state static LOW level of ground and zero current.

The sequence of levels at the load are indicated by numbers beside the dots on the V_{IN} static load line, with the corresponding numbers

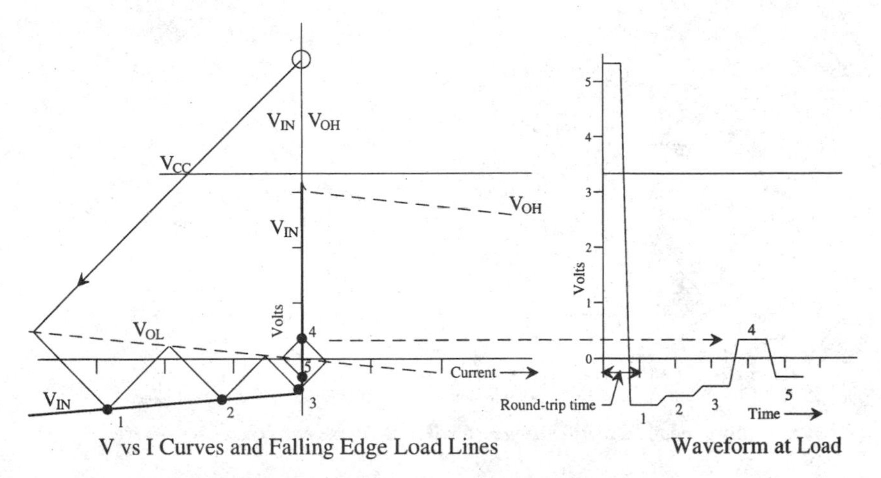

Figure 10.7 Falling edge reflection diagram with diode clamp.

on the signal levels on the waveform at the load diagram. It should be noted that none of the early reflection cycles result in conditions that are very close to the steady-state static level assumed for the starting point for the LOW-to-HIGH transition. Therefore, either the signal needs a long time to settle to the steady-state static level or the rising edge starts from a different point. This will result in a different quiescent HIGH level, which also changes the conditions for the HIGH-to-LOW transition. These characteristics illustrate that the high speed signal response may depend on signal history.

The previous illustration of the HIGH-to-LOW transition assumes that the static V_{IN} characteristics of the clamp diode at the load are the same as the dynamic response. However, if the diode to ground at the load is a large *p-n* junction semiconductor, the minority carriers must be swept out of the junction by reverse current before the diode can recover from forward conduction. Few devices are adequately specified for the dynamic characteristics of input clamp diodes, or even if diodes to ground or the power rail are present. FAST devices are described[12] as having Schottky clamps to ground with a dynamic impedance of about 10 Ω. FACT devices are described[13] as having diode clamps to both ground and V_{CC} on all CMOS inputs, and a schematic indicates that they are ordinary *p-n* junction semiconductor diodes. The discussion on latch-up suggests that adding Schottky diode clamps to the CMOS power rails is one measure to reduce the susceptibility to severe overshoot on signal inputs.

Figure 10.8 is an illustration of a falling edge transition from the previously defined HIGH level, but the diode clamp at the load is an ordinary *p-n* junction semiconductor diode with reverse recovery time

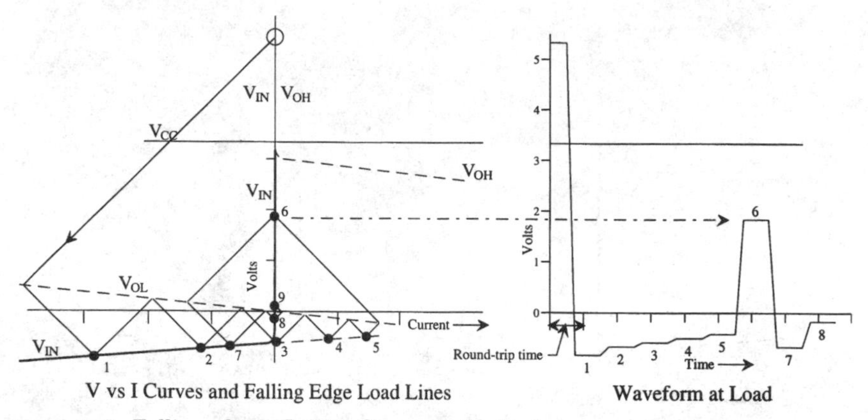

Figure 10.8 Falling edge reflection diagram with diode reverse recovery.

at reverse current. The model for the reverse recovery is an extension of the forward *V* vs. *I* curve into the reverse current region. This is similar to modeling a diode in conduction as a battery with some internal series resistance that defines the slope of the *V* vs. *I* line during conduction.[1]

The reflection diagram indicates that the clamp diode continues to conduct current during reflection cycles 4 and 5, then snaps to zero current at reflection cycle 6. It should be noted that the time at which the diode recovers to zero current depends on many variables which are not indicated by the reflection diagram. It should also be noted that although this reflection diagram indicates that the diode recovery is synchronous with reflection cycle 6, the actual recovery occurs at any time at which the minority carriers are finally swept out of the junction region and the diode snaps to zero current. When the reverse current flowing from the transmission line into the diode suddenly stops, there is a voltage transient that travels on the line equal to the change in current times the Z_0 transmission-line impedance. The effect is the same as if a signal with that reverse current arrived at an open-circuit end of a line; the current must change to zero, and the voltage changes in proportion to the change in current.

The voltage pulse that occurs when the clamp diode recovers is shown in the waveform at the load. The amplitude of this pulse is obviously a violation of LOW state noise margin requirements. The level illustrated is almost to the input HIGH threshold for TTL, and well into the threshold uncertainty region for 3.3-V CMOS. We would like to emphasize that this is not a hypothetical illustration that could only occur for unlikely circumstances. Rather, this is an illustration of the cause of an actual problem that was observed. The problem was corrected by substituting a driver with internal series resistors. This reduced the forward clamping current, so the reverse recovery current became insignificant.

One of the principles that we learned in trying to determine the cause for the problem is that the diode recovery pulse may occur much later than expected by the ringing of an ordinary linear device. The voltage waveform seemed to remain settled for many line delays after the initial falling transition, so we thought that it was settled to steady-state static levels. Of course, ground noise tended to obscure the observation that the signal was still below ground, and the actual measurements were not the clean steps of the illustration.

The recovery pulse occurred so late in this case that the problem occurred in a later bit time, so we spent much time trying to find why the conditions at that time could cause such a large pulse. Another principle that we learned is that when clamp diodes conduct on transmission-line overshoot, the diode has not recovered until the signal level pops above ground for ground clamps, or below the power rail for V_{CC} clamps.

10.4.6 Unterminated reflection diagrams

Figure 10.9 illustrates another case of a falling edge transition from the same HIGH level defined above, but this illustration has no clamp diode to suppress the negative overshoot and ringing. Although there

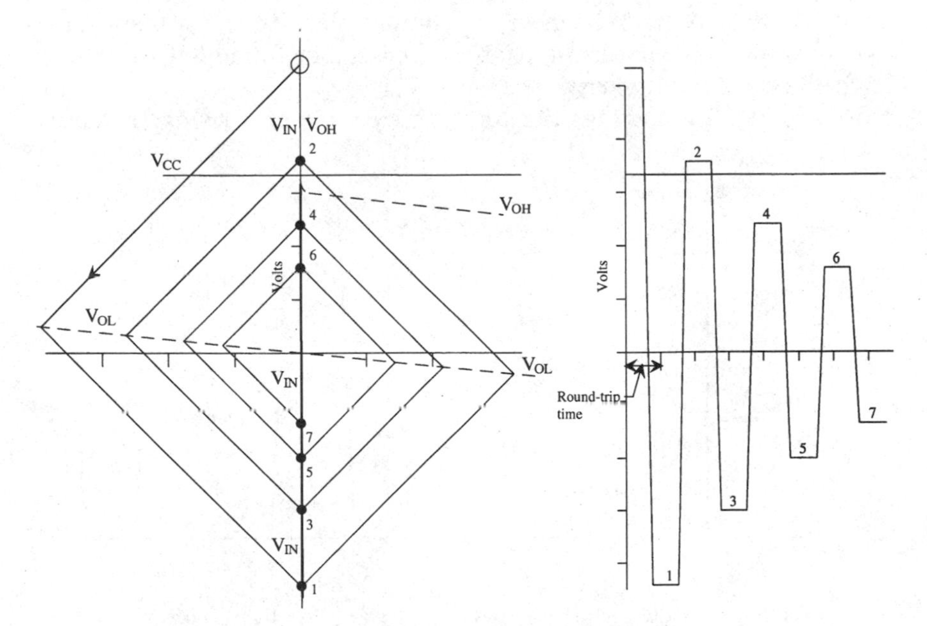

Figure 10.9 Falling edge reflection diagram without diode clamps.

is no diode recovery issue, the ringing is severe and significant levels remain even longer than when the clamp diode recovery pulse occurs. The response to the falling edge transition is linear, and the ringing levels may be predicted with the linear equations for the appropriate reflection factors.

The uncertainty issue here is what is the HIGH level from which the falling edge transition begins. Obviously, the LOW signal levels illustrated in Fig. 10.9 do not settle to anywhere near the steady-state static LOW level until after a very long settling time. The danger is that another rising edge transition begins when the source is conducting positive current after a reflection from one of the negative overshoot levels. For example, the source may be conducting the positive current on the V_{OL} line after the undershoot at the load marked by the number 1 at the dot below ground. If the rising edge transition at the source starts from this point, the load line intercepts the V_{OH} line much farther to the right than when starting at zero current. Therefore, the zero current intercept for the reflection of the rising edge transition will overshoot farther above the HIGH level assumed for these illustrations. The next falling edge transition then overshoots even farther below ground. A vicious cycle of larger and larger overshoot levels may progress to damage levels if the frequency of input transitions coincides with overshoot on the ringing cycles. The warning is to be aware of the likelihood of transmission-line problems when overshoot is not adequately terminated. The designer should verify that the parts to be used are compatible with the transmission-line environment or termination devices must be added to achieve high speed signal integrity.

Figure 10.10 illustrates the limiting extreme of a rising transition

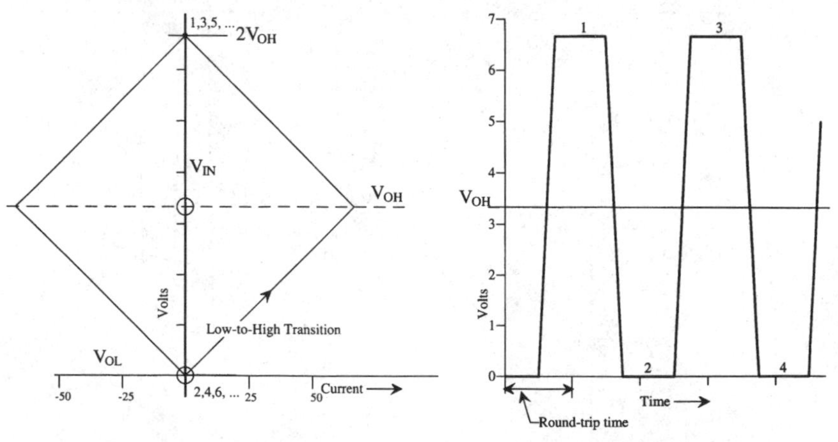

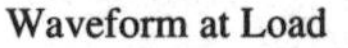

Figure 10.10 Rising edge reflection diagram with no damping.

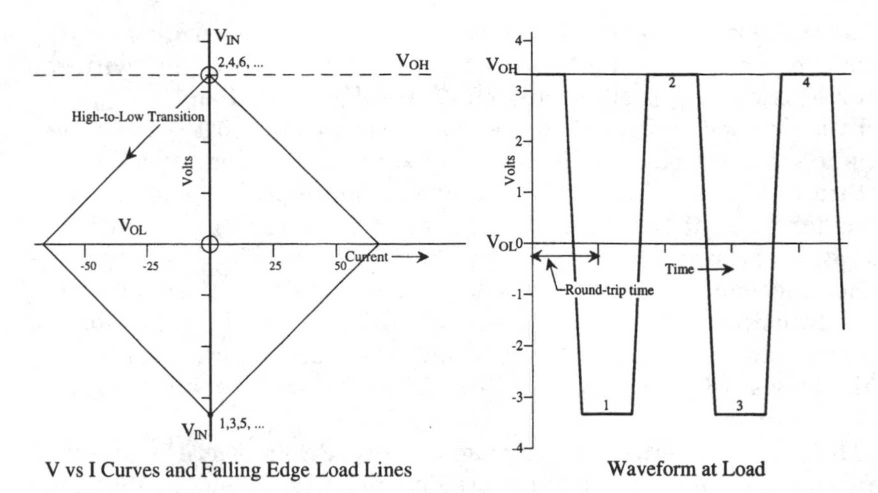

Figure 10.11 Falling edge reflection diagram with no damping.

on an unterminated line. The driver has no series impedance, so the V_{OH} and V_{OL} characteristics are horizontal at the V_{CC} and ground levels. The load is perfectly open over the entire range of signal levels, including overshoot above V_{CC} and below ground. There are only two steady-state static points, indicated by the circles in the figure. If the rising edge transition starts at the steady-state static LOW level at zero current, the reflection at the open load doubles to twice the V_{OH} level. If there are no further transitions, the voltage signal at the load oscillates between twice the V_{OH} level and ground. The average of the signal levels is V_{OH}, but that level is never reached. The voltage signal at the zero-impedance source remains at V_{OH}, but the current at the source oscillates between $\pm (V_{OH})/Z_0$. Of course, practical systems have some losses that will eventually reduce the amplitude of the oscillations.

Figure 10.11 illustrates the falling edge transition for the same case, assuming that the starting point is the steady-state static zero-current point at V_{OH}. For this case, the signal at the open load oscillates between the original V_{OH} above ground and V_{OH} below ground. Again, the response to a single transition is oscillation which is dampened only by the actual losses in the interconnection system. The voltage signal at the zero impedance source remains at V_{OL}, but the current at the source oscillates between $\pm (V_{OH})/Z_0$.

As bad as the previous responses to single transitions appear, they are not the worst-case response to *multiple* input transitions. The obvious problem with the limiting cases illustrated above is that the steady-state starting points are not consistent with the oscillations which (almost) never reach the steady-state static levels at zero current. For example, if another rising edge transition begins with the source conducting positive current on the V_{OL} line, then the V_{OH} current is doubled compared to the waveform in Fig. 10.10. The reflection to the open load will double to 3 V_{OH}, so the second reflection will return to V_{OH} *below* ground, for a total peak-to-peak swing of 4 V_{OH}. Furthermore, if the next falling edge transition begins with the source already conducting negative current on the V_{OH} line, the currents and voltage excursions will increase again (and again on the next transition). The signal levels would be far off the paper for the figures, and actual signal levels would continue to grow until the circuit "blows up" or some limit is reached, typically some nonlinear increase in losses.

The previous reflection diagrams indicate that there is a severe problem with high speed transmission lines without adequate termination to dampen ringing. These reflection diagrams illustrate the response to a single transition that begins at steady-state *static* levels. However, the graphical approach is not well suited to determine what the dynamic limits are for underdamped lines with new signal transitions that combine with previous transients that have not settled to static levels.

An open load is a linear element, so linear analysis can be applied to the dynamic response to repetitive waveforms. Since the VSWR of an open load is unlimited, there can be an unlimited ratio between the minimum and maximum signals on the line. The voltage will be maximum at the open end because $I = 0$. There may be a voltage minimum at locations on the line where the incident and reflected waves cancel. If the signal frequency and the line length between the source and the load result in a minimum at the source, the open end of the line appears as a short circuit to the source. Therefore, the maximum current at the source is limited only by the source resistance, and can become very large for a low source resistance. The maximum voltage at the open load is the maximum current at the source times Z_0. This maximum voltage can become much larger than twice the initial step for a single transition that begins from a steady-state static level. The potential exists for much higher dynamic signal levels than the designer ever imagined from analysis based on static conditions.

The zero-impedance source illustrated in Figs. 10.10 and 10.11 allows the current at the source to grow without limit, so the voltage at the load grows without limit. The maximum steady-state *dynamic* signal levels are determined by the line VSWR and the source impedance. Of course, actual driver outputs, receiver inputs, and the transmission line itself are not completely lossless. Therefore, there is always a practical limit to how high the currents can grow at a low-impedance node, how high the voltage can grow at a high-impedance node, and how long ringing can continue after an input transition launches energy into a low-loss line. However, the requirements for high speed interconnections are that at least one end of the line must be reasonably terminated so that repeated cycles of reflections cannot combine with new signal transition inputs. If not, dynamic currents may grow larger at low-impedance nodes, and voltages may grow larger at high-impedance nodes, compared to analysis based on static conditions.

Reflections will return to the source if the far end of a line does not draw current at steady-state static levels. Reflections are required to change the current on the line from the transient level (determined by the Z_0 line impedance) to the static levels at zero current. Transmission lines that do not draw static current include lines with "ac" termination, clamp diode termination, as well as lines that are high-impedance (more than a few multiples of Z_0) at the end farthest from the source. If the load does not draw static current, the source must be reasonably terminated to avoid repeated reflection cycles or signal voltage and current levels that vary with data patterns and frequency.

A quarter-wavelength line, or odd multiples of a quarter wavelength of the repetitive input signals result in reflections that are out of phase with the input transitions. However, even if the line is shorter than a quarter wavelength, input transitions may be out of phase with the ringing reflections. Any input transitions that are out of phase with reflections will increase the current at the source and the voltage at an open load. Therefore, lower-frequency signal patterns can also result in dynamic signal level growth until limited only by source impedance. Line lengths that are either odd *fractions,* or odd *multiples* of a quarter wavelength of a repetitive input frequency can result in signal level growth if the line is not adequately terminated so that steady-state static signal levels are nearly reached before each new input signal transition.

10.5 Reflection Diagrams for Clamp Diode Termination

The following reflection diagrams deal with the issues of lines with clamp diodes at the load to control overshoot and ringing, but no parallel load resistors that draw static current. These diagrams will include the associated signal voltage waveforms at the load and at midpoints. The reflection diagrams will indicate that high speed performance is affected by the dynamic impedance of the diodes as well as the source impedance of the driver. The results will indicate that high speed transmission-line performance can be achieved without a parallel load termination if the driver includes a reasonable source impedance (but not termination with Z_0) and the clamp diodes also include an appropriate dynamic impedance. Although Schottky diodes would offer superior performance, the proper use of series impedance can limit the currents to control the reverse recovery of ordinary *p-n* junction semiconductor diodes.

The conventions selected for the following illustrations are that only rising edge transitions are shown. This is chosen because it seems more conventional and convenient. Even the terminology of "rise time" and "overshoot" is oriented to the rising edge transition. Furthermore, the diagrams tend to be a little easier to draw with greater clarity because the signal level ringing is not near the horizontal axis.

The scales indicated and the conventions chosen for the diagrams are intended to be consistent with 3.3-V rail-to-rail CMOS technology, where diode clamps to the rail offer high speed performance improvement without the power dissipation of parallel resistor load resistor termination. However, it should be noted that with all CMOS technology with clamp diodes to the power rails there is the interface issue with multiple power supplies, live insertion in party lines with multiple drivers, or ground shifts. Any unpowered unit will clamp the signals to the lowest power supply. This will tend to load the clamped signal or to charge the power supply line of the unpowered device through the clamp diodes.

Illustrated in Fig. 10.12 is the ordinary case of a low-impedance source and a low-impedance clamp diode to the V_{CC} power rail. The initial signal at the source rises to nearly the power rail, so the reflection at the load causes heavy forward current in the clamp diode. This figure shows how the diode is supposed to work, without reverse

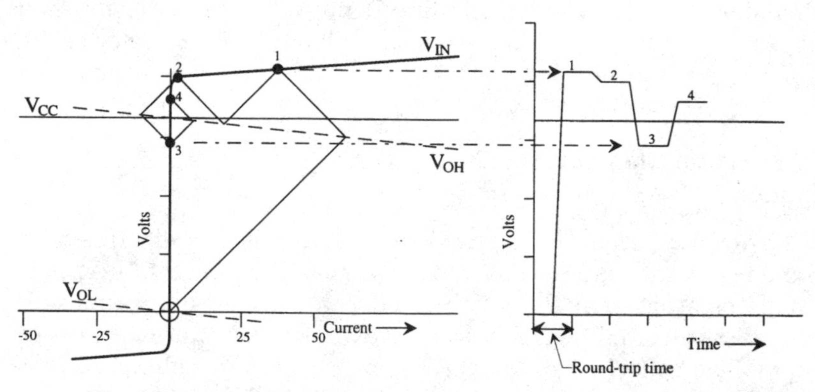

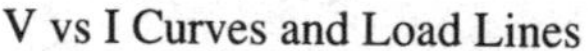

Figure 10.12 Reflection diagram for low-impedance source and clamp diodes.

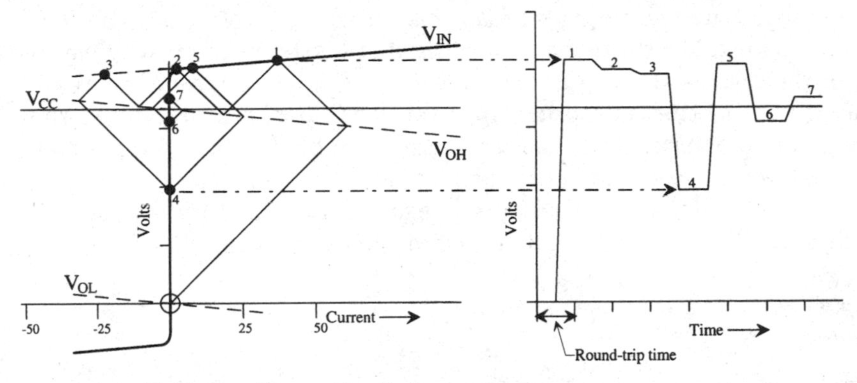

Figure 10.13 Reflection diagram for diode reverse recovery.

recovery after the heavy forward conduction. The waveform at the load shows that the signal level hangs above the V_{CC} power rail until the current finally drops to zero, then pops below the V_{CC} power rail to begin minor ringing about the zero-current steady-state static level at $V_{OH} = V_{CC}$.

Illustrated in Fig. 10.13 is the result if there is reverse recovery after the heavy forward conduction. This figure illustrates one additional reflection cycle of reverse current before the diode snaps to zero current. The change in current results in a negative pulse far into the threshold region after reflection cycle 4, with minor ringing about the zero-current steady-state static level after reflection cycle 5, or nine line delays.

The associated Fig. 10.14 shows the midpoint waveform for the same conditions. The midpoint of the line sees both the incident wave

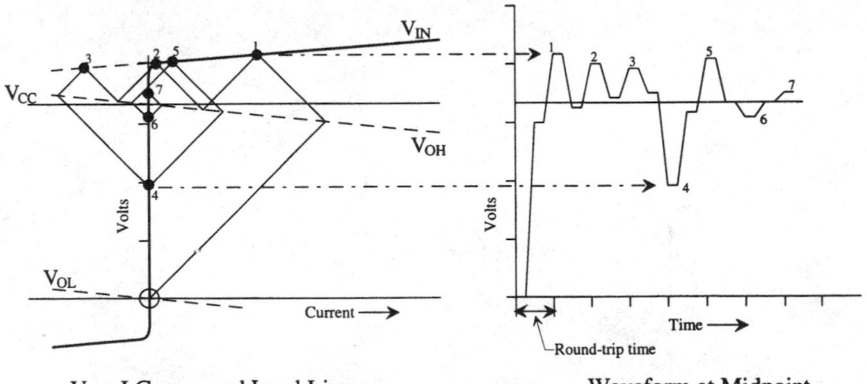

V vs I Curves and Load Lines
Low Impedance Driver Outputs
Low Impedance Input Clamp Diode

Waveform at Midpoint

Figure 10.14 Reflection diagram and midpoint waveform.

levels from the source as well as the reflected levels from the load. Since this is a low source impedance example, the incident wave levels from the source tend to be closer to the V_{CC} level, with the greater excursions on the reflections from the load. It should be noted that midpoint waveforms are generally more variable because the midpoint sees the worst levels of both ends. This is a reason why critical edge sensitive signals must be clustered near the far end of lines unless the ends of the lines have a parallel termination to reduce reflections.

10.5.1 Source termination and clamp diodes

Illustrated in Fig. 10.15 are the results of adding series termination resistance to the source. The initial level is reduced, so that the first reflection overshoot results in very little current in the clamp diode. There should be no reverse recovery current, so the load signal is very nearly exactly the steady-state static V_{OH} level after reflection cycle 2.

The associated Fig. 10.16 shows the midpoint waveform for the same source termination. The reduced initial level at the source results in a step at the initial level. Although the level is a little more than halfway to the HIGH level, edge sensitive critical signals should be clustered near the far end to provide more signal margin on the initial transition. The signal levels after the initial transition see the slight overshoot, with essentially no further ringing.

The significant reduction in the forward current in the clamp diode at the load should prevent any reverse recovery problems. However, the case where reverse recovery might occur is shown in Fig. 10.17 to illustrate that series termination at the source can nearly eliminate the negative undershoot pulse as the clamp diode snaps out of reverse recovery.

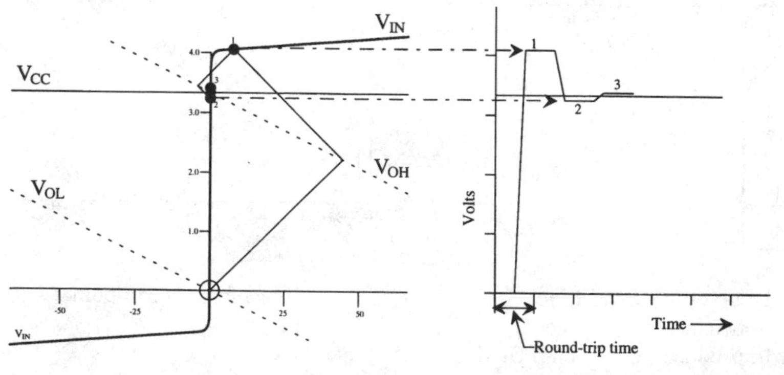

Figure 10.15 Reflection diagram for source termination.

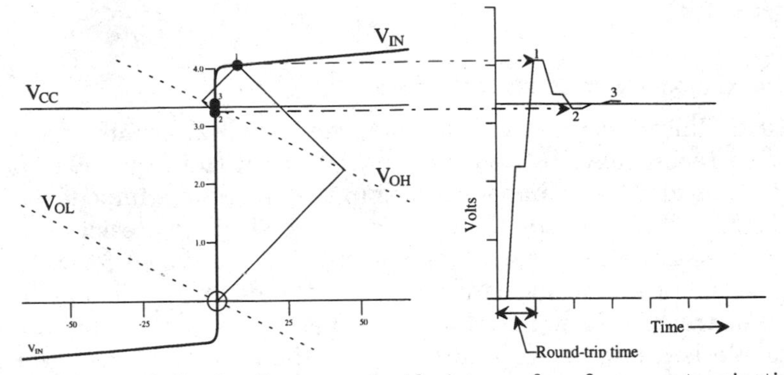

Figure 10.16 Reflection diagram and midpoint waveform for source termination.

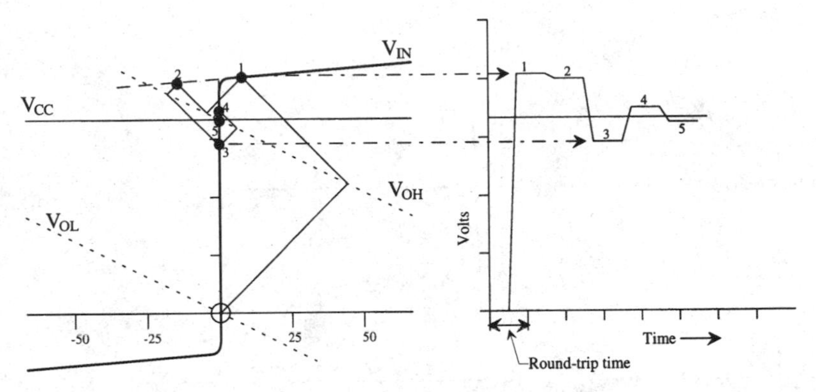

Figure 10.17 Source termination and diode reverse recovery.

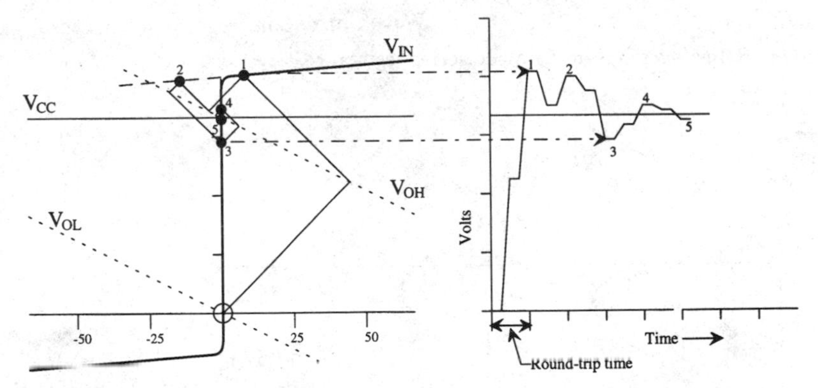

Figure 10.18 Reflection diagram and midpoint waveform for source termination and diode reverse recovery.

This is because the higher impedance of the source termination to reverse current limits the conversion of reverse current to undershoot.

Figure 10.18 shows the associated midpoint waveform for the case of possible reverse recovery. Even if the reverse recovery delay does occur, the ringing is well within typical signal level margins for the steady-state static HIGH level.

10.5.2 Clamp diodes with series resistance

Figure 10.19 illustrates the reflection diagram for an alternate series termination technology: resistance in series with the input clamp diodes at the load. This example returns to the low source impedance of Fig. 10.12, so it shows only the effect of the diode series resistance. The diode resistance allows a higher overshoot at reduced forward current levels on the first incident wave, but the next reflection cycle returns to nearly the static HIGH level, so there is no further ringing. This example is a low source impedance, so the midpoint waveform illustrated in Fig. 10.20 is very similar to the load waveform, with only a small initial step near the V_{OH} level.

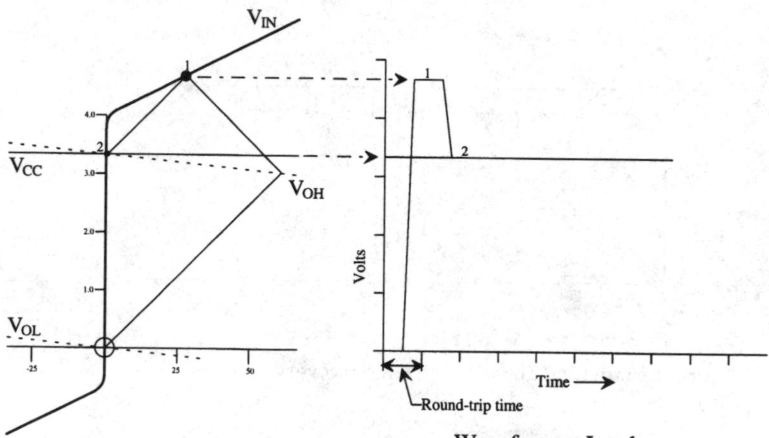

Figure 10.19 Reflection diagram for diode series resistance.

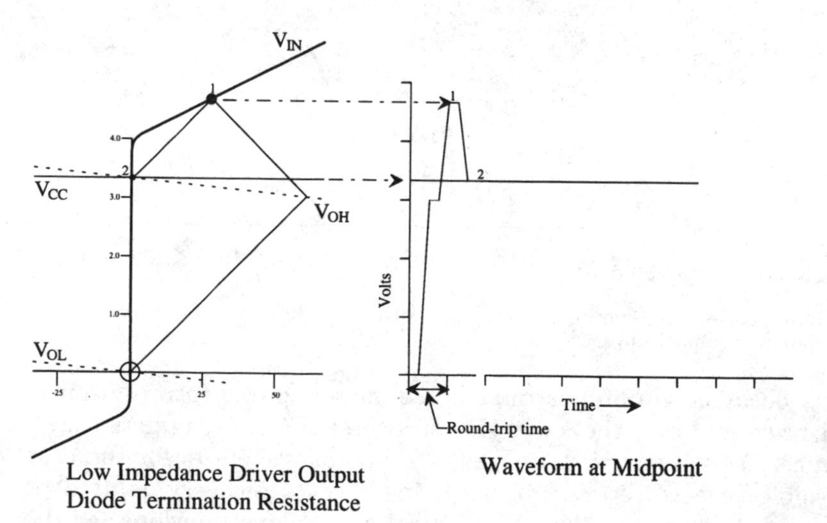

Figure 10.20 Reflection diagram and midpoint waveform for diode series resistance.

This seems like an ideal application for a Schottky diode clamp with sufficient series resistance so that the dynamic impedance is just a little less than Z_0. There would be no reverse recovery issues because the Schottky barrier diodes are primarily a majority carrier device. The low current threshold is lower than *p-n* semiconductor junctions, so the diode line is closer to the $V_{OH} = V_{CC}$ or V_{OL} = ground intercepts at zero current. Reflections from a lower-impedance source will always return to nearly zero current near the steady-state static LOW or HIGH level.

Figure 10.21 is an illustration of a Schottky diode forward characteristic. The load lines indicate that the nonlinear characteristics of this diode alone would terminate a transmission-line impedance ranging from 200 to 50 Ω at currents ranging from 2 to 10 mA, respectively. These current levels are appropriate for terminating the low current levels of emitter coupled logic (ECL) interconnections.[14] However, resistance should be added in series with the Schottky diode so that the total impedance more nearly matches the line impedance at the higher currents of CMOS or TTL interfaces. For the example of Fig. 10.19, the diode current is approximately 30 mA, although the current would be less for series termination of the source. The diode current is about one-half of the line current if the second reflection returns to nearly zero current at the steady-state static level. For the example of Fig. 10.19, the triangle from the steady-state static level indicated by the dot at the number 2, to the diode current indicated by the dot at the number 1, to the intercept of the V_{CC} axis with the reflection line from the V_{OH} intercept is an isosceles triangle.

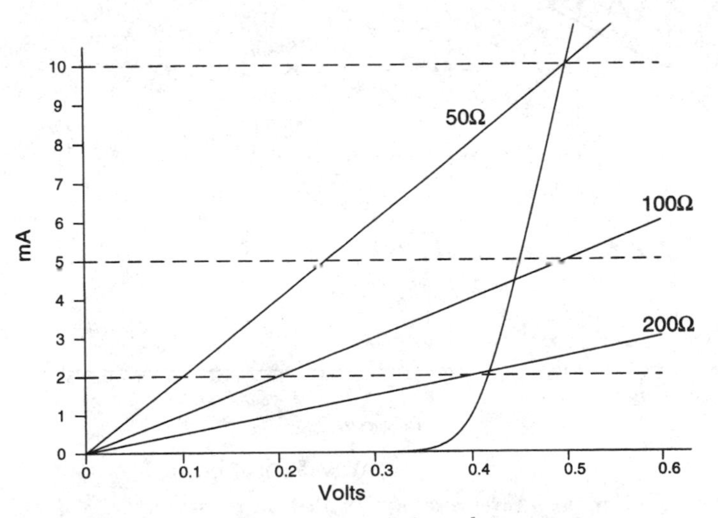

Figure 10.21 Schottky diode characteristics at low currents.

The comparison of ECL to CMOS or TTL diode termination is that the ECL termination is to a voltage between the signal levels, as shown in an example of ECL Schottky diode termination in Fig. 10.22. R_S is the series resistance for the emitter follower source, and R_P is the pull-down load resistor required for ECL. V_{CC} and V_{EE} are the ECL power supplies, which are typically ground and −5 V. V_{BB} is the reference bias voltage that is midway between the logic signal levels. The total ECL logic swing is about twice the Schottky diode forward voltage as either one or the other of the parallel diodes remains in forward conduction to avoid reflections. In contrast, the CMOS clamp diodes terminate overshoot beyond the power rails and do not remain in forward conduction for steady-state static signal levels.

Figure 10.23 illustrates the reflection diagram for resistance in

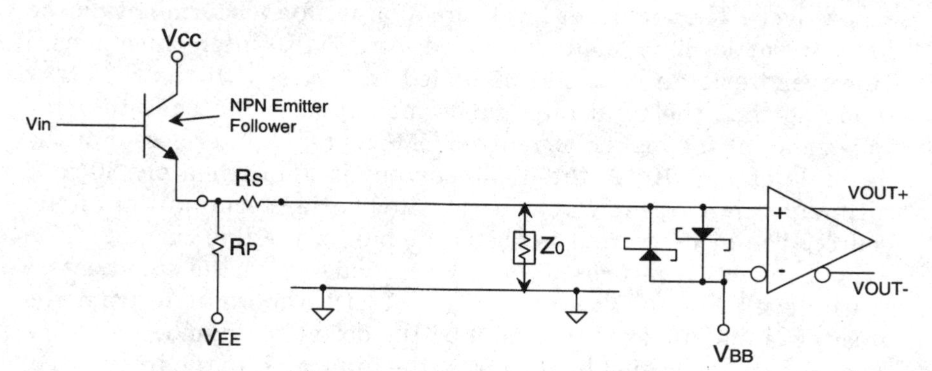

Figure 10.22 ECL interconnection with Schottky diode termination.

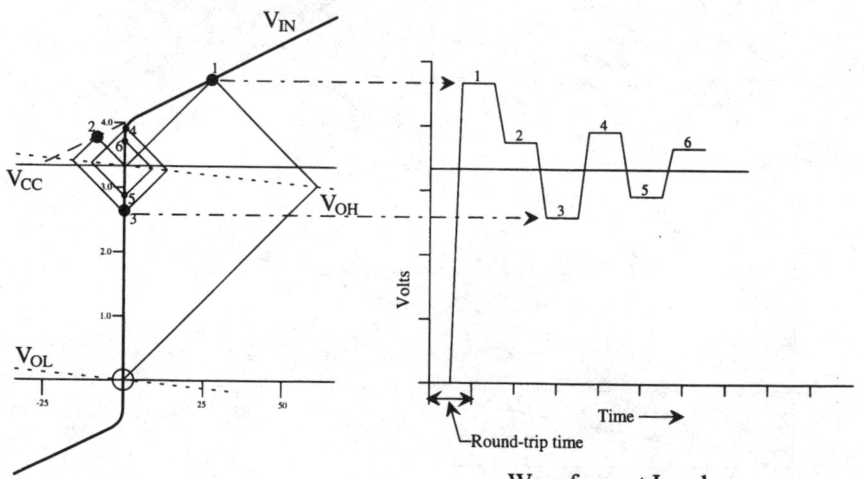

Figure 10.23 Reflection diagram for reverse recovery with diode resistance.

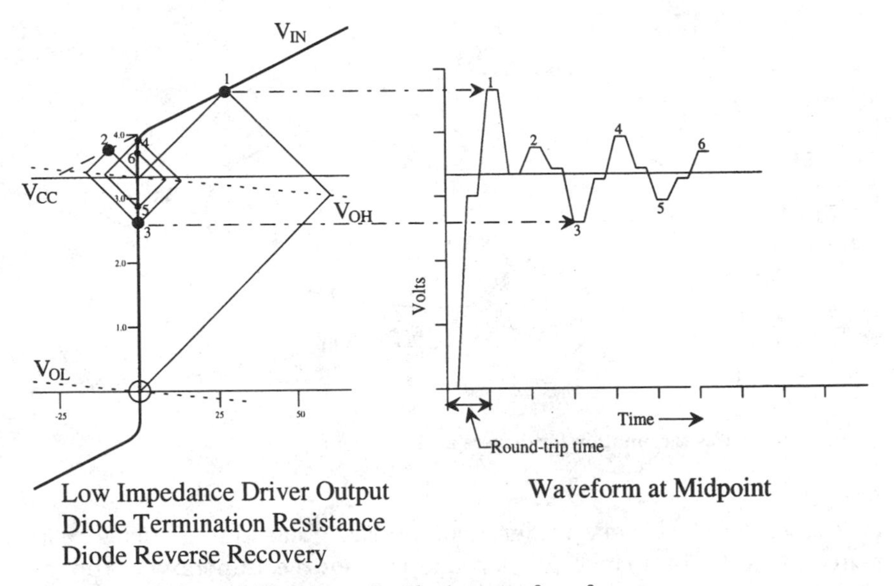

Figure 10.24 Reflection diagram and midpoint waveform for reverse recovery.

series with the clamp diode, but the diode has reverse recovery issues due to the forward current. Diode series resistance is not as effective as source resistance in reducing reverse recovery undershoot pulses. This is primarily because the extended diode line is not as close to the zero-current intercept with the V_{OH} point, and the low reverse impedance of the source converts the reverse current to undershoot. However, the undershoot pulse amplitude is significantly reduced compared to the low-impedance diode case of Fig. 10.13, and the recovery is completed one reflection cycle sooner than without diode resistance.

Figure 10.24 illustrates the midpoint waveform for reverse recovery with diode series resistance. The source impedance is low for this example, so the maximum excursions are equivalent to the corresponding excursions of the load waveform. Although there is some overshoot and ringing, there are no significant departures from the final steady-state static signal level at V_{CC}. The waveform is generally similar to the low-impedance diode case of Fig. 10.14, with the notable reduction in the reverse recovery undershoot pulse.

10.5.3 Both source and clamp diode series resistance

The previous reflection diagrams have shown that resistance in series with either the source or the clamp diode can control ringing and

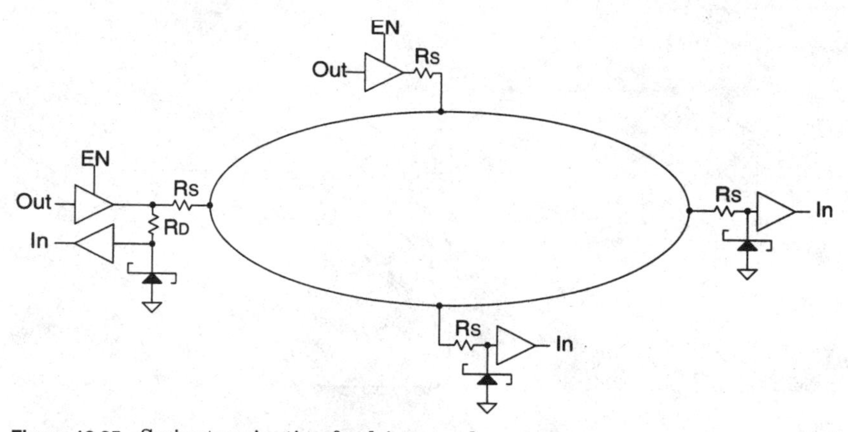

Figure 10.25 Series termination for drivers and receivers.

reduce reverse recovery undershoot pulses. Reasonable series resistance at both the driver to increase the source impedance and the inputs to increase the clamp diode ON impedance is even more effective at controlling overshoot and preventing line ringing. The concept is essentially equivalent to double termination with series termination at the source and parallel termination at the receiver. However, resistance in series with the input to increase the clamp diode impedance does not reduce the signal levels and consumes no dc power when the signals are at their steady-state static levels.

Bidirectional devices or transceivers may require only a single series resistor on the common input-output pin of the device to satisfy both requirements. Figure 10.25 illustrates a single series resistor and the Schottky input clamp diodes for a transceiver, driver, and receivers. The resistance designated R_D may achieve higher resistance for the clamp diode on the transceiver input than for driver output. However, this may only represent the inherent internal impedance of the diode, not a separate physical resistor. If an external series resistor is being added to an existing transceiver or receiver with internal clamp diodes, only the single R_S will be added. The important concept to understand is that a small resistance (less than Z_0) is important for inputs with clamp diodes. Although Schottky diodes are preferred for superior high speed reverse recovery performance, input resistance is important for *any* inputs that have clamp diodes because unterminated inputs on long lines will overshoot beyond steady-state static levels.

A word for the integrated circuit industry is that impedance control for high speed performance needs to be integrated into the devices and the performance in a transmission-line environment must be assured by specifications. The optimum source and load-impedance requirements for high speed performance require impedance control at both the driver output and the clamp diodes on the inputs (Ref. 10, p. 169). The value of the input resistance is not extremely critical. It is suggested that the industry practice of about 25 Ω for driver series output resistance is also about right for input series resistance. It is preferred that the resistance value be reasonably well controlled to minimize sensitivity to temperature and process variability. It is important that the series resistance be isolated from the circuit substrate so that transient input currents do not flow in the substrate. Input signal levels beyond the power rails should result in current only through the clamp diode to the ground pin.

Series resistance may be added to the inputs of all overshoot clamp diodes on all devices on a party line bus, as the added input resistance has little effect on normal operation. However, the designer is cautioned that the use of devices with clamp diodes always results in the issues of diode currents if there is any difference in power supply voltages or ground shifts. These clamp diode current issues are only limited, not eliminated, by series resistance.

An example of the addition of series resistance and a Schottky diode clamp to the negative (ground) power rail only is shown in Fig. 10.26. This is a modification to the driver with the Schottky diode in series with the HIGH level output as illustrated in Fig. 10.4. This example is intended to preserve the blocking of charging currents to V_{CC} when outputs are driven above the positive power supply. The series termination resistor on the input-output pin limits the excursions above V_{CC} on the rising edge transition and limits the currents and ringing on the falling edge excursion. It should be understood that the series resistor must be designed to operate at signal levels beyond the power and ground rails without inducing other diodes to conduct or to inject currents into the semiconductor substrate.

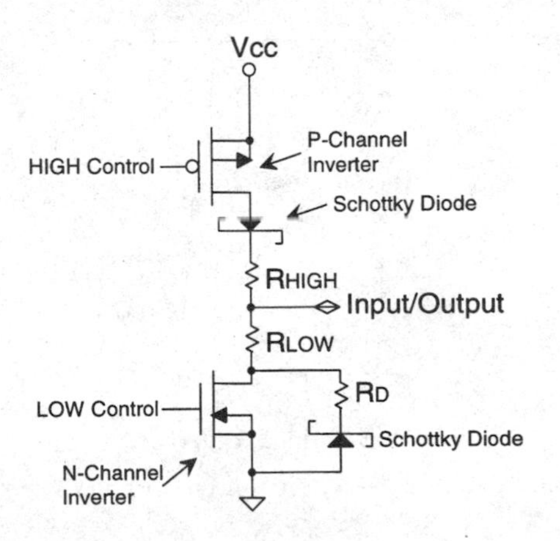

Figure 10.26 Series termination for CMOS input/output with series diode.

The Schottky clamp diode to ground is in parallel with the NMOS transistor and has no effect when the NMOS transistor is ON. However, if a falling edge transition is driven on the line by another driver, the Schottky clamp diode will tend to conduct current for any overshoot of the signal below ground. Although the Schottky diode is in parallel with the NMOS body diode in this example, the lower threshold of the Schottky diode should divert a significant portion of the overshoot current and reduce the potential of reverse recovery issues in the NMOS body diode. Additional resistance to limit the body current could also be connected in series with the body diode if the technology permits. However, if the device technology provides a body bias for the NMOS transistor that is more negative than ground or the body is otherwise isolated, then only the Schottky diode conducts the overshoot current.

An input only device would have only the series resistor and the Schottky clamp to ground to control falling edge overshoot and ringing. A reflection diagram for both rising and falling edges is illustrated in Fig. 10.27. The total impedance of the output and the diode clamp to ground is $Z_0/2$ for this example. For the combination input/output of Fig. 10.26, this would mean that R_D is zero because R_{LOW} is $Z_0/2$. The rising edge initial signal level is more than 50 percent of the power supply, so the reflection at the load overshoots beyond the power rail. Both the input and output are high impedance, so the signal level may remain at the overshoot level unless some other load tends to bring the level down. The falling edge is shown beginning at the overshoot level above the power rail. The initial sig-

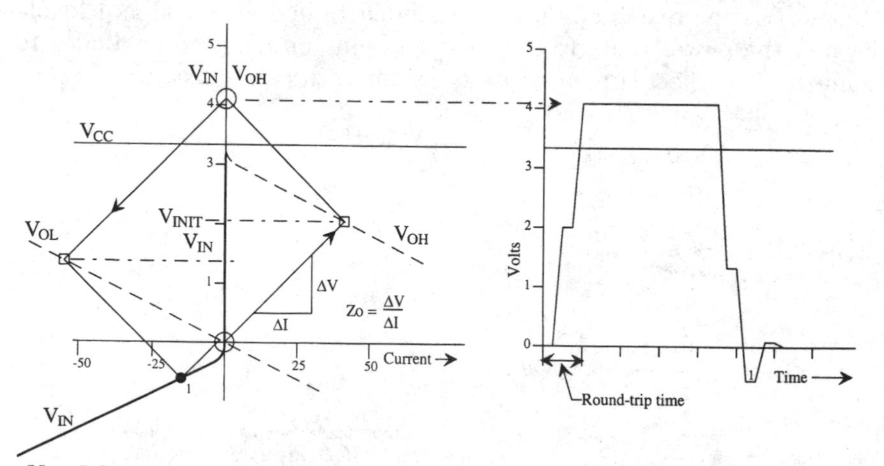

Figure 10.27 Reflection diagram and midpoint waveform for $Z_0/2$ series termination.

nal excursion is at a level about midway between the power rails. The reflection at the load falls below ground until the Schottky clamp diode begins to conduct. The impedance of the source and the clamp diode results in low currents, so that the negative overshoot recovers to almost exactly the steady-state static condition of zero current.

An important characteristic to note in the reflection diagram of Fig. 10.27 is that another rising edge may begin from either the steady-state static point at zero current, or at the initial transient level resulting from the condition at the load indicated by the number 1 to indicate the signal level after the first incident wave. The resulting rising edge gives essentially identical results for either case, so there is no concern that reflections could coincide with new driven transitions to increase the current and/or voltage levels. Although the midpoint waveform shows the initial levels about midway between the power rails that are characteristic of source terminated lines, the waveform near the far end of the line would show a smooth transition to the overshoot levels. However, it should be noted that this driver will not terminate crosstalk or other disturbances on the line when the series diode is reverse biased in the HIGH state.

Figure 10.28 illustrates a reflection diagram with an example of both series source resistance and series clamp diode resistance to both power rails. This diagram is essentially a combination of the source termination of Fig. 10.15 and the diode termination of Fig. 10.19, with the same impedance values chosen for convenience. However, other values may be more appropriate, depending on such

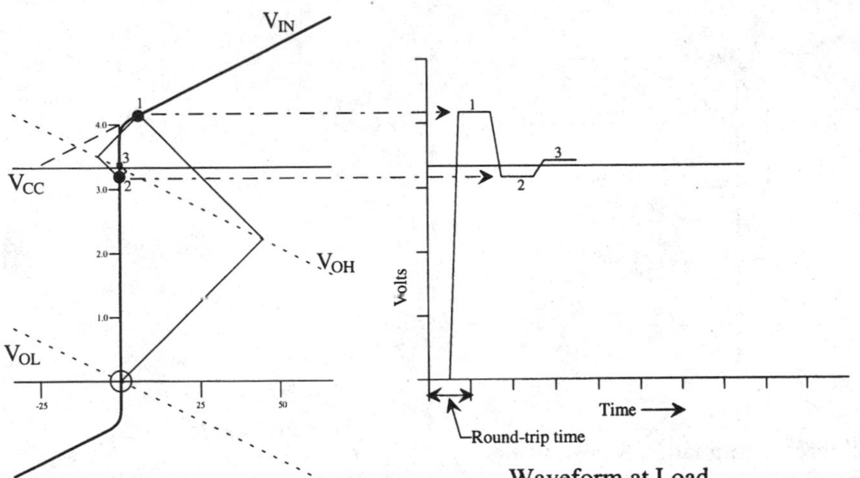

Figure 10.28 Reflection diagram for both source and diode resistance.

conditions as whether it is known if there is always resistance at both ends of the line, if the overshoot clamp is a Schottky diode, and whether it is more important to limit diode currents or to limit overshoot voltage. This figure illustrates almost an ideal underdamped waveform at the load. There is some overshoot to improve rise time and compensate for some high-frequency losses, with any subsequent ringing essentially eliminated. The driver is reasonably terminated in both logic states, so the driver terminates reflections, crosstalk, or other disturbances traveling toward the driver. This may be even better than parallel resistor termination which draws dc power, attenuates the signal levels, and achieves overshoot only with the result of damped ringing.

Even if the clamp diode is a *p-n* junction semiconductor, reverse recovery effects are almost impossible. The diode current is very low and the resistive extended diode line into the reverse current region intercepts the resistive V_{OH} line (or the V_{OL} line for the falling edge transition) very close to the steady-state static signal level. Furthermore, the limiting of the diode currents reduces the risk of latch-up due to injection of input overshoot currents into the substrate.

Figure 10.29 illustrates the midpoint waveform for both source and clamp diode series resistance. The initial voltage from the driver shows the initial step characteristic of source termination, but the remainder of the signal converges rapidly to the steady-state static

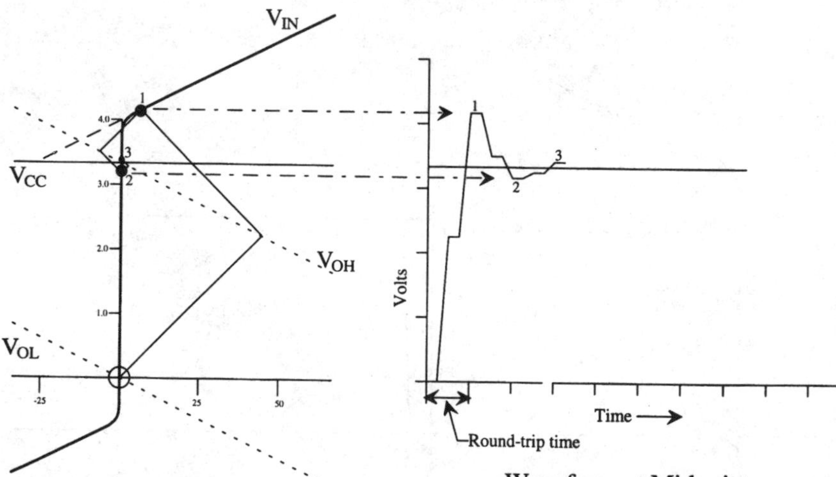

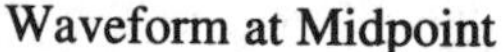

Source Termination Resistance
Diode Termination Resistance

Figure 10.29 Reflection diagram and midpoint waveform for both source and diode resistance.

signal level. Examination of this reflection diagram, the midpoint waveform, and comparison to Fig. 10.20 indicates that lower source impedance increases the initial voltage and increases the amplitude of the reflected wave. However, lower source impedance increases the diode current and the risk of increased settling time for any reverse recovery of the clamp diode. Increasing the clamp diode impedance reduces the diode current by allowing more overshoot voltage on the initial wave.

The integration of resistance in series with both the source output and the clamp diode inputs provides flexibility in selecting impedance values appropriate for the application, as well as controlling ringing on high speed interconnections without the dc power consumption of parallel termination. Typical values for the series resistance added to output drivers or input clamp diodes is about $Z_0/2$ as a compromise between limiting currents and ringing with minimal effect on transient performance. The industry practice for integrating the series resistance with logic drivers is about 25 Ω for the 2000 series of resistor output logic.[15] This is appropriate for most typical single-ended transmission lines with little dc loading (high input impedance). About the same resistance in series with each leg is usually appropriate for a balanced differential line with a differential receiver, even if the far end is terminated with a line-to-line resistor to control reflections and standing waves on long lines. Although the signal levels are moderately attenuated by the series resistance at the source and the parallel resistor at the load, a differential receiver should be capable of detecting the resulting differential signal in the presence of common mode voltages. If the parallel termination is $2 * Z_0$ and the total series termination is $Z_0/2$, the steady-state static differential signal level is about 80 percent of the open-circuit level. The initial signal excursion should overshoot to about 40 percent beyond the steady-state static level, ignoring any losses which will tend to reduce the signal excursions.

10.6 References

1. James E. Buchanan, *Signal and Power Integrity in Digital Systems: TTL, CMOS, & BiCMOS,* McGraw Hill, New York, 1996, chap. 3.
2. E. James Angelo, Jr., *Electronics: BJT's, FET's, and Microcircuits,* McGraw-Hill, New York, 1969.
3. S. M. Sze, *Physics of Semiconductor Devices,* Wiley, New York, 1969, pp. 378, 505.
4. W. J. Cunningham, *Introduction to Nonlinear Analysis,* McGraw-Hill, New York, 1958.
5. G. Massobrio and P. Antognetti, *Semiconductor Device Modeling with SPICE,* 2d ed., McGraw-Hill, New York, 1993.

6. James E. Buchanan, *Signal and Power Integrity in Digital Systems: TTL, CMOS, & BiCMOS,* McGraw-Hill, New York, 1996, chap. 8.
7. *INTERFACE: Line Drivers and Receivers Databook,* National Semiconductor Corporation, Santa Clara, Calif., 1992 ed., sec. 6, AN-807.
8. George V. Mueller, *Introduction to Electrical Engineering,* McGraw-Hill, New York, 1957, chap. 5.
9. Samuel Seely, *Electron Tube Circuits,* McGraw-Hill, New York, 1958, chap. 2-3, p. 63.
10. James E. Buchanan, *Signal and Power Integrity in Digital Systems: TTL, CMOS, & BiCMOS,* McGraw-Hill, New York, 1996, chap. 7.
11. *FACT Advanced CMOS Databook,* National Semiconductor Corporation, Santa Clara, Calif., 1989 ed., sec. 1.
12. *FAST Advanced Schottky TTL Logic Databook,* National Semiconductor Corporation, Santa Clara, Calif., 1990 ed., sec. 1.
13. *FACT Advanced CMOS Databook,* National Semiconductor Corporation, Santa Clara, Calif., 1989 ed., sec. 3.
14. *MECL System Design Handbook,* Motorola Semiconductor Products, Inc., Phoenix, Ariz., 4th ed., 1988, chap. 4.
15. David Wyland, "Resistor Output Logic Gives High Speed with Low Noise," AN-07, *Quality Semiconductor Databook,* Quality Semiconductor, Inc., Santa Clara, Calif., 1991, p. 6-61.

Chapter

11

Transmission Lines: Signal Levels

This chapter describes the signal levels on a transmission line as a function of the termination resistors and the reflection coefficients resulting from linear terminations. The duals of transmission-line terminations that result in the same output waveform at the load end with different source and load terminations are described. The increase in line current and loss of signal levels that results when the lengths of open lines match the signal frequency are discussed and illustrated. Transmission lines with the same ratio of termination resistance to the line Z_0 at both ends are analyzed, with charts to illustrate the results of increasing the mismatch ratio of low source resistance and high load resistance. This chapter presents the power series that defines the relationship of the final steady-state static signal level to the overshoot and undershoot ringing amplitudes, and the short-line scaling factor that reduces ringing on lines that are short relative to the rise time.

11.1 Basic Signal Level Definitions

The basic signal levels are the following:

V_S is the no-load, or open circuit voltage excursion between the two static signal voltages of the source of the signal transition. V_S may be ambiguous for TTL-compatible or other nonlinear drivers, so V_S is usually considered as the difference of the linear intercepts of the operating voltage versus current lines with the zero-current axis. This is illustrated in Fig. 11.1 for a typical Schottky TTL driver output characteristic.[1] The actual V_{OH} output intercepts the zero-cur-

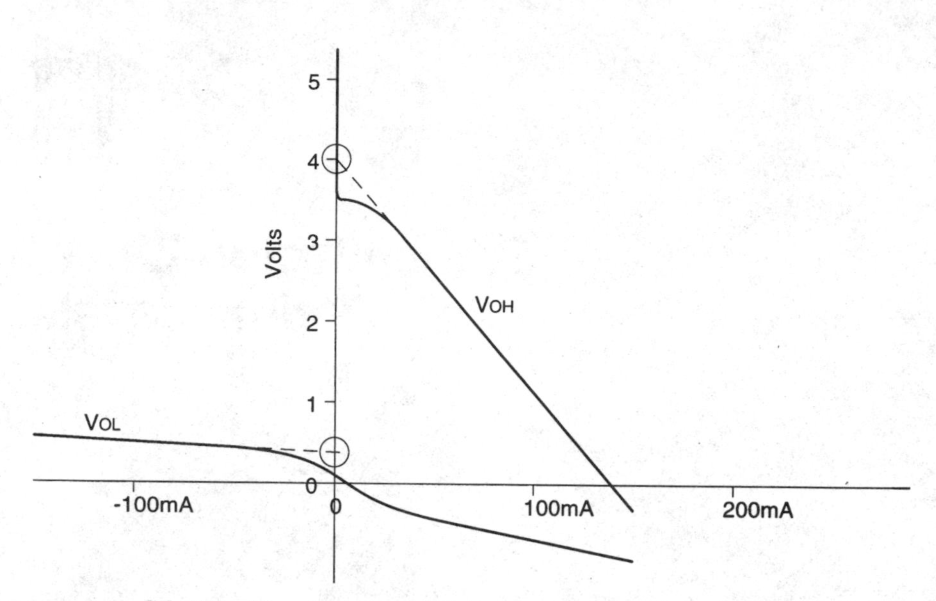

Figure 11.1 Schottky TTL output characteristics.

rent axis at a different level owing to nonlinear characteristics at low current. Many TTL Schottky line drivers do not conduct negative current, even for output voltages above the power supply, but do conduct positive current for output voltages below ground. When the transmission line is driven with two drivers in a balanced, differential configuration, V_S is double the open-circuit voltage of a single driver. For example, a Schottky TTL driver operated at 5.0 V has a typical no-load intercept for the HIGH state of approximately 4.0 V, and a LOW state of about 0.5 V, so V_S for a differential line is about 7.0 V; an ECL driver with a signal excursion of 0.8 V would have a differential V_S of 1.6 V.

V_i is the initial signal excursion launched into the transmission line when the driver changes logic state. This is the signal excursion from one static signal level onto the line when the driver is loaded with the transmission-line impedance. The total source impedance of the driver results in some reduction of the no-load signal excursion before the signal excursion is launched into a long transmission line. The total source impedance may include the ON resistance of the switching transistor and any resistance internal to the driver circuit, as well as external impedance added to control signal response. When the source impedance is nonlinear, the initial signal voltage excursion may be determined by drawing a Z_0 load line through the previous static signal level and the V_{out} vs. I_{out} characteristic curves.

V_f is the final (static, steady state, or quiescent) signal voltage level when source and load current(s) are equal and the signal level is equal at all points along the line connecting the source to the load(s) if line losses are ignored.

The notations for transmission-line analysis may vary. For example, the National Semiconductor FAST Advanced Schottky Logic and FACT Advanced CMOS Logic Data Books describe transmission-line effects in the design considerations sections. V_E or V_0 is used for the no-load source voltage, and V_{out} for the initial signal excursion into the transmission-line load.

The total voltage excursion for differential lines is double that for the individual complementary signals. The signal response of differential lines may be analyzed as if it were a single-ended line with all of the signal change on one line and the other line held fixed. Most analysts comfortable with TTL and CMOS signals can consider the one active line switching from ground to a positive signal at double the steady difference between the actual LOW and HIGH levels. This concept is adequate for analyzing the dynamic switching currents but does not properly model the steady-state currents. An equivalent concept is to consider a differential line as a dual polarity (plus or minus) signal on one line and the other line held at ground so the currents change direction when the signal switches. These concepts can be observed on an oscilloscope if the two differential signals are combined as $A - B$ (algebraic add with inversion) to observe the differential signals that exist on the line at any point. The total signal excursion is double the individual signal on each line; the concept of a reference can be considered as either the center of the trace or one of the extreme levels.

The signals may also be combined as $A + B$ to observe the differential output common mode. Exactly balanced complementary signals should add to a constant value, usually called the common mode voltage. If the true and complement signals do not switch simultaneously, there will be a signal "spike" of common mode signal during the unequal signal excursions. When a differential line is analyzed as a single-ended line with double the voltage excursion, the combined source impedance is also the sum of the HIGH state and LOW source impedance. However, one-half of the actual physical series termination impedance for differential lines should be placed in series with each line to maintain balance, so that any induced common mode signals (such as crosstalk) are not converted to differential noise.

11.2 The Initial Signal Excursion

The transmission-line impedance is the initial load on the driver, so the normalized initial signal level into a long transmission line is determined by the voltage divider ratio. The initial voltage excursion is $V_i/V_S = Z_0/(R_S + Z_0)$, where V_S is the no-load (open-circuit) source signal voltage excursion and V_i/V_S is the initial signal voltage transition as a fraction of the no-load voltage.

The difference between the no-load voltage and the initial voltage as a fraction of the no-load voltage is $(V_S - V_i)/V_S = R_S/(R_S + Z_0)$. This is how much the signal falls short of the no-load voltage as a fraction of the no-load voltage. For example, when $R_S = Z_0/3$, the initial signal is within one-fourth of the no-load voltage; when $R_S = Z_0/2$, the initial signal is within one-third of the no-load voltage; and when $R_S = Z_0$, the initial signal is one-half of the no-load voltage. In general, when $R_S = Z_0/n$, the initial signal level is within $1/(n + 1)$ of the no-load voltage. However, the final voltage is less than the no-load voltage if there is any significant load resistance. The final static voltage level may not be achieved if the interval between signal transitions is less than the signal settling time, including the round-trip delay of the transmission line.

11.3 The Initial Undershoot

V_i approaches V_S as the source impedance R_S becomes small compared to Z_0. The final steady-state static signal voltage depends on both the source and load resistance, but not on the dynamic transmission-line impedance. The final voltage is $V_f/V_S = R_L/(R_S + R_L)$. Therefore, the ratio of the initial voltage to the final voltage is $Z_0(R_S + R_L)/R_L(R_S + Z_0)$. The difference between the final voltage and the initial signal voltage as a fraction of the final voltage is $R_S(R_L - Z_0)/R_L(R_S + Z_0)$. This is called the *initial undershoot* because this is how much the initial signal falls short of the final steady-state voltage.

The obvious case of $R_L = Z_0$ for an exactly perfect load termination results in the initial voltage equal to the final voltage, independent of the value of R_S. Another common case is R_L approaching an open circuit, so that the final value is nearly the open-circuit source voltage. The initial undershoot as a fraction of the final steady-state voltage is the same as the difference between the open-circuit source voltage and the initial signal voltage. As described above, when $R_S = Z_0$ for exact source termination, the initial signal is at 50 percent of the final signal level. When $R_S = Z_0/2$ the initial signal is within 33 percent of the final signal level, and when $R_S = Z_0/4$ the initial signal is within 20 percent of the final signal level. In general, when $R_S = Z_0/n$, the initial signal is within $1/(n + 1)$ of the final signal level for a line with no dc load on the line. Although the initial signal is closer to the final signal level for larger n, the ringing increases for larger n. The average load current at the source increases in proportion to the length of the line, up to a maximum when the open line is a quarter wavelength of the signal frequency. Therefore, the average load current also increases for larger n.

11.4 Duals of Transmission-Line Terminations

The final voltage approaches the no-load source voltage when R_L is much greater than R_S. Either the source or load resistance (but not both) may be near Z_0, so the line may be terminated at one end to avoid ringing and achieve nearly full signal amplitude at the load. It is generally true that the same final static signal levels at the load can be achieved at the load with two different pairs of source and load resistance. However, the initial voltage is only a function of the source resistance and may be different for each of the two pairs of resistance. The two pairs of source and load resistance are called *dual* solutions. The duals are defined by the following relationships between the reflection coefficients, where numbers in parentheses indicate the respective dual solutions:

$$\rho_S(2) = -\rho_L(1) \qquad \text{and} \qquad \rho_L(2) = -\rho_S(1) \tag{11.1}$$

11.5 Three Special Cases of Exact Termination

There are three cases of exact termination on a simple ideal line with two ends. Case 1 is with absolutely zero resistance at the source and the load exactly terminated with $R_L = Z_0$. Case 2 is the dual of Case 1, with the source exactly terminated with $R_S = Z_0$, and the load end is a perfect open circuit. These two cases provide exact termination at just one end, with complete reflection at the other end, so both have the steady-state static final voltage equal to the no-load source voltage. Case 3 is the dual of itself, with exact terminations at both ends. The steady-state static final voltage for Case 3 is one-half of the no-load source voltage.

11.5.1 Case 1: Termination at the load only

Case 1 has $\rho_S = -1$ and $\rho_L = 0$, and the initial voltage is equal to the final voltage. Case 1 is often considered an ideal signal waveform because the final signal level is established as the signal travels along the line. Power is dissipated at all times, independent of data rate and line length. The impedance of the line as "seen" by the driver is always $Z_0 = R_L$. This line is called "flat" because the exact termination at the load end means that there are no reflections and the load impedance seen by the driver is a constant, independent of signal frequency and line length.

Although Case 1 may be an ideal concept, it cannot be achieved in actual circuits and has undesirable characteristics in practical applications. Zero source impedance (both resistance and inductance) is not only impossible to implement in actual circuits and packages, but a low-impedance source does not absorb crosstalk traveling toward the source. Differential drivers with low source impedance may not have identical impedance for both high and low signal levels, so common mode crosstalk traveling toward the source will not be reflected as purely common mode noise.

Many common TTL and CMOS drivers have higher impedance in the high state than in the low output state, typically to control currents during switching transients or a short circuit to ground. When the source impedance of differential drivers is not balanced for both signal levels, common mode crosstalk will tend to be converted to differential noise. Therefore, a low-impedance driver will tend to reflect common mode crosstalk as both common mode and differential noise traveling toward the receiver. Furthermore, the larger transient currents resulting from low source impedance may interfere with (i.e., crosstalk to) other circuits sharing power and ground connections.

Although the load termination does absorb signals (including differential noise) traveling toward the load, typical differential receivers have very high common mode impedance (to avoid ground loop currents). Therefore, common mode crosstalk is not absorbed by a differential termination at the receiver but is reflected toward the source. Therefore, common mode crosstalk is not absorbed at either end of differential lines for Case 1. Crosstalk and reflections of crosstalk may combine and exceed the common mode rejection capability of the receiver. Common mode conversion at the source may induce differential noise that competes with driven signals at the receiver and may result in data pattern sensitive errors.

Practical problems with termination at the far end of the line include that it may not be economical to locate a termination resistor at the receiver which exactly matches the line impedance. Lines with multiple loads and/or multiple sources may require a separate load termination at one or more ends of the line. Illustrated in Fig. 11.2 is a line with two ends. There are multiple drivers and receivers at various locations. The drivers are inactive to a high-impedance state unless they are enabled as indicated by the "EN" control input. The

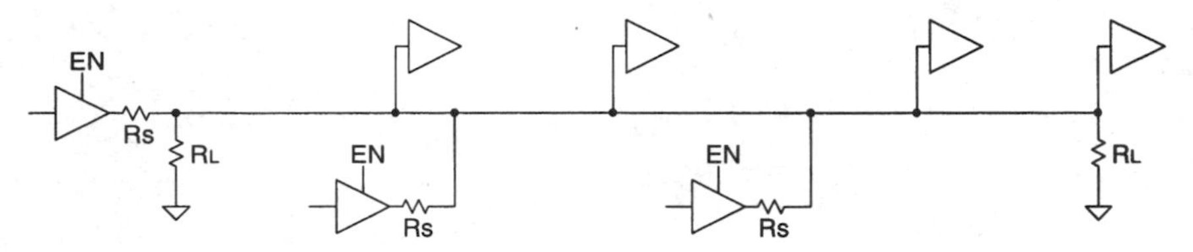

Figure 11.2 Party line terminated at both ends.

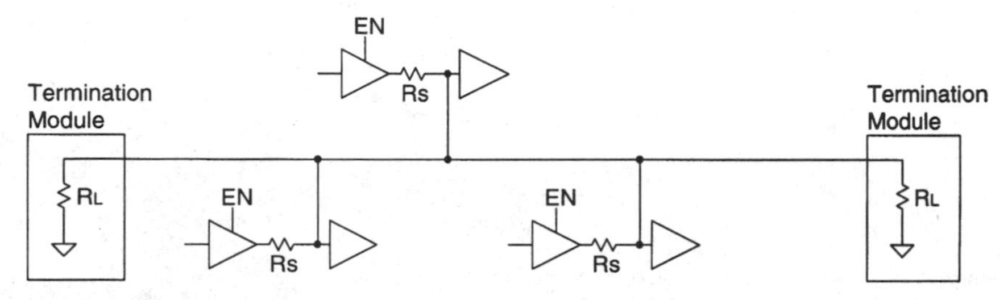

Figure 11.3 Party line with termination modules.

signal level will default to the termination voltage (shown here as ground) if no drivers are enabled. A parallel load termination is associated with a driver which is located at one of the ends, and another load termination is associated with a receiver at the other end.

Another example of a party line is illustrated in Fig. 11.3. Multiple driver and receiver units may be located along the line. Termination modules are located at the two ends. The termination modules must be relocated if the line is rerouted so that the location of the end is different.

11.5.2 Case 2: Termination at the source only

Case 2 has $\rho_S = 0$ and $\rho_L = +1$, and the initial voltage is one-half of the final voltage. There is no load at the end, so the signal level doubles at the load and the reflection travels back to the source to establish the final level. Current flows and power is dissipated during the time that the signal travels to the load and reflects back to the source, but no current flows during steady-state static levels.

Source termination allows greater flexibility in placement of multiple loads or multiple drivers (often called "party lines," from telephone terminology). Multiple load and/or drive points may be connected on a "loop line," where all connections are two-way, so there are no ends, as illustrated in Fig. 11.4. Only the active driver (if any) is at the line impedance because the inactive drivers are high-impedance. The drivers are connected to lines in both directions, so the line impedance seen by the driver is one-half the impedance of a single line. The dri-

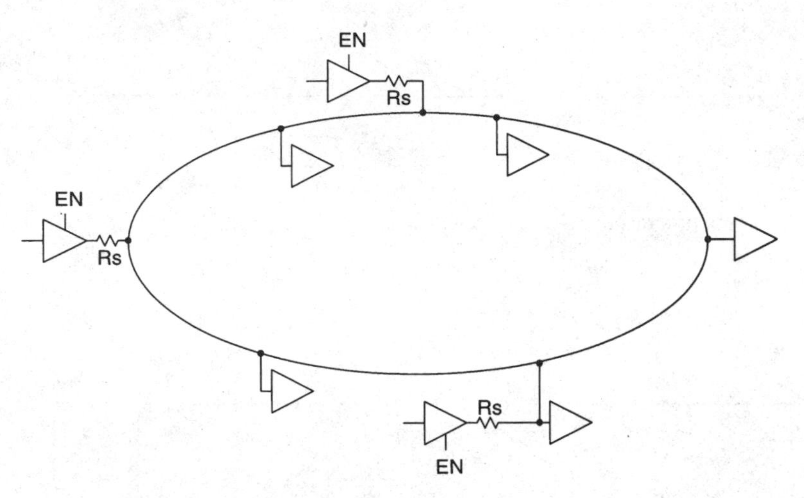

Figure 11.4 Loop line with source termination.

ver source impedance R_S is the termination appropriate for $Z_0/2$. The virtual "end" of the line is a point on the loop that is electrically an equal distance from the active driver in both directions. This figure does not indicate what level the line would default to if no drivers are active, but some bias would typically be provided by the receivers.

Although actual load circuits do not have perfectly open input impedance, most inactive drivers, single-ended logic inputs, and differential receivers have an input impedance that is so much higher than typical line impedance that load reflection factors of at least 90 percent are easily achieved. Practical applications of source terminations are usually underdamped rather than exactly terminated. This will tend to increase the initial signal level and provide some overshoot at the end of the line which may help to compensate for some of the rise time limitations caused by lumped capacitance loading. Either differential mode or common mode crosstalk traveling toward the source end of the line is usually absorbed by an appropriate source termination of typical signal drivers.

The logic input to the gate of a CMOS transistor has essentially no dc resistance, so the input impedance is primarily capacitance. However, most devices designed to receive digital signals include circuits to protect against excessive voltages that could damage the circuits (usually called ESD protection), as well as a pull-down or other bias resistance to establish stable levels when the input is not connected to a valid signal. Logic inputs may not receive a valid input signal under such conditions as when the receiving circuits are powered, but an interface cable is not connected, or when a part or assembly is being tested. The input circuits may also affect the operation of interfaces when both drivers and receivers are not powered identically. Receiving circuits that protect against input voltage excursions beyond the power rails may have a significant effect on transmission-line performance.

The signal levels on a line terminated only at the source are not constant along the line because the reflection arriving from the load end is required to establish the final signal level. If the data rate is low enough so that the interval between signal transitions is longer than the round-trip propagation delay of the line, the reflection arrives at the source before another transition is driven on the line. The initial signal level at the far end is established on the entire line length after one round trip on the line. If final signal levels are reasonably achieved at the driver before the next transition is driven, the line impedance seen by an exactly terminated driver is equivalent to a short circuit (zero impedance to a level midway between the two static signal levels) during the time that the one-half signal level travels to the end of the line and reflects back to the source. The current which flows during this interval is determined by the source no-load voltage and the source termination resistance of Z_0. When the reflection arrives at the source, the impedance seen by the driver is the open circuit at the load end, and no current flows until another transition is driven onto the line. For the typical case of source termination less than Z_0 for an underdamped response, there may be some overshoot and ringing before the signal levels achieve the static levels, and the currents may be higher.

If a fixed-frequency square wave is driven onto a line terminated only at the source, the average impedance seen by the driver depends on the relationship of the line length to the signal frequency. The minimum impedance seen by the driver (a short circuit for an open load) occurs when a reflection of the opposite polarity arrives just as the new signal transition is driven onto the line. The lowest frequency at which this occurs is when the round-trip propagation delay of the line is the interval between transitions of opposite polarity (one-half of a full cycle of the signal), or the one-way propagation delay is one-quarter of the wavelength. This line is called an open quarter-wavelength stub, which inverts the impedance at the load because the reflection for that frequency is opposite phase (180° phase shift) to the driven signal. This pattern repeats for all longer lines that are an *odd* multiple of a quarter wavelength. Conversely, the maximum impedance seen by the driver is equal to the open load when a reflection of the same polarity arrives at the same time as a new signal transition is driven onto the line. This means that at least one transition of the opposite polarity is still traveling on the line. The lowest frequency at which this occurs is when the round-trip propagation delay of the line is the interval between transitions of the same polarity (one full cycle of the signal), or the one-way propagation delay is one-half of the wavelength. This pattern repeats for all longer lines that are a multiple of a half wavelength but is commonly referred to as *even* multiples of a quarter wavelength.

Lines longer than a quarter wavelength have locations where a valid signal level may never be achieved for repetitive signals. The signal at the open end of the line is always valid, but the signals at odd multiples of a quarter wavelength from the open end are not. For an example of a line exactly one-quarter wavelength long, the signal at the driver will always remain at an intermediate level midway between the two valid levels if the reflected signal is exactly the same as the driven signal and losses or distortion are ignored. However, the signal at other locations along the line will be at the intermediate level for a time which is proportional to the distance from the open end of the line. For another example of a longer line exactly one-half wavelength long, both the open end and the driven end achieve valid signal levels, but the midpoint (one-quarter wavelength from the open end) does not.

In terms of standing waves, the location that does not achieve a valid signal level is called a *null* or minimum location. Illustrated in Fig. 11.5 is the percentage of time that a valid signal exists as a function of the distance from the open end of a line. The data in the figure is based on a continuous square-wave signal. This indicates that a signal that is far from the open end of a high speed line may be seriously degraded, but the signal at the open end of the line is not affected. This is of little concern for lines with the only load at the end of

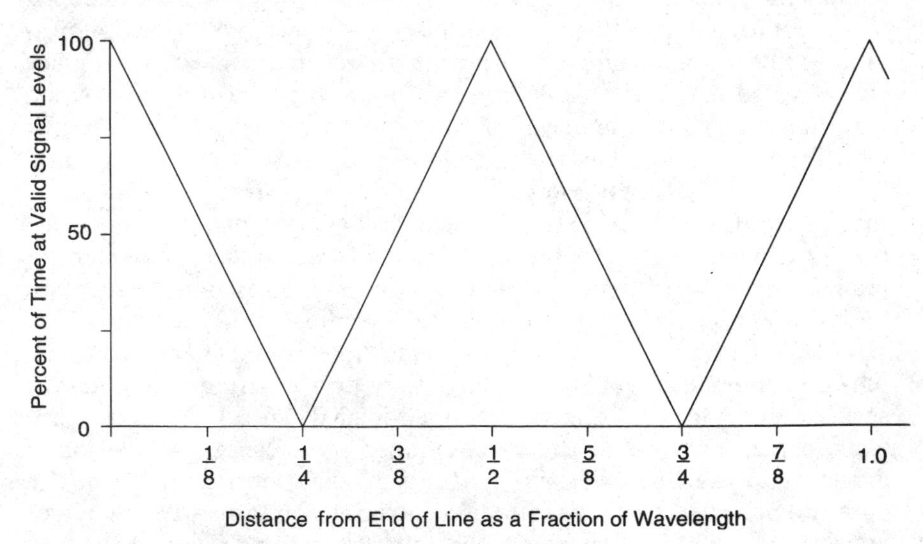

Figure 11.5 Signal percentage for source-terminated lines.

the line. However, troubleshooting a long line from the series-terminated source may be very difficult. The author can speak from personal experience of a long differential line with one wire failing to contact at a connector part way along the line. It was not practical to distinguish between reflections from the fault and reflections from the far end at the operating frequency. However, checking the signal at the far end clearly showed a problem. End sections were disconnected one at a time and checked for correct signals. The longest section with a correct signal isolated the problem.

Of course, losses and distortion result in reflections that are not exactly equal and opposite to the driven signal, so the null location does not have exact cancellation of signal transitions. Furthermore, if the driven signal waveform is not exactly 50 percent duty cycle, the null is at a different line location for the different signal transitions.

11.5.3 Case 3: Terminations at both ends

Case 3 of special interest is *both* ends terminated. This double-terminated case is the dual of itself, with both $\rho_S = \rho_L$. The double-terminated case may be referred to as *fully matched* when the termination at both ends exactly matches the line impedance Z_0.[2] The initial and final signal levels are the same, but only one-half of the no-load source voltage. The initial signal excursion is the same as for Case 2 with source termination only, but there is no reflection to increase the final level. Because the load end of the line is exactly terminated, Case 3 is a flat line with no reflections, so the load impedance seen by the driver is a constant, independent of signal frequency and line length. This is the $R_S = R_L = Z_0$ case shown in Fig. 11.6.

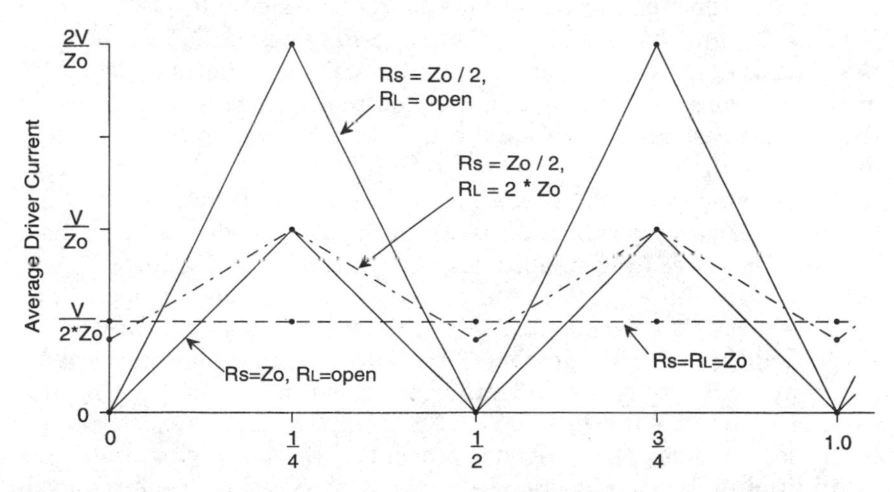

Figure 11.6 Average driver current for source-terminated lines.

Crosstalk traveling toward the source would be terminated, as in Case 2. Signals traveling toward the load would be terminated, as in Case 1 (but a line-to-line differential load termination does not absorb common mode crosstalk at the load end). This third case provides essentially perfect terminations, but at only one-half of the no-load source voltage signal level. Current flows, and power is dissipated at all times, but the current is one-half that for Case 1. The Case 1 current would be a flat line at V/Z_0 in Fig. 11.6. The total loop resistance is doubled, so the total power is one-half that for Case 1. The power dissipated is split equally between the source and load terminations, so the power dissipated in each termination resistor is one-fourth of that in the load resistor only for Case 1. Figure 11.6 indicates the variation in average driver current for a few cases of source termination and load termination. Driver current is minimum at zero line length and even multiples of a quarter wavelength for an open load. It should be noted that although the curves indicate zero current at even multiples of a quarter wavelength, practical lines with some loading and losses will require some nonzero current. This figure indicates the various driver *currents*; power consumption is proportional to the square of current. Therefore, each doubling of current as indicated on the vertical axis increases power dissipation by a factor of 4.

Although it is clear that the average current is high when the line length is an odd multiple of a quarter wavelength, this is only the *average*. The transmission-line driver draws the *peak* current at each transition until any reflection returns from the higher-impedance end of the line to reduce the line current. Therefore, the driver must supply this peak current during the round-trip time to the end of the line. There is no reflection when the load is exactly terminated with Z_0, so there is no reduction in current for steady-state signal levels.

Two cases are shown with the source termination at $Z_0/2$ to increase the initial signal amplitude. The open-load case shows a dramatic increase in current at a line length of odd multiples of a quarter wavelength. The case shown with a $2 * Z_0$ load termination has a modest increase at the quarter wavelength, but not greater than an exact source termination with an open load. The $2 * Z_0$ termination at the load reduces current compared to an open load for all but very short lines (or lines close to even multiples of a quarter wavelength).

Figure 11.7 illustrates several cases of signal levels for open lines as bar charts. The groups are the initial undershoot, the first overshoot, and the second undershoot. The bars within each group are for different values of source-termination resistance, indicated by the ratio to Z_0, or VSWR. The VSWR values vary from 1.0 (exactly equal to Z_0) to 5, which represents 20 percent of Z_0. The figure indicates that initial undershoot decreases as the source resistance is reduced,

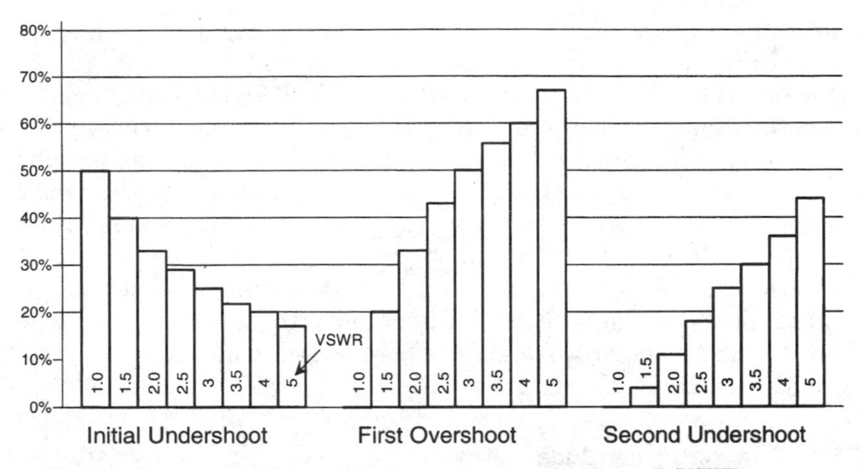

Figure 11.7 Signal levels for open lines vs. source termination = Z_0/VSWR.

but the overshoot and resulting second undershoot increase with lower source resistance. At VSWR = 2 (source resistance = $Z_0/2$), both the initial undershoot and the overshoot are 33 percent. The second undershoot is the square of overshoot, or a little over 10 percent. At VSWR = 3 (source resistance = $Z_0/3$), both the initial undershoot and the second undershoot are 25 percent. This ringing is higher than usually recommended for high speed interconnections. However, it may be suitable for some applications, such as shorter lines where there is sufficient time for signals to settle, or if diode clamps control overshoot beyond steady-state levels.

11.6 Constant VSWR Termination Duals

Case 3 can be considered as one member of a family of termination cases that are duals of itself, defined by $\rho_S = -\rho_L$. When the termination is not exact, there will be partial reflections and ringing that depend on the mismatch ratio (VSWR). The $R_S = Z_0/2$ and $R_L = 2 * Z_0$ case shown in Fig. 11.6 is an example of a constant VSWR dual because both ends have the same mismatch ratio.

Increasing the load resistance and/or decreasing the source resistance will increase the steady-state final voltage. A family of termination cases that are duals of itself can be defined by $\rho_S = -\rho_L$. This family of cases has resistance at both ends of the line that has the same ratio to the line impedance, and may be regarded as constant VSWR lines. Signals or crosstalk traveling in either direction are equally well terminated at each end. Although any value of ρ_L from -1 to $+1$ may be included in a complete definition, only $\rho_L \geq 0$ is usually considered to avoid the severe loss of signal amplitude if the load resistance is low (and the source resistance high).

If we consider the VSWR at both ends to be defined as the variable n, then the source resistance is $R_S = Z_0/n$, and the load resistance is $R_L = n * Z_0$. Resistance ratios of about 2 to 4 at both ends provide results which would be considered practical for many applications. However, the increase in load current for quarter-wavelength lines may be significant for a VSWR at the load of greater than about 2. A load resistance higher than Z_0 and source resistance less than Z_0 is an underdamped line which will overshoot the steady-state level when the first incident wave arrives at the load, with alternate ringing on the subsequent waves.

11.6.1 Final signal amplitude

$V_F = n^2/(n^2 + 1)$ is the ratio of the final static voltage to the no-load source voltage.

11.6.2 Initial undershoot on the line

$US_{\text{init}} = (n - 1)/n(n + 1) = (V_F - V_{\text{init}})/V_F$ is the ratio of the initial undershoot to the final voltage. Initial undershoot is of concern for inputs located along the line, but not near the far end of the line.

11.6.3 First incident overshoot at the far end

$O_S(N = 1) = (n - 1)^2/(n + 1)^2 = -\rho_S * \rho_L$ is the ratio of the first incident wave overshoot at the load to the final voltage. Some overshoot is desirable because it increases signal amplitude. However, excessive overshoot at the far end results in undershoot and may increase power consumption and signal jitter on long lines if the signal level does not stabilize between signal transitions.

11.6.4 Second incident undershoot at the far end

$U_S(N = 2) = (n - 1)^4/(n + 1)^4 = (\rho_S * \rho_L)^2$ is the ratio of the second incident wave undershoot to the final voltage. This undershoot is usually of most concern because this phase of ringing results in a reduction of the signal amplitude toward the receiver threshold for the usual cases of $n \geq 1$. This signal reduction toward the receiver threshold reduces the signal noise margin. The amplitude of the second undershoot is the square of the first overshoot.

Table 11.1 lists the equations for, and values of, signal levels and driver currents for double-terminated lines for several values of VSWR from 1.0 (exact termination at both ends) up to 6.0 at both ends of a long transmission line. The initial undershoot, overshoot,

TABLE 11.1 Signal Levels and Currents for Double-Terminated Constant VSWR Lines

VSWR	n	1	1.5	2	2.5	3	3.5	4	5	6
US_{init} (initial undershoot)	$\frac{n-1}{n(n+1)}$	0.000	0.133	0.167	0.171	0.167	0.159	0.150	0.133	0.119
$OS_{(N=1)}$ (first overshoot)	$\frac{(n-1)^2}{(n+1)^2}$	0.000	0.040	0.111	0.184	0.250	0.309	0.360	0.444	0.510
$US_{(N=2)}$ (next undershoot)	$\frac{(n-1)^4}{(n+1)^4}$	0.000	0.002	0.012	0.034	0.063	0.095	0.130	0.198	0.260
V_{final}/V_S (final voltage)	$\frac{n^2}{n^2+1}$	0.500	0.692	0.800	0.862	0.900	0.925	0.941	0.962	0.973
$I_{dc}/(V_S/Z_0)$ (dc current)	$\frac{n}{n^2+1}$	0.500	0.462	0.400	0.345	0.300	0.264	0.235	0.192	0.162
$I_{max}/(V_S/Z_0)$ (max current at $\lambda/4$)	$\frac{n}{2}$	0.500	0.750	1.000	1.250	1.500	1.750	2.000	2.500	3.000

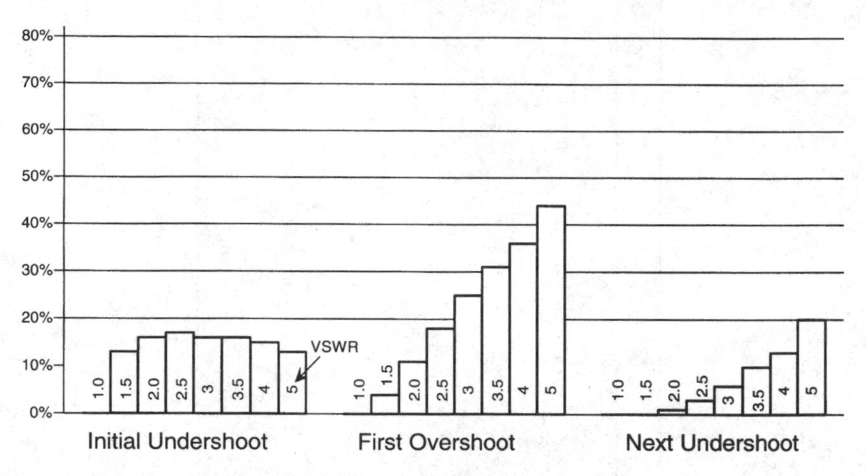

Figure 11.8 Signal levels for double-terminated lines vs. VSWR at both ends.

and second undershoot are the ratio to the final steady-state signal level. Of course, undershoot and overshoot are zero for a line exactly terminated at both ends.

Figure 11.8 illustrates the values for initial undershoot, overshoot, and second undershoot for the values of VSWR from 1.0 to 5.0 from Table 11.1. A comparison to Fig. 11.7 for the same signals indicates that double termination significantly reduces the deviations from steady-state levels. Unlike the open line cases, the initial undershoot is zero for VSWR = 1.0, increases to about 17 percent for VSWR = 2.5, then decreases slowly for increasing VSWR. The overshoot is also about 18 percent for VSWR = 2.5, with second undershoot less than 4 percent.

Figure 11.9 illustrates the values for the final signal level, dc current, and maximum current for the values of VSWR from 1.0 to 5.0 from Table 11.1. The final signal level compared to the open-circuit voltage reached for open lines is 50 percent when the line is exactly terminated at both ends. The final signal level approaches the open-circuit voltage for increasing values of VSWR. The dc current relative to V_S/Z_0 decreases as the load resistance increases. However, the maximum current for a quarter-wavelength line increases in proportion to VSWR. This is an illustration of the need for load termination to limit current on long lines operated at high data rates.

The ultimate limit as n increases toward infinity is an indefinitely ringing line with perfectly zero source impedance and no load resistance. Since there is no resistance at the source, it remains at the new signal level regardless of the effect of any reflections from the load. However, the reflection of the signal voltage at the load results in twice the source voltage at the first incident wave at the load. The reflection is reversed at the source, so the signal voltage at the load undershoots

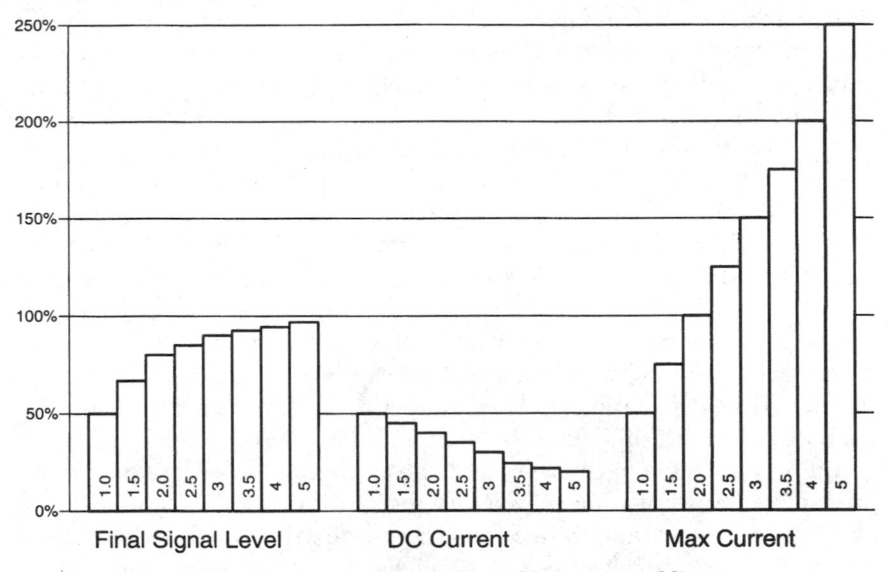

Figure 11.9 Final signal levels and currents for double-terminated lines.

all the way to zero (the level prior to the first incident wave). If there are no losses at the ends or on the line itself, the oscillations at the load will continue indefinitely. The average of the two signal levels is the source voltage, but the load only settles to that value as losses dampen the oscillations. The lossless line may be considered as a combination of Case 1 and Case 2, with $\rho_S = -1$ and $\rho_L = +1$.

Of course, continuous oscillation is a theoretical case, because all real systems have finite losses. However, many high speed interconnections with no terminations would tend to approach a response with severe ringing unless limited by such factors as source impedance, rise time and short line length, lumped capacitance, or clamping diodes to limit overshoot at the receiver. However, low-loss interconnections are relatively easy to achieve and may be ideal for very short interconnections. It is important to recognize that the lossless, unterminated line is *not* an ideal interconnection system for high-frequency networks of any significant length.

The specifications for common TTL and CMOS circuits even encourage low-impedance outputs because the propagation delay is usually specified at a lumped capacitance load, so lower output impedance results in faster specification times. However, practical operation on long lines usually requires a source impedance that is within a reasonable ratio of (but not exactly matched to) the line impedance. Source termination only, with no termination at the receiver, may perform adequately for transmission lines that are long compared to the rise time, but not long compared to the quarter wavelength of the highest frequency to be transmitted. As described above, very long lines may result in standing waves and significantly increased current flow if the frequency is not low enough to allow the signal level on the entire line to settle before each new transition is driven onto the line.

When a line is so short that the two-way propagation delay of a line is less than the rise time, the amplitude of reflections and ringing is reduced because reflections begin to arrive from the load before the initial signal rise time has been completed. This can be considered as the reflection (typically from a high-impedance load) assisting the driver, so less energy is driven onto the line and there is less reflection. Alternately, it may be considered that the driver begins to adjust to the actual load resistance as the reflection begins to arrive from the load during the initial signal transition.

If it is reasonable to consider that the circuits involved are linear, the principle of superposition may be used to estimate the response to the overlap of signals and reflections. Superposition for linear circuits means that the response to signals (reflections in this case) is not affected by the signal level, so that the response to overlapped signals may be superimposed, or added. Most CMOS drivers are reasonably linear, and most logic inputs remain high-impedance if the signals remain within the range so that clamp diodes do not begin to conduct. However, TTL and ECL drivers may not maintain a linear response to some signal conditions. Graphical techniques called *reflection diagrams* are useful for estimating nonlinear response to transmission-line conditions.

11.7 Ideal Transmission-Line Waveforms

Shown in Fig. 11.10 is an example of various line signals for an underdamped source-terminated line. This figure illustrates signals at the load for two cases of long line lengths. This example assumes a linear rise time, an ideal lossless line, a source impedance of $R_S = Z_0/3$, and a perfectly open load. Since there is no load resistance, the final signal level is the no-load signal at the source.

The upper waveform is the ideal no-load source signal excursion with a linear rise time. This would represent the signal at the ideal source, before any source internal resistance. The second waveform is the incident wave launched onto the line. The finite source impedance of $Z_0/3$ means that three-fourths of the no-load signal is driven as the initial signal into the long transmission line with an impedance of Z_0. Although this voltage divider reduces the slope of the signal transition (the volts per unit time slew rate), it does not reduce the rise time of the transition.

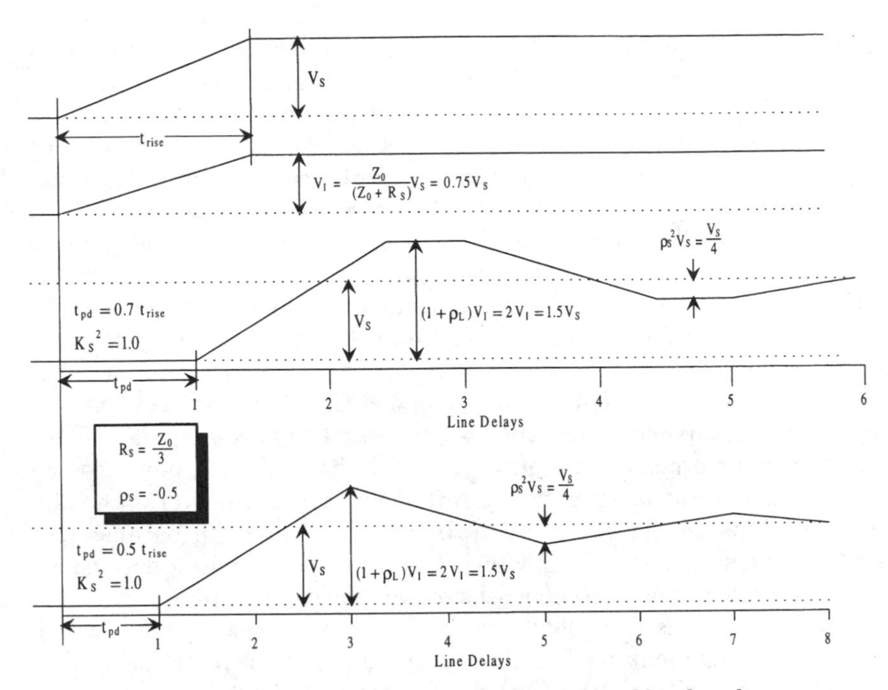

Figure 11.10 Incident and reflected waves for long and critical line lengths.

The third waveform in the figure is the signal at the perfectly open load after traveling on a line with a propagation delay of 70 percent of the linear rise time. The initial transition signal amplitude will double at the load owing to the 100 percent reflection coefficient at the open load. Therefore, the maximum signal at the load will be double the initial signal, which will be 2 * 3/4 = 1.5 times the no-load signal. This is an overshoot of 50 percent of the final signal level. Because this amplitude is larger than the no-load source amplitude, the volts per unit time slew rate at the load is actually greater than the no-load slew rate, as well as the incident wave launched onto the transmission line. Of course, any line losses or lumped capacitance loading will tend to increase the rise time at the load.

The reflection at the open load begins to arrive at the source at the two-way propagation delay, or 140 percent of the rise time for this example. The source reflection coefficient is −50 percent, so only 0.5 * 3/4 = 3/8 of the no-load signal is returned from the source as a negative signal transition. This second incident wave begins to arrive at the load at the two-way delay after the first incident wave, or three times the one-way delay after the driver launched the initial signal. This negative 3/8 reflected transition doubles at the open load, for a total negative transition at the load of 3/4 below the previous level of 1.5. This results in a net signal level after the second complete incident wave of 1.5 − 0.75 = 0.75 of the no-load signal level. This is an undershoot of 25 percent relative to the final signal level. These overshoot and undershoot cycles continue to repeat, with the amplitude of the excursion reduced by the source reflection coefficient of 50 percent for each cycle. These levels are achieved if the round-trip propagation delay on the line is long enough that the reflections are separated and not attenuated. This first case illustrates a propagation delay more than twice the total rise time, so the waveform at the load remains at each level until the next reflection arrives from the source.

The lower waveform in Fig. 11.10 illustrates the case where the line propagation delay is exactly 50 percent of the total rise time. The maximum levels of 50 percent overshoot above the steady-state level and 25 percent undershoot below the steady-state level are the same as for the longer line propagation delays. However, the maximum excursions are just reached as the round-trip reflection arrives to reverse the signal level. This line length is called the *critical line length* because this is the shortest line length at which the maximum ringing excursions occur on an ideal line with no losses or waveform distortion.

Figure 11.11 illustrates load waveforms for the same incident waveforms and reflections as Fig. 11.10, but for line propagation delays less than the critical line length. Assuming that the components involved are essentially linear, the principle of superposition may be

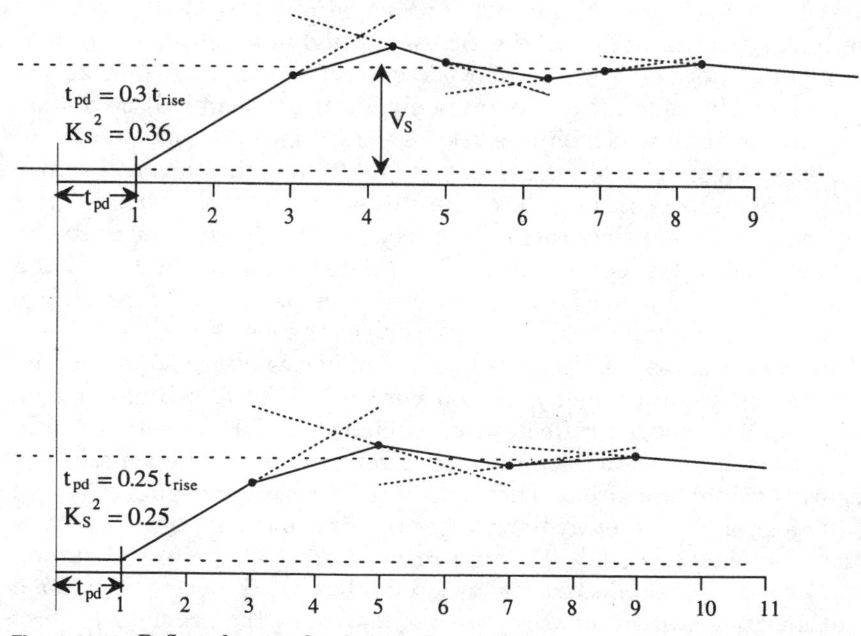

Figure 11.11 Reflected waves for short and very short line lengths.

applied to determine the resulting waveforms. This means that the separate incident and reflected waveforms shown previously for the long-line cases may be superimposed for the shorter-line cases. The reflected waveforms begin before the end of the previous incident wave, so they overlap in time. During the overlap times, the resulting signal is the linear sum of the overlapped waveforms.

The signal waveforms shown in Fig. 11.11 indicate the superimposed linear signal components as dashed lines. The combined waveform is shown as the solid lines connecting the dots at points where linear sections begin and end. The beginning and ending of the linear sections coincide for the case of the one-way propagation delay only one-quarter of the linear rise time. This case is the lower waveform in Fig. 11.11.

The signal components for an underdamped line alternate in polarity, so the sum of the overlapped components is closer to the final signal value. Actual signals with some skin-effect or other line losses, and/or waveshape distortion typically have even further reduction of worst-case overshoot and undershoot excursions. The waveforms in this figure illustrate one-way propagation delays of 36 and 25 percent of the total rise time. When the one-way propagation delay is 25 percent of the total rise time (or one-half of the critical length), there is just complete overlap of incident and reflected waves. The undershoot is typically negligible if the one-way propagation delay is less than about 25 percent of the rise time.

This figure also illustrates that the times at which the maximum and minimums of the line ringing occur are a function of both line propagation delay and rise time. The ringing frequency is determined primarily by the line propagation time. As can be noted from the waveforms, the maximum overshoot is reached at the one-way propagation time plus the total rise time. The maximum level is maintained for lines longer than the critical length. Overshoot recovery begins after three times the one-way propagation time. The minimum level (maximum undershoot) is reached at three times the one-way propagation time plus the total rise time.

The second overshoot maximum occurs at five times the one-way propagation time plus the total rise time. This is the additional settling time that may be required for undershoot recovery to valid logic levels if the ringing is so severe that the first undershoot level is not valid. Of course, higher speed operation can be achieved if the ringing can be controlled to acceptable levels. High speed operation requires that adequate signal levels are achieved quickly at the first incident wave and any undershoot is small enough to maintain adequate signal levels. Five times the one-way propagation time alone is not sufficient because the signal level is still at the maximum undershoot at that time for long lines, and the maximum undershoot occurs later than five times the one-way propagation time for short lines. For the example in this figure of the one-way propagation time 25 percent of the total rise time, the maximum undershoot occurs at seven times the one-way propagation time.

11.7.1 The short-line scaling factor

Although not analytically exact, it is convenient to scale each reflection factor by the ratio of the round-trip propagation delay to the initial rise time. The square of this ratio is a reasonable approximation for scaling the overshoot and undershoot of a line with less round-trip delay than the signal rise time. The short-line scaling factor is designated K_S and is essentially the same factor as used for the coupled length for crosstalk. When K_S is used for scaling the ringing of lines with shorter round-trip propagation delay than the linear rise time, $K_S = 2T_0'/t_{\text{rise}}$ for $K_S < 1$, or unity for $K_S = 1$.

$2T_0'$ is the total two-way loaded propagation delay, including any increase due to lumped loading. The square of K_S is indicated on the waveforms of Fig. 11.11. It may be noted that the reduction in the overshoot amplitude is reduced by about K_S^2. The definition for rise time is based on the total (0 to 100 percent) rise time of a linear transition. However, most practical waveforms of high speed transitions have significant rounding, so it is usually more practical to define the rise time of high speed signals as the 20 to 80 percent levels as this central 60 percent of the signal transition is a more linear representation of the signal slew rate, or slope.

Illustrated in Fig. 11.12 is an overlay of the linear waveforms from the previous figures for the long line and the very short line. These waveforms illustrate the effect of short-line scaling. The first overshoot at the load is $V_S/2$ for the long-line case, but only $V_S/8$ for the very short line case with $K_S = 0.5$. The reduction by a factor of 4 for this case is exactly K_S^2, although for other cases the reduction may not be exactly K_S^2. The overlay of the waveforms also indicates that the greater overshoot crosses the steady-state level sooner, but the initial slew rate is the same before the reflection arrives for the very

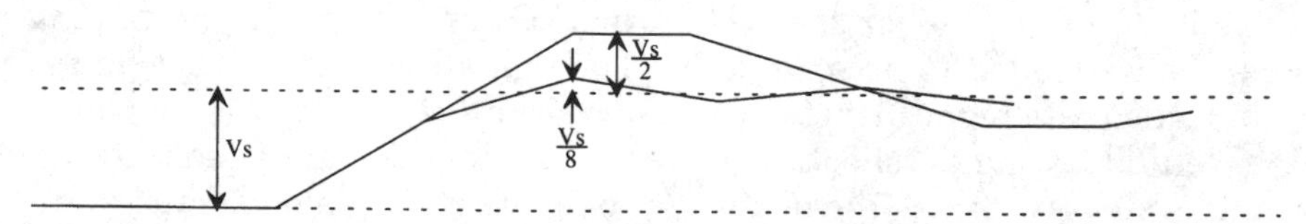

Figure 11.12 Overlay of long and very short line response.

short line case. The time to reach the maximum overshoot level is the same, but the recovery for the very short line is much faster, so the time to the first undershoot is much sooner. This illustrates that the ringing frequency for a linear system depends on the line propagation delay. The ringing excursions on the very short line are so small that the waveform is not significantly different from the no-load source signal transition, although the rise time at the load is slightly faster.

11.8 The Power Series for Signal Levels

The consecutive reflection cycles of the signal levels on a transmission line are exponentially damped, with a damping ratio between cycles that defines a power series for the sequence of signal levels.[3]

Setting the sum of the series of the incident signal waves at the load equal to the known final signal level of $R_L/(R_S + R_L)$ results in the partial sum of the first N terms of the series as follows:

$$\frac{V_N}{V_F} = 1 - K_S^2(\rho_S * \rho_L)^N \tag{11.2}$$

V_N is the signal level at the load just after the Nth incident wave transition is complete. V_F is the final steady-state static voltage level. K_S is the short-line scaling factor, with $K_S = 1$ for long lines. The series converges to the final value for lines with finite damping, so that $(\rho_S * \rho_L)^2 < 1$.

The general form of a power series is

$$a + ar + ar^2 + \cdots + ar^N \tag{11.3}$$

The signal level after the first incident wave is

$$a = (1 + \rho_S)V_{\text{init}} = \frac{2Z_0 * R_L}{(R_S + Z_0)(R_L + Z_0)} \tag{11.4}$$

The final value of the series sum is $a/(1 - r) = R_L/(R_L + Z_0)$, so $r = \rho_S * \rho_L$.

Equation (11.2) for the signal level after the Nth incident wave indicates that the $K_S^2(\rho_S * \rho_L)^N$ term represents the deviations from the final signal level. Since overshoot and undershoot are defined as deviations from the final value, $K_S^2(\rho_S * \rho_L)^N$ is an estimate of the sequence of deviations from the final signal level. It is convenient to use a single definition for both overshoot and undershoot. The convention selected is that positive values of undershoot represent deviations of the opposite polarity to the driven transition. Conversely, overshoot will be considered as negative deviation values.

Since undershoot is defined as the deviations that are opposite polarity to the driven signal transition in either direction, they are of most concern because the deviation is in the direction toward the receiver threshold level. This reduction of desired signal level results in reduced noise margin, and increased risk of erroneous operation.

It should be noted that although signal transitions are often illustrated as positive-going edges, the corresponding definitions and conditions apply as well to negative-going edges. Therefore, undershoot resulting from a positive transition is a negative deviation from the final signal level, and undershoot resulting from a negative transition is a positive deviation from the final signal level. Typical signals from TTL and TTL-compatible CMOS outputs have more severe ringing on the negative-going transition because the source impedance of outputs without series resistors is lower in the LOW state than for the HIGH state.

11.8.1 Ringing signal levels on long lines

The undershoot as a deviation from the final signal level after the *N*th incident wave arrives at the load as a proportion of the final signal level is estimated as follows:

$$\frac{US_N}{V_F} = \frac{V_F - V_N}{V_F} = K_S^{\,2}(\rho_S * \rho_L)^N = K_S^{\,2} * K_R^{\,N} \tag{11.5}$$

The short-line scaling factor $K_S = 1$ for long lines. $K_R = \rho_S * \rho_L$ is the line reflection factor.

The usual case of $R_S \leq Z_0 \leq R_L$ results in a negative line reflection factor because the reflection factor for the source is negative. When the line reflection factor is negative, the first and all odd-numbered incident waves arriving at the load result in overshoot because the value of the undershoot is negative. Even-numbered incident waves always result in undershoot. A negative reflection factor results in underdamped ringing of the signal.

Some ringing is desirable because the final signal level is increased, and the first incident wave overshoot crosses the input threshold faster and tends to compensate for lumped loading and line losses, especially skin-effect distortion on high speed lines.

11.8.2 The exponential damping envelope

Figure 11.13 illustrates an underdamped step waveform for a line reflection factor of $K_R = -0.50$, with the corresponding exponential envelope of the lower excursions. A line with a positive reflection factor will have positive undershoot at every incident wave, so the waveform at the load approaches the final value in a monotonic, stair step

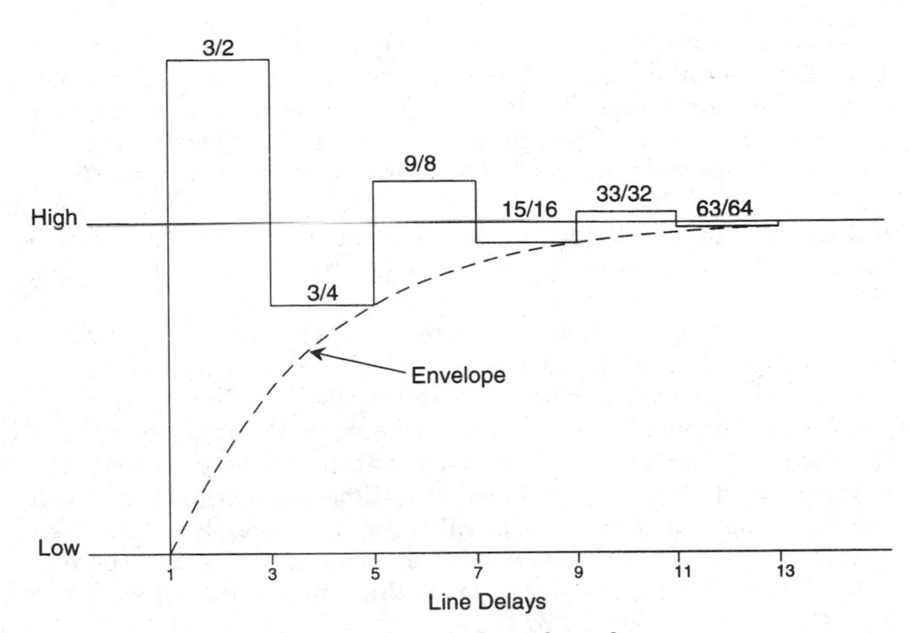

Figure 11.13 Exponential envelope for underdamped waveform.

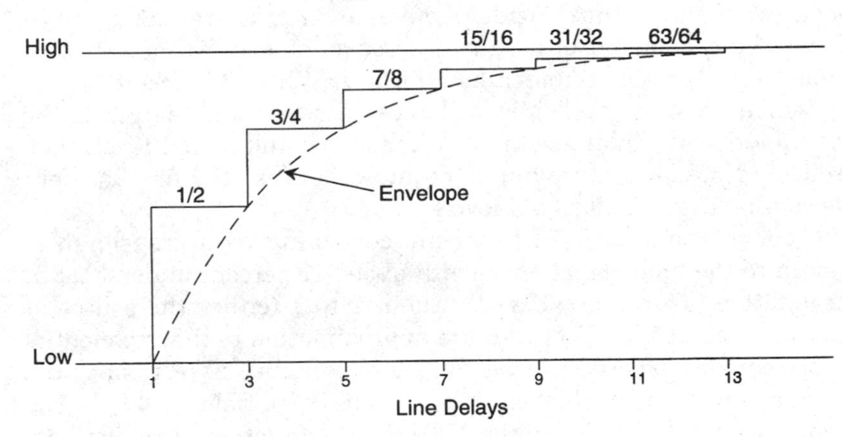

Figure 11.14 Exponential envelope for overdamped waveform.

fashion. This is generally undesirable for practical lines, but it is of interest to note that if two lines have the same *magnitude* of reflection factor, both lines will have the same undershoot at the even-numbered incident waves, so the damping envelope is the same.

Figure 11.14 illustrates an overdamped step waveform for a line reflection factor of $K_R = +0.50$, with the same exponential envelope of the lower excursions. The exponential damping envelope defines an equivalent lumped circuit with an RC charging time constant for an imperfectly terminated long line. The exponential envelope does not include the short-line scaling factor and must use only the *magnitude* of the reflection factor. The exponential begins (traditionally as $t = 0$) when the first incident wave begins to arrive at the load ($t = -T_0'$ at the beginning of the driven signal at the source). The lower exponential envelope passes through the signal level at the beginning of each incident wave for long lines, based on a positive reflection factor. The time constant of the exponential envelope is $-2T_0'/\ln|K_R|$.

When a long line does not have a termination that closely matches the dynamic line impedance, there is a reasonable linear approximation to the natural logarithm of the reflection factor, which is $-2/\mathrm{VSWR}$. Therefore, a reasonable estimate of the time constant of the line is $T_0' * \mathrm{VSWR}$. This time constant can be used to solve for the envelope of the ringing as a function of time to determine how well the signal has settled to the final value. For example, if two time constants of an exponential waveform are considered as reasonably achieving the final signal level (more than 80 percent), then $2T_0' *$ VSWR is a reasonable settling time.

Two round-trip delays are required from the beginning of the first incident wave to the third incident wave, or $4T_0'$. Therefore, a VSWR of 2 may be considered to result in a reasonable two-time constant settling time. The actual undershoot for a VSWR of 2 is about 11 percent, which most logic devices will receive as a valid signal. If the worst undershoot signal level is a sufficiently valid signal level, then no settling time for line ringing is required because the first incident wave establishes a valid signal level.

The above example used two time constants as a "reasonable" approach to the final signal level, with about 11 percent undershoot at a termination mismatch (VSWR) ratio of 2 to 1 (either the source at $Z_0/2$ or the load at $2Z_0$). However, the approximation to the exponential envelope can also be used to solve for the maximum VSWR to meet the maximum undershoot allowed at two round-trip delays ($4T_0'$). The approximation for the maximum VSWR as a function of undershoot (US) is

$$\mathrm{VSWR} = \frac{-4}{\ln(US)} \tag{11.6}$$

This approximation is conservative (it specifies a lower VSWR than an exact solution) and can be used to estimate the termination mismatch ratio at one end when the other end of the line is not terminated. The resulting equation for the minimum source resistance for an open load is $-0.25Z_0 * \ln(US)$. For example, if the maximum undershoot allowed is 25 percent, the exponential approximation is a minimum source resistance of about $0.35Z_0$. This compares to the exact solution of $Z_0/3$, which can be determined from

$$\text{VSWR} = \frac{1 + \sqrt{US}}{1 - \sqrt{US}} \tag{11.7}$$

The above equations for the maximum VSWR at one end of a line will limit the undershoot (US) so that an adequate signal level is maintained after the first incident wave arrives at the load, even if there is some ringing due to an underdamped termination. For example, if a maximum undershoot of 10 percent is substituted into Eq. (11.7), the maximum VSWR at one end is 1.9, or nearly 2.0. Either a source resistance of at least 50 percent of the maximum line impedance with no-load impedance, or a load impedance at the end of the line of less than twice the minimum line impedance should limit the undershoot to within about 10 percent to support incident wave switching. For a loop line with source terminations as was illustrated in Fig. 11.4, the source resistance R_S is at least 25 percent of the maximum Z_0 because the two-way impedance seen by drivers in the middle of a line is $Z_0/2$.

It should be noted that the above equations for VSWR are for one end only. For example, they apply to the VSWR at the source for an open load, or to the VSWR at the load for a very low impedance source. The undershoot on a double-terminated line is reduced by the square of the product of the reflection factor at each end, as defined for Eq. (11.5). Therefore, the square root of undershoot in Eq. (11.7) should be replaced with the fourth root for the maximum VSWR at *both* ends of a double-terminated line.

The VSWR at both ends for a double-terminated line is defined as follows:

$$\text{VSWR}\,(2x) = \frac{1 + \sqrt[4]{US}}{1 - \sqrt[4]{US}} \tag{11.8}$$

For example, for a maximum undershoot of 10 percent, the fourth root of 0.10 is about 0.562, so the maximum VSWR is about 3.5. Therefore, a source resistance of at least 30 percent of Z_0 and a load resistance of less than 3.5 times Z_0 at the end of a single line should limit the undershoot to within about 10 percent to support incident wave switching.

For this example, consider a party line bus with multiple drivers along the line and terminations at both ends as was illustrated in Fig. 11.2. Each driver should have at least 15 percent of the line Z_0 as the total source impedance (about 30 percent of the $Z_0/2$ impedance in the middle of a line). Both ends should have a parallel load termination of less than 3.5 times Z_0 to limit ringing at both ends to less than 10

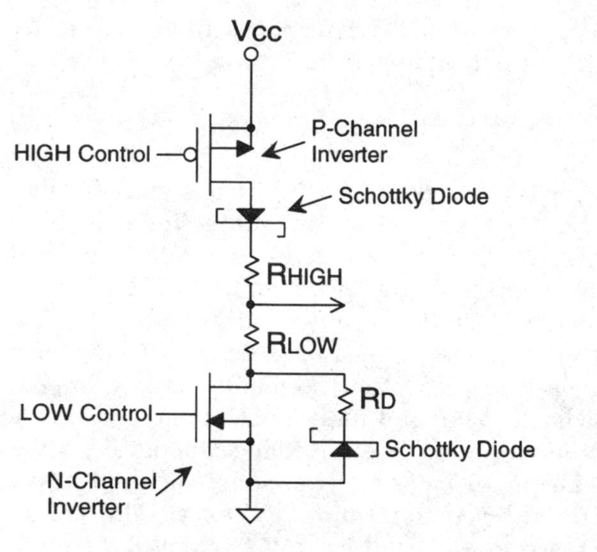

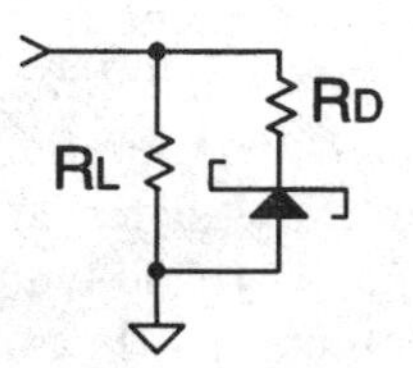

Figure 11.15 CMOS driver and termination module with series diodes.

percent undershoot. This means that if enough time is allowed for recovery after the first overshoot, the line should settle to within 10 percent of the final steady-state static signal level.

The final signal level is over 90 percent of the open-circuit voltage, and the dc current is about one-half of the dc current for a line exactly terminated with $Z_0/2$ at each driver and Z_0 at both ends. The signal values for a party line terminated with the same VSWR at each driver and at both ends are the same as in Table 11.1, although the magnitude of the currents is doubled owing to the terminations at both ends.

Diode clamps with a series impedance appropriate for the line impedance Z_0 may be incorporated into the R_L terminations at each end to control overshoot, which will further reduce undershoot. Figure 11.15 illustrates a driver and termination module with diode clamps intended to be suitable for a party line bus such as was shown in Fig. 11.3. The termination module has a resistor to ground, so the steady-state static LOW level is at ground. This is also the static level when all drivers are disabled, or if power is off, so the party line bus remains in a static LOW state if drivers are disabled during power up. The bus should be driven to a LOW before all drivers are disabled so the level does not drift through the threshold region as the load resistor to ground would discharge a HIGH level. A LOW level also reduces power dissipation during periods of inactivity. The termination module also has a Schottky clamp diode to ground in series with another resistor to terminate negative overshoot at a lower dynamic impedance than the load resistor alone.

The resistive load to ground also means that the steady-state static HIGH level is a single value, about a diode drop below the V_{CC} power supply level. This eliminates the ambiguous HIGH level, reducing the signal excursions and the negative overshoot on the falling edge signal transition. The termination module requires no power but must be connected to the common ground reference for all devices on the party line bus. The circuits illustrated in Fig. 11.15 do not include any diodes that would conduct forward current if the output is driven above the V_{CC} power rail. This illustration may be suitable for multiple drivers on a party line bus with multiple power supplies if there are no additional isolation or protection diodes to the V_{CC} power rail other than shown in this schematic. The output signal levels are essentially consistent with TTL-compatible signal levels and may be compatible with operation on a party line bus with TTL-compatible bus drivers powered by 5 V. This example is intended to be compatible with single-ended operation on a common ground reference to illustrate the integration of both series resistance and series diodes at both ends of a party line bus. The driver output may also be connected to the input side of a bidirectional transceiver (not shown).

Figure 11.16 is a reflection diagram that illustrates the operational characteristics of a line with the output driver and termination module example from Fig. 11.15. The R_{HIGH}, R_{LOW}, and R_L resistors were

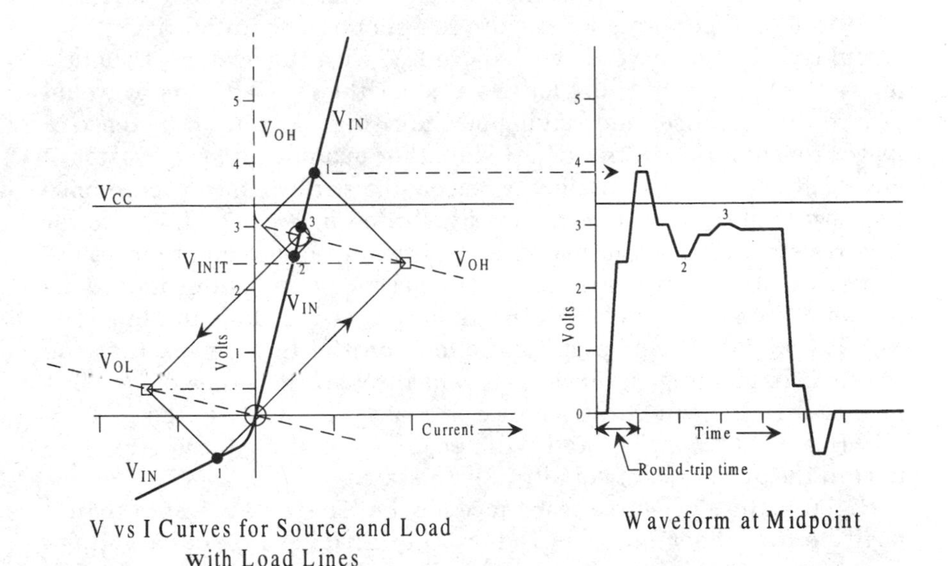

Figure 11.16 Reflection diagram and midpoint waveform for $4 * Z_0$ load termination.

chosen for convenience in this illustration to result in a dynamic impedance of $Z_0/4$ for the driver outputs and $4 * Z_0$ for the load terminations at each end of the line. This relatively high impedance ratio is not necessarily optimum to meet performance objectives for all applications, but the reflection diagram can be evaluated for potential changes to improve performance for specific applications.

The R_D resistors for the overshoot clamp diodes to ground were maintained at the same dynamic impedance of $Z_0/2$ which was previously shown to be an effective overshoot clamp impedance. The actual values of the separate physical resistors may be lower because the transistors and diodes have internal impedance. However, it is generally recommended that the physical resistors be at least one-third of the total dynamic impedance to limit the effects of process tolerances, supply voltage, and temperature range on the transistors and diodes.

The reflection diagram of Fig. 11.16 is drawn as a single active driver in the center of a line, launching current into both arms of the line. This means that the driver impedance is actually one-half of the value for the Z_0 of one line. The two-load termination modules are considered to be at equal distances from the active driver. The negative overshoot clamp diode on the active driver has no effect on the signal. This reflection diagram ignores the effects of the negative overshoot clamp diodes on any other inactive drivers, or the different waveforms that may result if the active driver is not centered in the middle of the line and the ends are not terminated with exactly Z_0.

Even though this example illustrates a rather high ratio of terminations to Z_0, the performance is quite reasonable. The initial rising edge signal level at the driver is well above 2 V, with the overshoot slightly above the V_{CC} power rail. A lower value for the R_L load resistor would reduce the overshoot and ringing on the rising edge, at an increase in power consumption in the HIGH state. For example, a load resistor of $2 * Z_0$ at each end is possibly a reasonable performance compromise for most applications. This case is illustrated in Fig. 11.17. The lower load resistance is a closer match to the line impedance and reduces the overshoot to what is nearly an ideal level for typical interconnections to improve rise time and compensate for lumped loading and line losses.[4] The higher steady-state static load on the line means that the source is conducting significant current through the series diode. This means that the source impedance is reasonably terminated in both HIGH and LOW states and will terminate disturbances traveling toward the driver about equally in both states.

Higher values of source resistance for the HIGH state would reduce positive overshoot but also reduce the initial rising edge voltage excursion at the driver. The initial falling edge signal level at the driver is about 0.5 V, with some diode current conduction below ground.

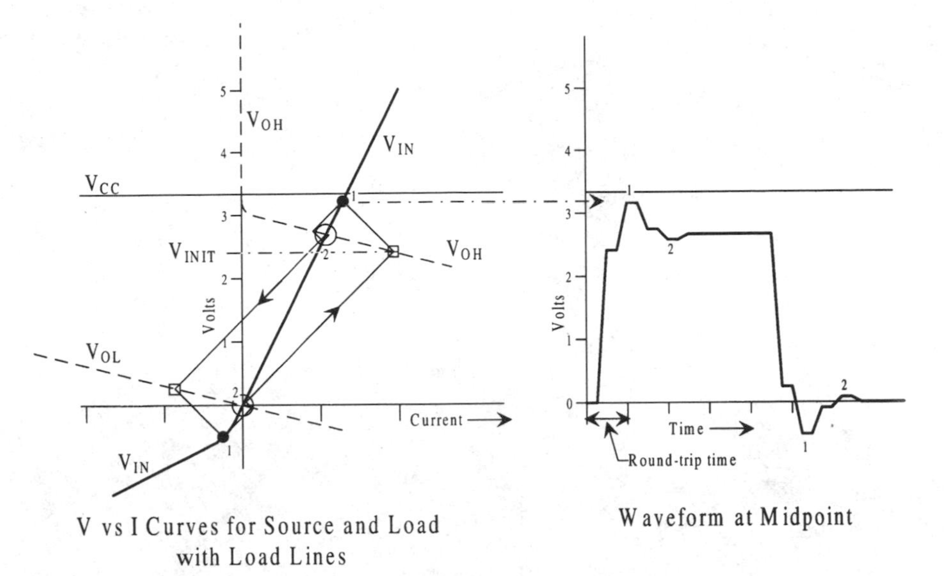

Figure 11.17 Reflection diagram and midpoint waveform for $2 * Z_0$ load termination.

The reflection back to the driver results in almost exactly the steady-state quiescent conditions, so there is no further ringing on the falling edge transition for this example.

This example illustrates that series resistors and diodes can be combined to achieve reasonable high speed performance without requiring an exactly matched termination load resistor at the ends of either a simple two-ended transmission line or at both ends of a multiple driver party line bus. The higher impedance of the parallel termination load resistors to a common ground reduces the high switching currents and power dissipation compared to double-ended exact terminations and does not require a separate termination power supply.

11.9 References

1. *FAST Advanced Schottky TTL Logic Databook,* National Semiconductor Corp., Santa Clara, Calif., 1990 ed.
2. *INTERFACE: Line Drivers and Receivers Databook,* National Semiconductor Corp., Santa Clara, Calif., 1992 ed., sec. 6, AN-807.
3. Robert E. Canright, Jr., and Arden R. Helland, "Reflections of High Speed Signals Analyzed as a Delay in Timing for Clocked Logic," *24th ACM/IEEE Design Automation Conference,* Miami Beach, Fla., June 1987, paper 7.4.
4. James E. Buchanan, *Signal and Power Integrity in Digital Systems: TTL, CMOS & BiCMOS,* McGraw-Hill, New York, 1996, p. 169.

Chapter

12

Interconnections: Routing of Critical Signals

This chapter discusses the distribution and routing of critical signals in high speed interconnections. The characteristics of loop routing are described and illustrated. The general loop routing configuration locates the source in the middle of a signal net that is connected in a closed loop without ends. It is shown that a loop-routed line is equivalent to two single lines in parallel. The termination of twin waves returning to a series-terminated source in the middle of a line is illustrated. Loop routing is applicable to source-terminated lines, multiple sources on a party line bus, and nets with terminated clamp diodes to terminate overshoot to the power rails.

12.1 Simple Lines

If there is only a single load at the end of a simple line, the total propagation delay and settling time is the one-way propagation delay (including lumped loading effects) plus the rise time. This is usually slightly conservative because any initial overshoot will cause the signal to cross the threshold before the trailing end of the rise time. If multiple loads are distributed arbitrarily along a line that is terminated only at the source, then additional settling time may be required for the reflection at the end of a line to return to all the loads along the line. If the loads are clustered near the unterminated end of a line, the total propagation delay is the one-way delay from the source to the far end of the line, plus the time for the reflection to travel from the far end and return to the load *nearest* the source (the load farthest from the far end).

If the loads are clustered near the far end of the line so that the maximum propagation delay between all of the loads is within about one-half of the 20 to 80 percent rise time of the signal, then all of the loads within the cluster will receive a waveshape which is nearly the same as at the end of the line. Although the loads farther from the end of the line will have a slower rise time than those loads very near the end of the line, there should be no significant step at the midlevel of the waveform. For example, if the propagation delay for 10 cm (about 4 in) is about 1 ns, then loads clustered within 5 cm of the end of a line will receive about the same waveshape if the signal rise time is at least 1 ns. This is an important rule for distribution of clocks or other edge sensitive signals. These signals must be distributed to multiple loads without any distortion near the threshold region which might result in double clocking or other processing errors, but load termination is not practical and not necessary.

12.2 Clock Routing

It is important to consider the routing, load placement, and number of loads for critical signals. Critical signals such as clocks often need to be buffered to a rather large number of loads, but there are often conflicting goals to maintain good signal integrity at high speed. Loads should be kept close together on each line to avoid steps or other waveshape distortion. The number of loads on each line should be minimized to reduce loading on the buffers, but the number of loads should be similar on each line to equalize loading to reduce timing skews. Parallel load terminations should be avoided to reduce power supply currents and loading on the buffers, but load terminations should be used to avoid standing waves on very long lines. Clock fanout trees should be kept to a minimum number of levels (two levels is generally recommended), but the number of buffers used in a single package should be limited to control ground bounce.[1] Reducing the number of loads and keeping the loads close together may be consistent goals because a smaller number of loads may be more convenient to locate together. Keeping the number of clock tree levels down is usually more important than reducing the number of loads, especially if the loads are high input impedance (especially for low capacitance) and located close together.

Paralleled input line receivers are usually acceptable, especially when parallel load terminations are used because the input impedance of the receiver is much higher than the termination resistor. The receivers should be in the same device so the input pins are close together. Each of the paralleled receivers should respond alike if they are on the same semiconductor chip and the input signals transition cleanly through the threshold region with adequate timing margins. However, the designer is cautioned that asynchronous inputs to *clocked* devices should be received and registered at only *one* device to avoid ambiguous operation. The placement of termination resistors is important for high speed signals with critical timing. It is preferred that the appropriate termination resistors be located internal to the integrated circuits or on the substrate of hybrid packages. Terminations should be located properly for each critical signal. Designers are cautioned to be very careful in the use of "termination packs" which include a number of terminations sharing common reference pins. It is often too tempting to use available "spare" terminations in a package not properly located with respect to the signals. Several critical signals clustered together with common package reference pins is also an invitation to crosstalk, ground bounce, or power droop.

Divided parallel terminations can be used for multiple receiver devices if a termination resistor is located near (compared to the signal rise time) each package. Even if the packages are located a significant distance apart, divided terminations may be used if all the terminations are properly located and the response is carefully analyzed for proper performance. For example, receivers may be located on two different boards if the boards are sufficiently close together and each board is terminated near the connector with double the appropriate single load termination. The source must be reasonably terminated to prevent ringing from crosstalk and minor discontinuities such as connector pin fields, line stubs, and divided load terminations. Source termination tends to relax the requirements for routing accuracy, termination placement, and resistor tolerances.

An example of a clock fanup tree is illustrated in Fig. 12.1. A typical recommendation for an appropriate source resistor R_S is about $Z_0/4$ for each side of the differential driver, with the receiver R_L at least $2.5 * Z_0$ at the end of each of the *two* lines. The example illustrates a fanout of four for each parallel termination because four loads is a typical rule of thumb for fanout of critical signals and many receivers provide four devices per package. Greater fanout may be considered if the need for higher fanout in two levels is justified. For example, the xxE116 and xxE416 ECL differential receivers provide five devices with true and complement outputs.[2]

Strict attention should be paid to high-frequency fanout buffers. All devices on each input line must be in the same package; the parallel terminations must be located very close to the loads. It is important that *both* the true and complement outputs of each critical ECL signal must be equally loaded, even if only one of the outputs is actually

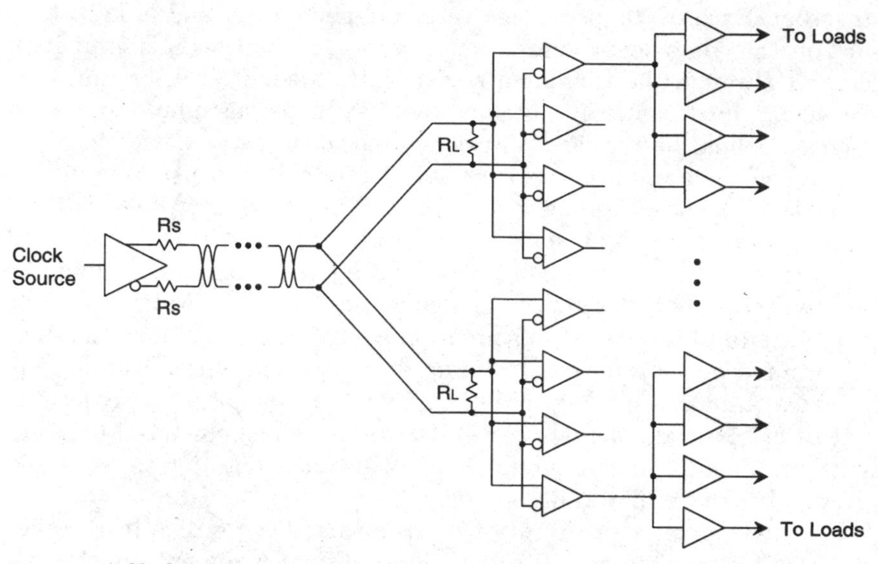

Figure 12.1 Clock fanup tree.

used. The power and ground must be distributed as solid planes on multilayer printed circuit boards or substrates with adequate high-frequency and bulk decoupling capacitors adjacent to each part on all power supply voltages.

12.3 Loop-Routed Lines

The total distance between clustered loads may be nearly doubled if the interconnection net is routed in a loop rather than a single line with two ends, and the two lines in parallel are often a better match for typical bus drivers. When series termination resistors or shield beads are used for waveshape control, the termination value is reduced to drive the two lines, so lumped loading is driven faster. If the driver can be placed in the center of a cluster of loads, the distance to the most remote load usually can be reduced.[3] Serpentine length adjustment may be inserted in a shorter arm of center-driven or loop-routed nets so that the two arms connected to the source are the same length. Length adjustment of a loop-routed line to center the load cluster at the midpoint of the loop is illustrated in Fig. 12.2. Adjacent serpentine lines should be spaced apart so that the coupling between them is not significant, so that little of the signal can "jump" across the serpentine. An open space between conductors of at least 3 times the conductor height above ground is reasonable for less than 10 percent coupling, which is essentially the same issue as for crosstalk.

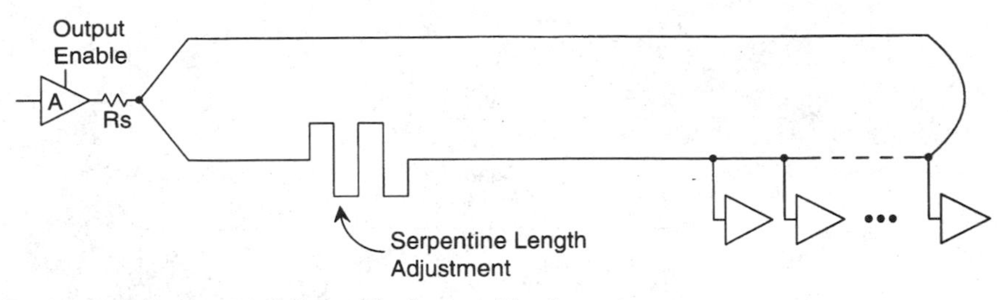

Figure 12.2 Loop-routed line with clustered loads.

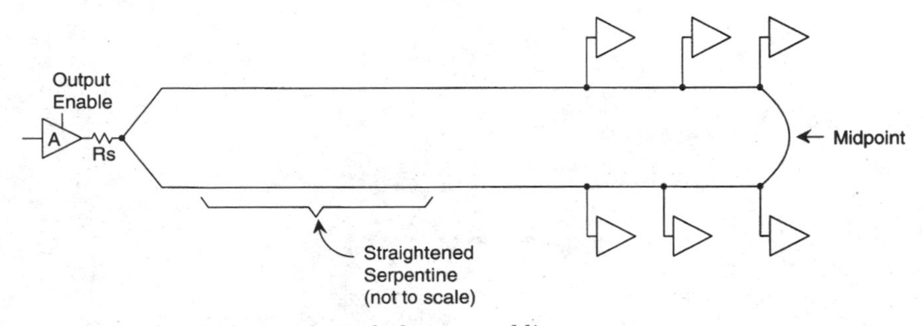

Figure 12.3 Electrical equivalent of a loop-routed line.

It is desirable (but not critical) that the loads be reasonably evenly divided between the two arms so that there are about the same number of loads on either side of the midpoint of the loads, which should also be the point farthest from the source. The midpoint of the loads is the point equally *distant* from the loads nearest the source on each arm, not necessarily the point at which there are an equal number of loads on either arm. However, balancing of the number of loads on each arm may help to equalize any waveshape distortion due to lumped loads, although this should not be significant for loads very close to the midpoint. Figure 12.3 is an illustration of the electrical equivalent of the loop-routed line. The serpentine is straightened to show that the loads are clustered about the midpoint.

About one-fourth of the rise time propagation delay in either direction from the midpoint does not result in any significant difference in waveshape, especially if the signal threshold is reasonably centered between the static signal levels. The signal transitions from each arm overlap near the midpoint and combine by superposition. The overlap of signals from the two arms of a loop connection is illustrated in Fig. 12.4. The signal traveling clockwise is illustrated on the "inner track"; the equivalent signal traveling counterclockwise is illustrated on the "outer track."

The signals combine by superposition as they meet at the midpoint. The combining of the signal levels is illustrated in Fig. 12.5. The inci-

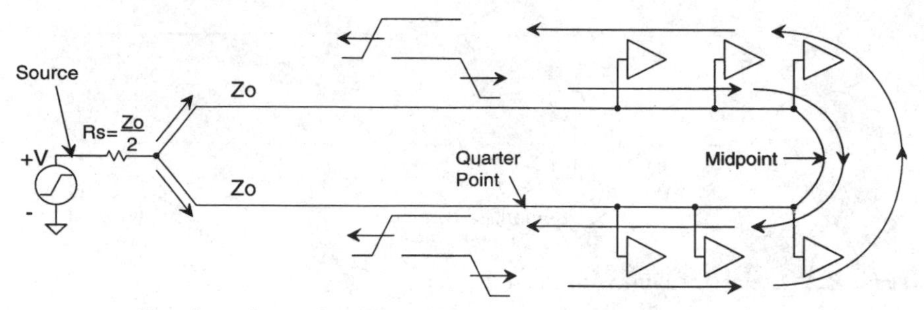

Figure 12.4 Signal overlap on loop-routed lines.

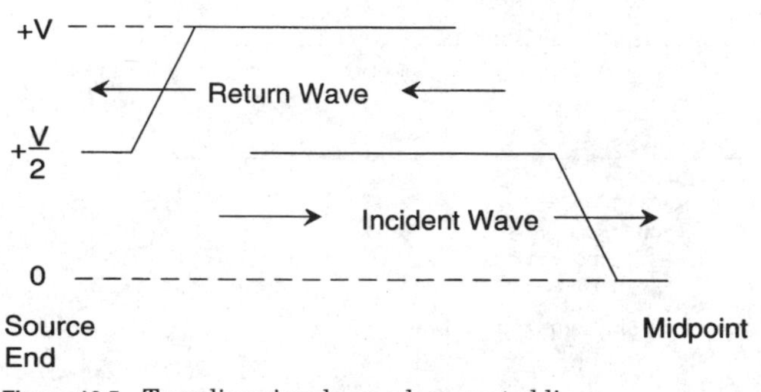

Figure 12.5 Traveling signals on a loop-routed line.

dent wave of one-half amplitude is launched from the source onto each arm of the loop. The return wave of the same amplitude returns to the source on the other arm to establish the final steady-state level.

Figure 12.6 illustrates the waveforms at the source, the quarter point around the loop (indicated on Fig. 12.4), the 3/8 point (near the midpoint), and the midpoint of the loop where the signals traveling around the loop meet to combine. The signal at the source rises to one-half of the steady-state level, which is characteristic of exact source termination. The signal rise time is greater than twice the round-trip propagation around the loop, so there is no overlap at the quarter point. However, at the 3/8 point (within 1/8 of the midpoint), there is overlap and no intermediate step for this example. The total rise time is slower than the source no-load voltage, or at the midpoint. However, the overlap near the midpoint results in a voltage slew rate in the threshold region between the static levels that is equivalent to both the source no-load voltage and the signal at the midpoint. The midpoint has complete overlap, so the rise time is minimum at that point.

As the individual signals continue around the loop, the combination brings the signal level up to the steady-state static level. There are no reflections at the source if the source impedance is exactly $Z_0/2$.

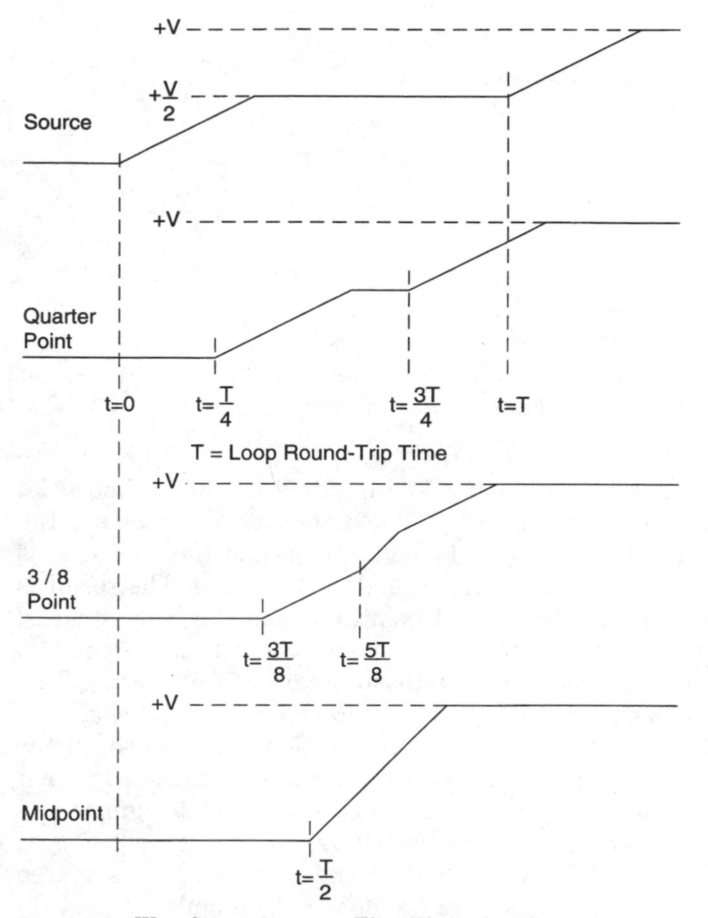

Figure 12.6 Waveforms at source, $T/4$, $3T/8$, and midpoint.

However, practical lines should use a slightly underdamped source termination to increase signal levels to compensate for lumped loading and losses, with some overshoot. The diagram is drawn to illustrate a single-ended signal, but the concept applies equally well to differential lines.

Loop routing of critical signals can present some difficulties in automatic signal routing. If there are only a few critical signals, each with a limited number of loads, the routing may be done manually. The final layout routing must be carefully checked to assure that it is proper and that any automated processing has not deleted one link of the loop as redundant duplication. Automatic routers can be controlled to define a single string with two ends and no branches (this is sometimes called ECL rules), so that a driver can be placed at one end and a load termination at the other end.

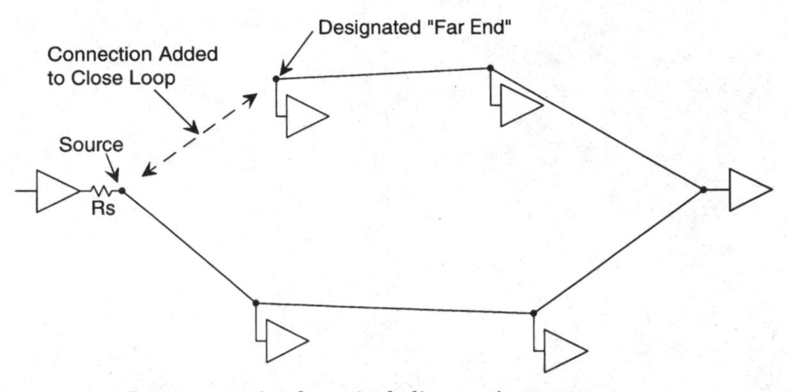

Figure 12.7 Loop connection from single-line routing.

One potential method for using a single string router for initial routing is illustrated in Fig. 12.7. The approach is to assign the source as one end in the normal fashion but assign the load that is *nearest* to the source as the far (load) end of the string. The result is checked for reasonable routing and balance. Then the loop is closed manually with a connection from the source to the load assigned as the end, inserting a serpentine length adjustment if necessary. This approach is equivalent to closing lattice grids for power and clock distribution. A slightly different naming convention may be used if the router does not accept the additional connection with the same net name. The designer checking layout rules may need to ignore any warnings that different nets are connected together.

Figure 12.8 illustrates conventional in-line routing of loads spaced at regular intervals with the source at one end. The multiple loads in this illustration represent significant loading for a single line. The following figures illustrate techniques for breaking this line into two parts. The loop routing source termination resistor is reduced to a value appropriate for $Z_0/2$. This launches more energy into the two lines to increase bandwidth and charge lumped loading more quickly.

Figure 12.9 illustrates that the line can be converted to a loop line by dividing the line into two separate lines. Each separate line is connected to approximately equal loading, with the two ends connected together. If the loads are spaced at approximately equal intervals, the loads may be assigned alternately to the separate lines, as shown. The loading on each arm of the line is reduced by about one-half, and the distance to the midpoint is unchanged, so the loaded propagation delay should be improved.

Figure 12.10 illustrates that a line can be converted to a broken loop line by locating the driver in the middle of the loads and providing connections in both directions. This does not reduce the line load-

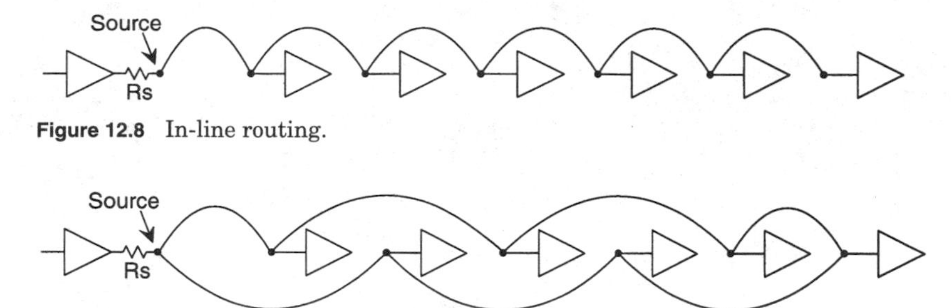

Figure 12.8 In-line routing.

Figure 12.9 Loop routing of in-line loads.

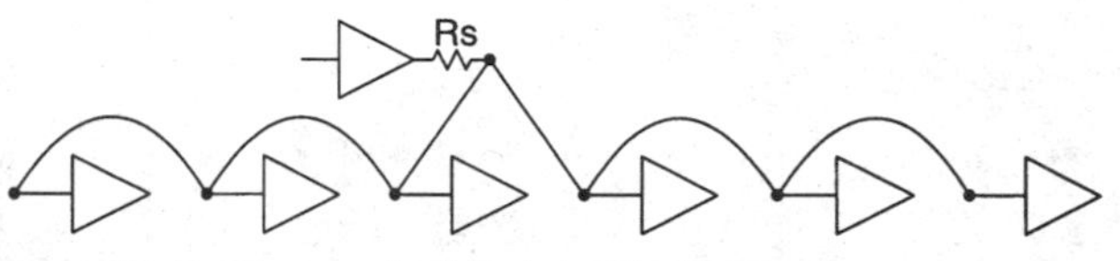

Figure 12.10 Source at midpoint of in-line loads.

ing, but the ends of the line are closer to the driver, so the propagation delay should be lower than the original in-line routing.

12.3.1 Single-line equivalence

A line connected in a loop can be considered as equivalent to a single line. The loop line has two arms of Z_0 impedance that meet at some midpoint around the loop that is equally distant from the source. The impedance seen by the source is $Z_0/2$. One of the advantages of the loop routing is that the lower line impedance is usually a better impedance match for most high speed line drivers. If series impedance is used to match the source impedance to the line, reducing the source impedance means that lumped capacitance loading will be charged faster. Since only about one-half of the loads are connected on each arm, the ratio of lumped load capacitance to intrinsic line capacitance is reduced, so the K_L loading factor is lower, which also reduces the propagation delay of the loaded line.

Another potential advantage of loop line connections depends on the layout of the multiple loads. As was illustrated above, a single line with multiple loads in a reasonably straight line can be converted to a loop line; the distance from the source to the midpoint of the loop is no longer than for the single line. Therefore, loads that are naturally located in a straight line can normally be converted to a loop line with no increase in line length to the farthest load. However, more arbitrarily located loads often can be connected in a loop that reduces the distance from the source to the farthest load. For example, loads

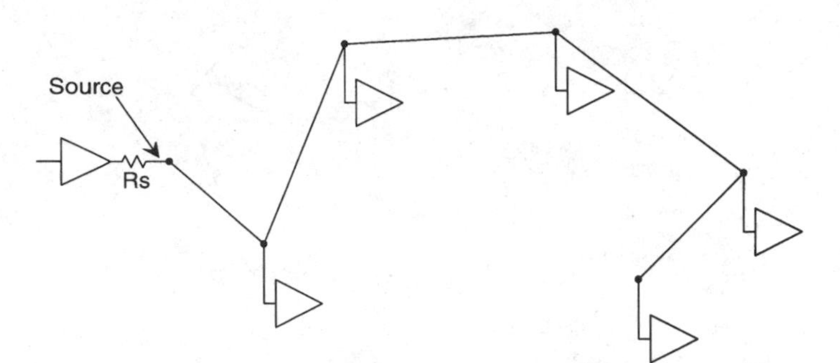

Figure 12.11 Single-line routing.

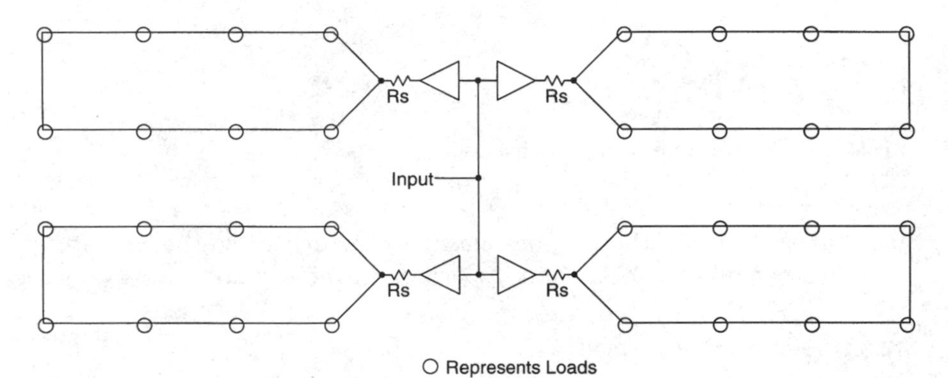

Figure 12.12 Example of multiple loops.

that are naturally located along an elliptical or rectangular perimeter can be connected with less distance from the source to the farthest load with loop routing, compared to a single line that zigzags between sides. Single-line routing of the loop that was shown above in Fig. 12.7 is illustrated in Fig. 12.11. The zigzag length to the end of the single line is greater than the length to the midpoint of the loop.

When a large number of critical loads must be driven (for example, clock distribution), the loads may grouped into multiple loops. This is illustrated in Fig. 12.12, with the loads divided into a loop in each of four quadrants.

Figure 12.13 illustrates a loop-routed network with series-terminated driver and/or receiver units that can tap into an established loop. The connection may be made to or removed from the loop without breaking the loop, as with a coax "tee" connector. New loop segments can be inserted into the loop as illustrated in Fig. 12.14. The source terminations are not affected by changes to the loop, although the maximum length of the loop limits the data rate on the line. The midpoint of the loop is always at the point equally distant from the active driver.

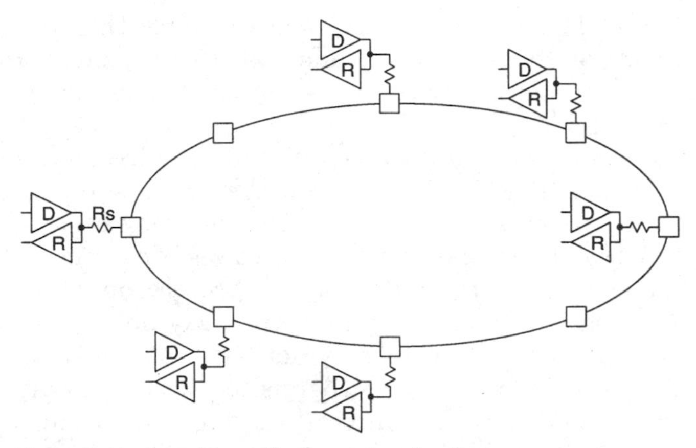

Figure 12.13 Loop line with source termination.

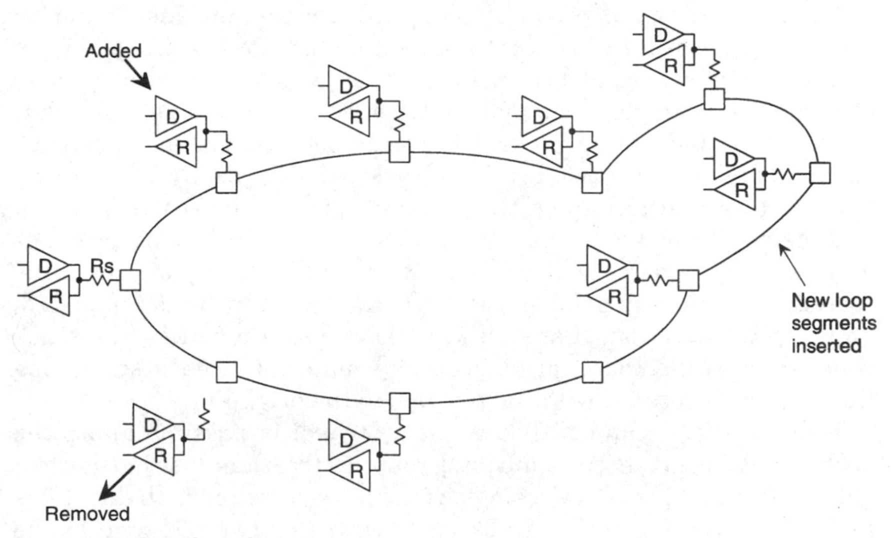

Figure 12.14 Loop line alterations.

12.3.2 Connecting looped lines

The determination of the shortest loop routing to connect multiple loads is an application of a classical mathematical problem called *The Traveling Salesperson Problem.*[4,5] The traditional definition of the problem is that a traveling salesperson has n stops to call on before returning to the origination; the problem is to find the route which will visit all n stops exactly once with the minimum distance traveled. In other words, the problem is to determine the sequence of n stops that minimizes the length of the route. This is similar to a delivery truck which has n delivery stops or a bus which has n stops to make on a school bus route. These traveling problems assume that the route begins and ends at the same point and that the sequence of stops is otherwise unimportant, except for the objective to minimize the travel length of the route.

The general sequencing problem is very difficult to solve as the number of stops becomes large because the number of possibilities grows as $n!$. When n is reasonably small, the solution may become obvious by comparison of a few reasonable alternatives. The general problem may be satisfied with a route that may be traveled in either direction. Of course, real vehicle routing problems may have other restrictions (e.g., the truck or bus yard may not begin at the same location as the loading point which must be visited first, one-way streets may restrict travel direction, or aircraft routing may consider prevailing winds), and doubling back on part of a physical transportation route is usually permissible.

An electrical signal traveling along a transmission line should be routed past each load in a continuous line, but electrical signals can divide and/or reflect at branches or stubs which a physical vehicle does not. An example of an analogy to reflections on a single line that is unterminated at the far end is to consider that the same truck route is divided into two separate tasks: delivery only on the outbound travel and pickup on the inbound return. The truck is loaded with deliveries at the origin (typically a sorting center) and proceeds along the shortest route to make deliveries at each stop on the route. At the last stop at the end of the route, the last delivery is completed, and any pickups from that stop are loaded into the truck. The truck then reverses the route, picking up any material at each stop along the route as it returns with the pickups to the origin.

This analogy assumes that two-way travel is possible along the route, so that the shortest inbound route is the same as the shortest outbound route. Therefore, a single route is traversed in both directions to return to the origin. The truck visits each stop twice, except for the turnaround at the last stop. The time between delivery and pickup at each stop is the round-trip travel time from the end of the route.

The analogy to electrical signals is that outbound delivery is like the first incident wave; the inbound pickup is like the return wave reflected back to the source. Only the end of the line "sees" the incident and reflected wave simultaneously; all other points see the incident wave first, followed by the reflected wave. These two signals are separated by a time which is proportional to the distance from the end of the line. Of course, if the separation time between the incident and reflected signals is small compared to the signal rise time, the central portion of the combined signal is the sum of the trailing portion of the incident wave and the leading portion of the reflected wave. When the round-trip propagation delay time from the end of the line is longer than the signal rise time, the separation between the incident and reflected signals results in a "step" in the signal level before the reflected signal begins.

Although loop routing is described as analogous to one traveling salesperson who travels the loop route in only one direction, the electrical signals travel in *both* directions simultaneously on a loop-routed line. The delivery truck analogy to loop routing from a source connected to both sides of the loop is that *two* trucks are loaded at the origin and proceed in opposite directions around the loop. Each truck makes deliveries only along the outbound portion of the route. The two trucks have completed their deliveries to their portion of the route when they meet at the point farthest from the source (ignoring any idle time at stops). It may be observed that for typical routes, two trucks can finish their deliveries faster than a single truck could. When the two trucks meet, each continues in the same direction, stopping to pick up any packages at each stop on the other portion of the route as they return to the origin. Note that the meeting point where the trucks pass is not necessarily at a stop. Since each truck travels around the same route and stops once at each stop, the two trucks return to the origin at the same time.

It is interesting to consider an alternate analogy which is indistinguishable from the two trucks that travel completely around the route in opposite directions. Suppose that just when the two trucks meet at the point farthest from the source, they stop and turn around; so that each retraces inbound the *same* portion of the route they traveled outbound. If the two trucks are indistinguishable and return to the origin at the same time, the results are the same as if each truck continued around the loop instead of reversing direction when they meet.

In practice, of course, each truck would begin pickups at the predetermined last stop and return to the origin in the reverse direction. If each truck arrived at their last stop at nearly the same time, they will also return to the origin at nearly the same time. This alternate avoids travel on the portion of the route between the two last stops. However, the route must be evenly divided between the two parts of the route, and adding or deleting stops on one part may require some reassignment of the stops at the ends so that the two trucks return to the origin at nearly the same time.

12.4 Broken Loop Routing

The electrical equivalent of the two delivery trucks that reverse their direction halfway around the loop is that the loop does not need to be connected at the midpoint. Performance is indistinguishable if the loop is broken at exactly the midpoint of the loop. Again, if the connection is broken reasonably close to the midpoint (within about one-half of the 20 to 80 percent rise time is usually reasonable), the reflections arrive back at the source nearly at the same time. When the loads are connected in a complete loop, the two signals naturally meet at the farthest point and return to the origin at the same time. The important concept for loop routing is not that the loads *must* be connected in a complete loop but that the source may be connected in the center of two arms of nearly equal length. Clusters or lines of loads that are driven from the center may be referred to as *center-fed* or *center-driven* routing.[3] Of course, it is not necessary that the two ends of a broken loop be located closely together, so that the source may be located near the midpoint of the loads.

The practical difficulty of a broken loop routing is determining the location of the midpoint of the loop, or equivalently, placing the source at the midpoint of a string of load points. However, if the source can be placed reasonably close to the midpoint of the loads, the distance from the source to the two ends should be less than if the source is at one end.

12.4.1 Loop source termination

It should be fairly obvious that if a source is connected to two lines of Z_0 impedance each, the source "sees" a line impedance of $Z_0/2$, so the exact source termination is $Z_0/2$. However, the analysis of the response to signals traveling toward a source also connected to another line may be a little more subtle. The following paragraphs discuss the response to signals traveling toward a source-terminated driver in a loop line.

If the source is terminated for an exact match to two lines, the source impedance matches the $Z_0/2$ line impedance. Therefore, if a signal travels toward the source from one side only, that signal "sees" an impedance of $Z_0/3$ at the junction of the source and the other line. This is the parallel impedance of the $Z_0/2$ of the source and the Z_0 of the other line (this may be more obvious if $Z_0/2$ is considered as one Z_0 in parallel with another Z_0, so the third Z_0 in parallel results in $Z_0/3$). This is shown in Fig. 12.15, where the only returning signal arrives from the right. The driver is shown with a switch closed to indicate that it is the active source.

The condition just as the wave arrives at the active driver is illustrated in Fig. 12.16. The active driver across the line is an impedance mismatch. This mismatch causes some of the signal to be reflected back to the right, but the difference continues to travel to the left.

12.4.2 Reflected and transmitted waves

The reflection factor for a $Z_0/3$ mismatch is -0.5, so there is a negative one-half reflection (to the right) and the remaining one-half of the

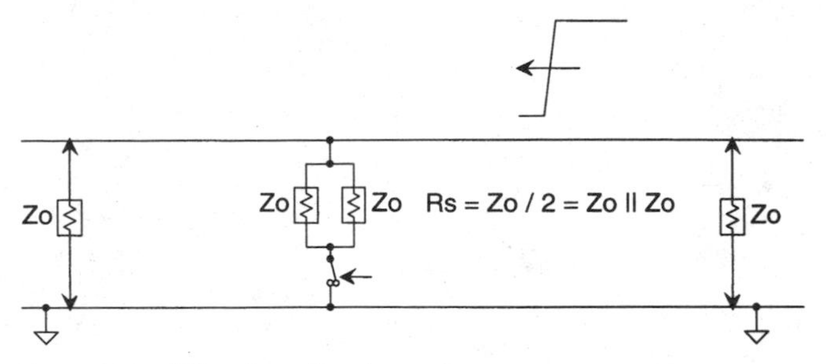

Figure 12.15 Wave traveling toward source.

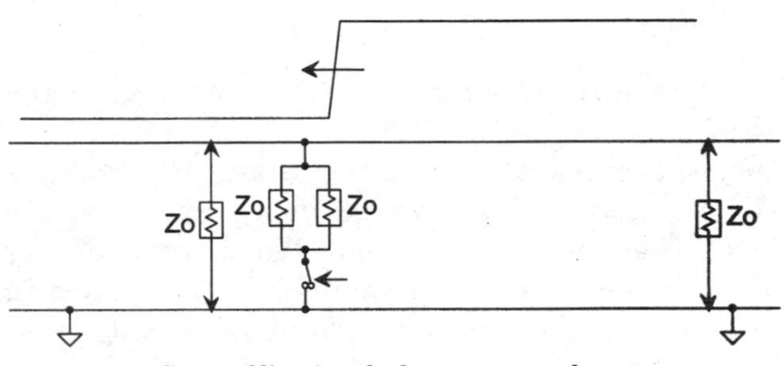

Figure 12.16 State of line just before wave reaches source.

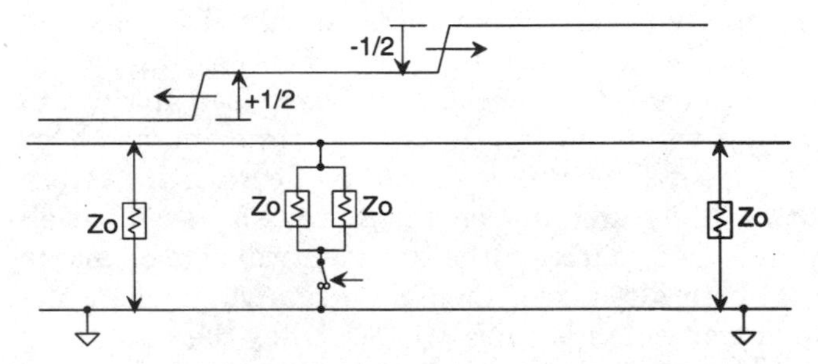

Figure 12.17 Incident and reflected waves travel away from source.

signal continues traveling along the other line (to the left). In other words, a signal that returns along only one arm to a perfectly terminated, loop-connected source will be reduced to one-half of the amplitude of the traveling wave. The signal levels as the waves travel away from the driver are illustrated in Fig. 12.17.

It should be noted that the amplitude of the traveling wave is *not* necessarily the same as the signal level. For this example of a loop-connected source terminated with exactly $Z_0/2$, the initial incident wave signal that travels from the source in both directions has an amplitude of one-half of the steady-state signal level. The two traveling incident waves combine for a total signal level equal to the steady-state signal level when they meet at the midpoint, but the

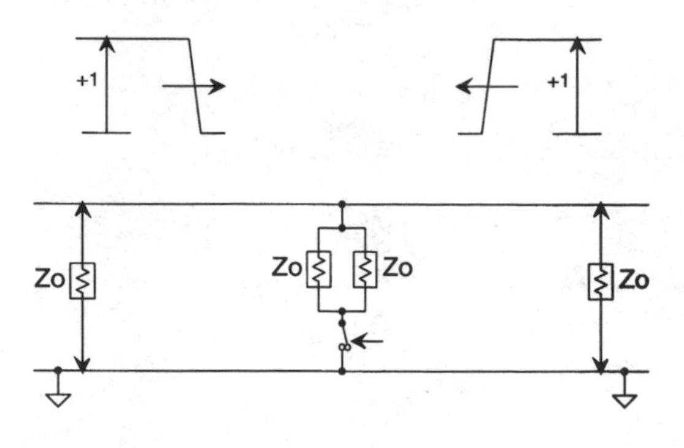

Figure 12.18 Two waves traveling toward a source.

incident waves that continue to travel are still only one-half of the steady-state signal level.

When one of the incident waves returns to the source, the negative reflection is one-half of this traveling wave, or one-fourth of the steady-state signal level. The net resulting signal level is three-fourths of the steady-state level. However, if the remaining signal traveling in the other direction also adds another one-fourth of the steady-state level, the final result is exactly the steady-state signal level.

The analysis and illustrations above show the results from a signal arriving at a loop-connected source from one side only. However, signals return to a loop-connected source from *both* sides simultaneously. Therefore, the signals from each side respond as described above, and the results of each of the two signals can be superimposed (or summed, if the response is linear, as it should be for resistive terminations at the source). Illustrated in Fig. 12.18 are two waves traveling toward a loop-connected source. If the two identical signals arrive simultaneously from both directions, then the negative one-half signal reflection from one direction is exactly canceled by the remaining one-half signal that flows through from the other direction.

Figure 12.19 illustrates the result when two identical waves arrive

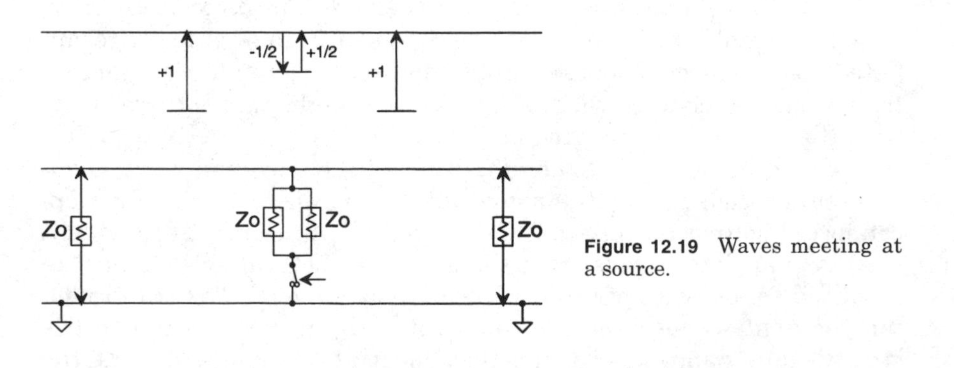

Figure 12.19 Waves meeting at a source.

simultaneously at a loop-connected source. To the extent that the reflections and remaining signal exactly cancel from both directions, the net result is no signal change in both directions. The steady-state static signal level is reached when the signals arrive from both sides simultaneously. Therefore, a source terminated in exactly $Z_0/2$ will exactly terminate two signals which arrive from both sides simultaneously. This also means that a loop-routed line with impedance Z_0 is exactly equivalent to a single line with impedance $Z_0/2$ and a length equal to the length from the source to the midpoint on the loop.

12.5 Underdamped Loop Terminations

When the source impedance of a loop-routed line is reduced to less than $Z_0/2$, this will increase the incident wave signal amplitude to more than one-half of the steady-state level. Therefore, when the signals from each direction combine when they reach the midpoint around the loop, the combined result will be overshoot to more than the steady-state level. When the larger incident wave signals continue around the loop and return to the source, the reflection factor at the source is more negative than -0.5. Therefore, there is more than one-half negative reflection and less than one-half of the incident wave signal amplitude remaining to continue past the source. Figure 12.20 indicates that the difference of the reflection and the incident waves travels away from the source after two waves meet at the loop connected source. These exactly cancel for the exactly terminated case.

However, if the negative reflection from one direction is not completely canceled by the remaining signal that arrives from the other direction, there is some negative reflection that travels around the loop again. The negative reflection results in some undershoot, which is line ringing as in an ordinary underdamped line. Again, the loop-connected source responds in the same manner as a single line with the impedance values reduced by a factor of 2. Significantly under-

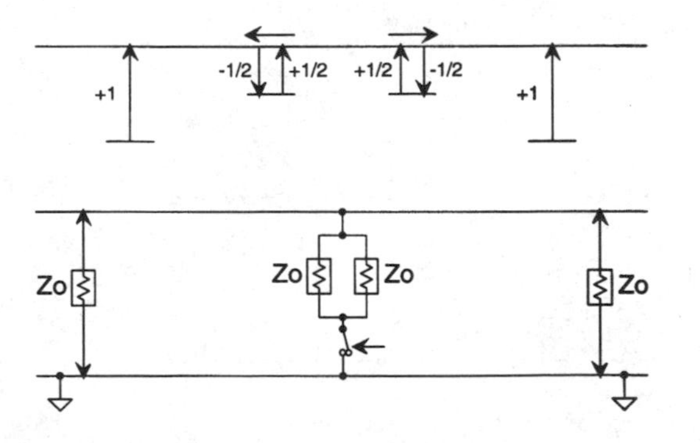

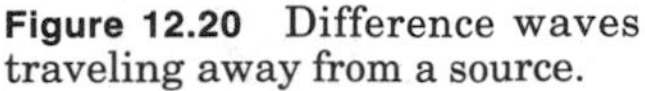
Figure 12.20 Difference waves traveling away from a source.

damped is usually recommended to increase initial signal amplitude and the capability to drive lumped loads, and to compensate for high-frequency losses such as skin-effect distortion.

If the source termination is one-half of an exact match to the loop impedance, or $Z_0/4$, then the reflection factor is 0.333, which results in an initial signal excursion of about 67 percent of the final value (or an initial undershoot of about 33 percent), about 33 percent of first incident wave overshoot, and only about 10 percent of second incident wave undershoot. These levels are usually very acceptable waveforms for signal distribution to multiple loads, especially for lines operated at low enough frequency so that signals settle to steady levels between each signal transition.

12.6 Off-Center Reflections

When a source is connected to two lines in a broken loop, but the two arms are not nearly the same length, the reflections from the ends of the two arms do not arrive back at the source at the same time. If the difference in the round-trip delay for each arm is greater than the rise time, the reflection from the shorter arm returns to the source before the reflection from the longer arm. This is illustrated in Fig. 12.21, where the first wave travels toward the source from the right.

The first signal to return to the source will result in a negative reflection, as was shown above in Fig. 12.17 for a single wave. The negative reflection from the first wave is not canceled until the later signal returns from the far end of the longer arm. The signal levels just after the first wave has passed the source are illustrated in Fig. 12.22. The reflection and the remaining incident wave from the first wave are traveling away from the source; the later wave is still traveling toward the source from the left.

Figure 12.23 illustrates the results just after the two incident waves have passed each other on the left side of the source. The later full-

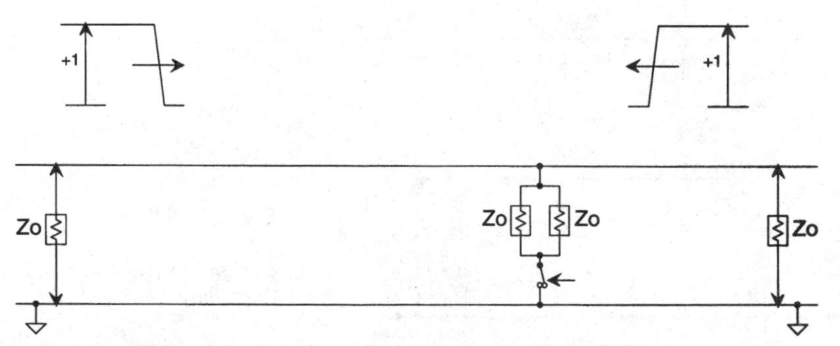

Figure 12.21 Waves traveling toward a source not at the midpoint.

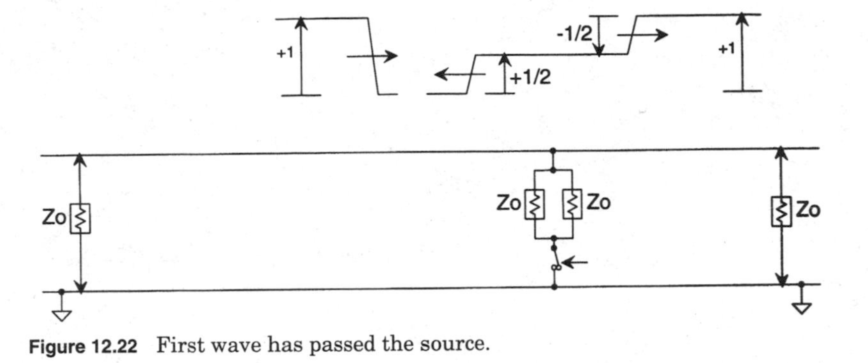

Figure 12.22 First wave has passed the source.

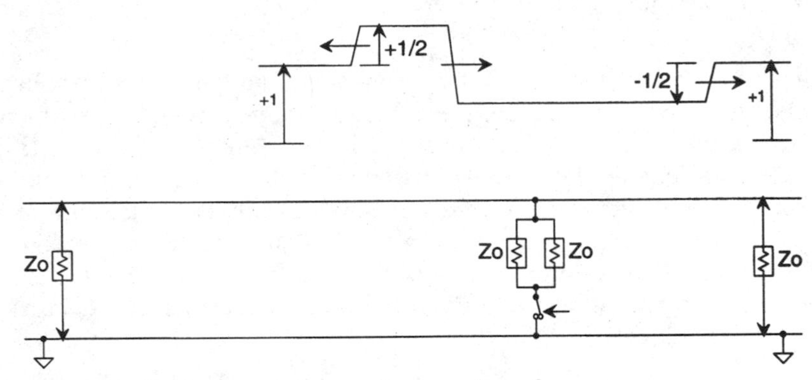

Figure 12.23 First and second waves have overlapped.

amplitude incident wave adds to the remaining incident wave of the earlier wave to overshoot the steady-state level. At this time, there is overshoot traveling to the left, toward the end of the longer arm, as well as the earlier undershoot from the early reflection traveling toward the end of the shorter arm.

Figure 12.24 illustrates the results just after the later incident wave has passed the source. Now the incident and reflected waves finally cancel at the source, and the steady-state levels travel away from the source. However, there is an overshoot pulse that travels toward the end of the longer arm. There is also the corresponding undershoot pulse that travels toward the end of the shorter arm. The width of these overshoot and undershoot pulses is the difference of the propagation delay between the longer and shorter arms.

The earlier negative reflection (undershoot) will be doubled by the reflection at the end of the shorter arm. Therefore, there may be a significant signal disturbance unless the lengths of the two arms are nearly the same. Reducing the source termination resistance below $Z_0/2$ will increase the initial incident wave amplitude but will also

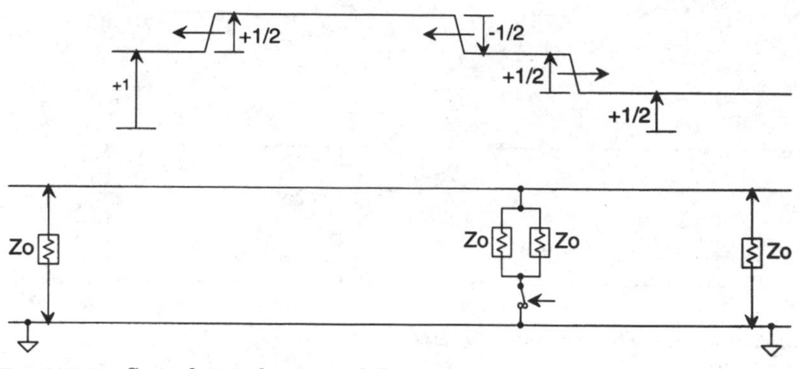

Figure 12.24 Second wave has passed the source.

increase the undershoot caused by the negative reflection on the shorter arm. These waveforms on a broken loop line with unequal arms illustrate what has been observed on a party line bus with series terminations at the sources but open at the two ends. The signals from active drivers near either end settled quickly, as did signals from active drivers located near the middle of the line. However, signals from active drivers neither close to the middle nor near the ends caused some ringing on the line that took more time to settle toward the steady-state levels.

12.7 Party-Line Bus Routing Summary

These examples indicate that a reasonably terminated source may be at the end of a line, as in conventional routing, or nearly at the midpoint of a broken loop, or anywhere in a closed loop. The signal levels along the entire line should settle within the round-trip propagation delay of the line with reasonable source termination. However, if a source is located significantly off the midpoint of a broken loop, there is significant signal distortion and additional time will be required for the signal to settle to valid levels. This may be a significant consideration for "party-line" bus configurations where sources are at various locations, and possibly subject to change as users are added or removed from the line.

A traditional configuration for a high speed, multiple user bus is to locate parallel load terminations at both ends, so reflections at the ends are adequately controlled. However, the parallel termination at both ends results in high static current loading, and the terminations may have to be moved for any bus reconfiguration that extends either end of the bus.

A source at any location along the "party-line" bus sees a load impedance of about $Z_0/2$. An internal source sees the transmission-line impedance Z_0 in both directions. If there is an active source at either of the ends, that active source sees a line impedance of Z_0 in one direction, in parallel with the load termination at that end. If the ends are terminated with greater than Z_0 to reduce the line loading and provide underdamped response, the signal response will vary with the position of the active source. The deviation of the initial signal level from the final static signal level has been defined as the initial undershoot. A source-terminated party-line bus may be connected in a complete loop, so that any active source is always in the "center" of the loop.

A closed loop functions the same as a single line with a single source at one end. The source termination is reduced to 50 percent of the termination appropriate for a single line. The equivalent of the one-way length is about one-half of the length around the loop. The round-trip line delay is the propagation delay once around the loop. There are no physical "ends" to the loop line. One of the two equivalent ends is at whichever source is active; the far end is at a point on the loop that is equidistant from that active source, or about halfway around the loop. No parallel load terminations are used, and the bus can be reconfigured to maintain a complete loop by connecting or removing users. All units in a loop are connected in both directions. Additional units may tap onto the bus without disturbing the connections.

For the general case of loads distributed arbitrarily on a long ideal line that is exactly terminated only at the source, the time required to settle to 100 percent of the final value is the time for the reflection to return to the source. The time for the traveling wave to return to the source is the two-way delay to the farthest end and back for a single line, or the one-way delay around the loop for a loop-connected line. Loads that are close to the far end of a single line receive a nearly full amplitude wave because the incident and reflected waves combine at the far end. Similarly, the two incident waves from each arm of a loop-connected line meet and combine at the midpoint and continue to travel around the loop as return waves.

12.8 References

1. James E. Buchanan, *Signal and Power Integrity in Digital Systems: TTL, CMOS & BiCMOS,* McGraw-Hill, New York, 1996, chap. 7.
2. *Synergy Product Data Book,* Synergy Semiconductor Corp., Santa Clara, Calif., 1992.
3. James E. Buchanan, *Signal and Power Integrity in Digital Systems: TTL, CMOS & BiCMOS,* McGraw-Hill, New York, 1996, chap. 6.
4. C. H. Chen (ed.), *Computer Engineering Handbook,* McGraw-Hill, New York, 1992, p. 18.22.
5. E. W. Felten, S. Karlin, and S. Otto, "The Traveling Salesman Problem on a Hypercube, MIMD Computer," *Caltech Report* C3P-093b (St. Charles, Ill.), *Proceedings of 1985 International Conference on Parallel Processing,* 1985.

Chapter

13

ECL Transmission Lines and Terminations

This chapter considers the relationship of driver circuits for high speed transmission lines with various line configurations and the terminations of the lines. This chapter emphasizes the characteristics of high speed emitter coupled logic (ECL) interfaces. ECL is typically the technology of choice for very high speed logic functions and wire interconnections at rates which can be significantly greater than practical with TTL, CMOS, or BiCMOS technology.[1] The characteristics of parallel Schottky diodes for line terminations are described.

13.1 ECL Comparisons

ECL signals are much lower amplitude signal levels than the typical TTL levels powered by 5 V, so the transmission-line currents are more reasonable. Almost all of the ECL outputs are capable of driving transmission lines, with a notable exception of the V_{BB} reference pin available on most differential receivers. Schottky TTL drivers have been observed to cease switching at about 10 to 25 MHz when loaded with only a few meters of an open transmission line. Although newer CMOS line drivers such as the 26C31 (available from National Semiconductor Corporation) may continue to function at data rates above 25 Mbits/s, the power dissipation of the driver rises very rapidly above about 10 Mbits/s.[2] Although ECL draws much higher dc current than TTL or CMOS, the internal current switching results in a much lower increase in current for higher data rates. This is partly a result of the lower ECL signal levels, and partly due to the characteristic of TTL and CMOS inverters that some transient current flows through both of the active pull-up and pull-down transistor switches every time a signal node changes logic state. This current is called *flow-through* or *simultaneous conduction* current. ECL output drivers operate in a linear emitter follower configuration, so the output rise time can be controlled by internal circuit parameters.

Emitter coupled logic (ECL) was originally introduced in the 1960s (Ref. 3, p. vi) as a high speed integrated circuit logic family based on silicon *npn* bipolar transistor technology. Early logic devices were capable of operation at over 10 MHz. ECL signal levels are low compared to TTL and CMOS, with typical peak-to-peak logic swings of less than 1 V, or about the forward voltage of one silicon diode. The logic transistors switch between cutoff and current-limited states to avoid saturation, which would otherwise result in reverse recovery problems. By about 1970, devices were introduced which were capable of operation at well above 100 MHz in a transmission-line environment. Since about 1980, more general-purpose families have been introduced which provide higher levels of integration with controlled rise time, as well as reduced power consumption and reduced sensitivity to variations in temperature, power-supply voltage, and input signal levels. ECL interfaces are the traditional technology for wire transmission-line interfaces operating above about 10 MHz. With device data rate capability at about 1 GHz by the mid-1990s, the data rate for long wire lines tends to be limited primarily by the bandwidth of the line itself.

13.1.1 Transmission-line limitations

The designer of high speed transmission lines should consider the limitations of the transmission line as well as the characteristics of the drivers and receivers. The performance of wire cables designed for transmission of high-frequency digital signals may vary significantly. However, the following is a rough order of magnitude guide for the approximate upper length limit for transmission of variable-frequency clocks or arbitrary NRZ (non-return-to-zero) data over twisted pair wires of 24 to 30 AWG equivalent wire size:

- Less than a few meters for up to 1 GHz.
- 10 m of cable length for up to 100 MHz.
- 100 m of cable length for up to 10 MHz.
- The practical limit for wire lines below 10 MHz remains at near 100 m because the total one-way dc resistance of 30 AWG wire is

about 100 Ω per 1000 ft, for nearly 70 Ω of two-way loop resistance for 100 m.

If the total line resistance is comparable to the typical line impedance of about 100 Ω for these wire lines, there is a significant loss of signal level even at low frequency. However, resistance approximately doubles for each increase in 3 AWG numbers, because higher gage numbers refer to the smaller die sizes through which the wire is drawn. The resistance ratio is also a little over a factor of 10 for each increase in 10 gage numbers. Therefore, if the wire size is increased by a reduction of 10 gage numbers from 30 to 20 AWG, the wire length can be increased by a factor of 10 for the same resistance. However, this does *not* mean that just increasing wire size will reduce line losses, because the ratio of line resistance to dynamic impedance determines losses. From Eq. (6.6), the loss $a = r/2Z_0$, so one-half of the signal is lost when the total loop resistance is equal to the Z_0 line impedance. Therefore, the line-to-line spacing must be increased in the same proportion as the increase in wire diameter to maintain Z_0. High-frequency components of a waveform will not propagate between conductors spaced far apart. If the wire size is increased by a factor of 10, then the line-to-line spacing must also increase by a factor of 10 to maintain the same Z_0. This would not be practical for most digital cables, so the length of wire lines remains limited by line losses even at low data rates. Longer runs of line will require that receivers be placed in the line so the signals are redriven at full signal levels before the signal is degraded below the levels that can be accurately received.

For a historical perspective of transcontinental telephone lines in the United States, typical open wire telephone line pairs with a 12-in (30-cm) line-to-line spacing on crossarms on poles and 10 AWG wire have a line-to-line impedance of about 600 Ω above 1 kHz and a resistance of about 1 Ω per 1000 ft. This is about 10 Ω two-way loop resistance per mile, or about 6 Ω of loop resistance per 1000 m.[4] Therefore, an open telephone line length of about 100 km has a dc resistance about equal to the line impedance. The dc signal output of a line with line resistance equal to the dynamic impedance is about 50 percent of the input, or a 6 dB line loss. If there is a 50 percent loss of signal for each 100 km, then an open line length of about 320 km (about 200 mi) would have only about a 10 percent output signal, or a 20 dB loss at audio frequency or carrier frequencies below 1 MHz.

Most digital differential receivers can recover signals at 50 percent of the normal signal level if the signal-to-noise ratio is reasonable. Since each increase in 10 gage numbers increases resistance by a factor of about 10, reducing the telephone wire size from 10 to 20 AWG would reduce the maximum length from 100 to 10 km for the same 600 Ω dynamic impedance. Reducing the wire size from 20 to 30 AWG reduces the maximum length to 1 km, or 1000 m. Reducing the dynamic impedance an order of magnitude, from 600 Ω to in the range of 60 Ω, also reduces the maximum length another order of magnitude to a maximum of about 100 m.

Skin effect has little effect on line resistance below about 10 MHz for 30 AWG wire because the skin depth penetration is comparable to the wire radius.[5] Increasing wire size has more effect on dc resistance than high-frequency losses because skin effect for larger wires becomes significant at lower frequency. This means that high-frequency losses are not proportional to wire size. However, it should be noted that although losses may be acceptable for recovery of data at these rates, the effect on the high-frequency components of digital signals may significantly alter the waveshape of fast rise time signals.

13.1.2 ECL characteristics

ECL provides several characteristics in addition to high switching speed that make it suited to digital transmission lines, including the following:

1. The very low output impedance of the bipolar *npn* transistor emitter follower output transistors to drive the high-transmission-line currents.
2. Low signal swing to limit the power loss in charging and discharging lumped capacitance loading.
3. Simultaneous true and complement outputs to drive balanced differential lines.
4. Nearly constant power-supply current as a function of both data rate and logic state when true and complement outputs are equally loaded.
5. Open emitter outputs for transmission-line terminations and to allow multiple drivers on a "party line" bus.
6. Low pin-to-pin skew between outputs to control timing skew between multiple output signals.

13.2 ECL Interfaces

This section describes and illustrates examples of various interface configurations for high speed ECL interfaces. ECL is typically the most appropriate choice for wire transmission-line interfaces for digital signals at data rates over about 10 MHz, depending on the particular applications. ECL offers the characteristics of tight timing accuracy, well-balanced differential signals, low-level signals with low generated noise, and power consumption that remains nearly constant with data rates.[6] However, the relatively high dc power consumption and the need for additional components usually do not justify ECL interfaces at lower rates for many ordinary applications.

There are special applications that may justify the use of ECL interfaces even though most of the logic processing within a physically small area is performed by lower speed TTL or CMOS logic. Examples of ECL interfaces between TTL or CMOS logic include clock rates or timing accuracy requirements beyond the capability of TTL or CMOS over long transmission lines, or when a large number of parallel signals can be converted to ECL and time-multiplexed to operate at higher rates with fewer wires.

Figure 13.1 illustrates that a single ECL output stage is simply an *npn* emitter follower with the collector connected to the positive power supply V_{CC}. The V_{in} signal is the internal signal, generally the collector of the differential current switch.[7] The output is typically very low impedance to more negative loads. The output may have a clamp diode from the output to V_{CC} which would conduct large currents if the output is driven more positive than V_{CC}. Even without a clamp diode, outputs driven above V_{CC} would cause the *npn* transistor to conduct in the reverse direction. Although this is of little concern for normal operation, it must be considered in some applications if power-supply faults could overload a powered driver, or more positive outputs could drive reverse currents into an unpowered driver.

The positive power supply is usually the ground reference for ECL, so signals are below ground potential. Since the typical ECL output is only a simple emitter follower, it can only source current to a more negative load. Therefore, some source of negative current is required to permit the output signals to function. Some higher-impedance ECL interfaces may incorporate an internal current source or load resistor to a more negative voltage. However, these currents may not be suffi-

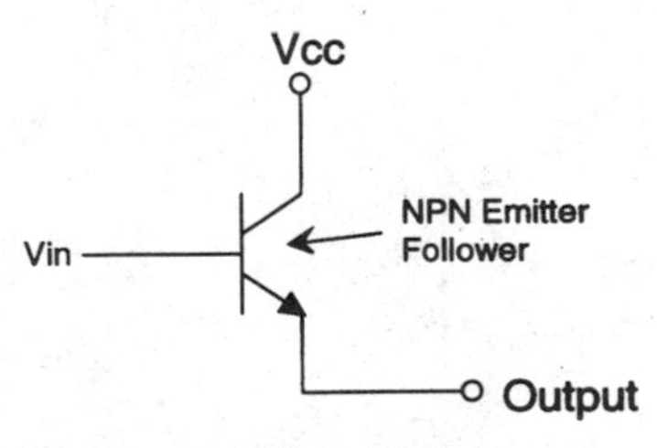

Figure 13.1 ECL output stage.

cient to supply the currents necessary to provide full switching currents into a low-impedance transmission line. For example, if the ECL signal excursion is considered to be 1000 mV into a 50-Ω impedance line, about 20 mA of switching current is required for that signal excursion. The dc bias current that is required to maintain current through the emitter follower under worst-case conditions is in addition to the switching current. If the driver is not at the end of a line, double the switching current is required to launch the signal in both directions from the middle of a transmission line, as well as to support the dc current for all of the termination and pull-down resistors on the line. The switching currents also double for a differential line if the line impedance remains the same as for a single-ended line. However, the differential impedance is usually greater than for the corresponding single-ended line, so the transient current for a differential line is less than doubled.

13.2.1 Load terminations

High speed transmission-line ECL interfaces often combine the termination and pull-down resistor in some fashion. Figure 13.2 illustrates an ECL source at one end of a line with a termination at the far end of the line. An exactly matched line (needed to minimize any reflections) requires an R_L as close to Z_0 as possible. Essentially the same signal level results as the signal travels along the line, so loads may be placed along the line with little regard for signal waveshape. R_L may be higher than Z_0 if the line is reasonably short compared to the signal repetition frequency, so sufficient time is available to allow any ringing to settle. Higher R_L reduces the static current, slightly increases signal levels, and tends to reduce rise time to the switching threshold.

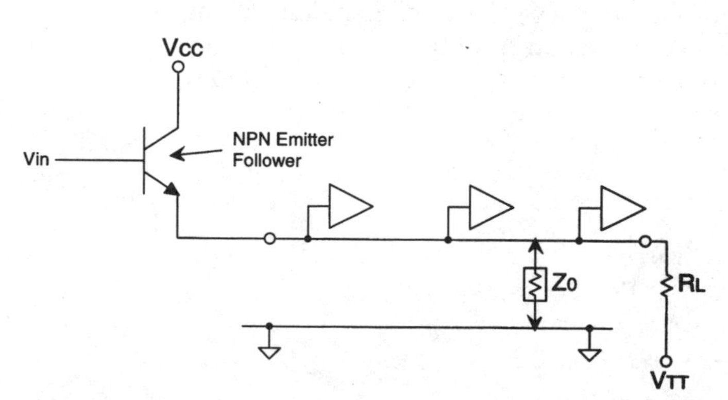

Figure 13.2 ECL output with load pull-down to V_{TT}.

The termination voltage V_{TT} must be at least as negative with respect to V_{CC} as the lowest LOW output signal, or V_{OL} min. However, at this minimum voltage, there is no LOW state bias current in the emitter follower output, so the output impedance and noise margin are degraded compared to the HIGH state. The traditional ECL termination voltage is −2.0 V. However, the performance of some ECL outputs with a more negative V_{OL} minimum is generally improved if the termination voltage is more negative. A nominal V_{TT} of about −2.4 V is usually sufficient to allow for temperature and power-supply distribution and regulation tolerances. The greater value for V_{TT} increases the LOW state noise margin. This may be important for high speed systems because an ECL driver converts negative-going overshoot to undershoot at the load. Although the load current increases slightly for a more negative V_{TT}, the termination resistance value can be increased above an exact match to Z_0 for many interconnections. A higher load resistance also increases the static HIGH state noise margin if the load current is less than for the specified load at $V_{TT} = 2.0$ V.

13.2.2 Signal return currents

It should be noted that interconnections with a load termination as illustrated in Fig. 13.2 result in both dc bias and signal return currents flowing through the ground reference interconnection. This results in ground loop currents, which must be controlled to limit interference. Also, the transient signal current that flows behind the signal wave as it travels along the transmission-line also flows in the transmission-line reference, which is not necessarily the same as either V_{CC} or V_{TT}. If the transmission line is coax, the coax shield is the transmission-line return and is typically grounded to the chassis at both ends. If the transmission line is printed circuit or other signal over "ground" interconnection, the ground or dc voltage plane next to the signal is the signal return line.

A typical ECL application provides a direct connection from the driver and through the ground system as the signal reference. When the signal waveform traveling on the transmission line arrives at the termination resistor, the return current from the transmission-line current begins to flow in the termination resistor. Therefore, it is important that the return current makes a low-impedance transition from the transmission line to the termination resistor supply. This is typically achieved by locating low-impedance bypass capacitance between V_{TT} and the signal return as close to the terminating resistor as practical.

There are common applications where the signal return line is not at chassis ground potential. For example, a multiple-layer printed circuit board may use dc voltage planes as the nearest signal return. This is acceptable if there is an adequate bypass capacitor at all signal return transitions. The dc paths shown in Fig. 13.2 are shown incomplete to emphasize that the transmission-line interface is incomplete until the voltages, signal returns, and decoupling (bypass) capacitors are properly located to complete the circuit. The power supply and interconnection analyst and the system designer should draw and understand the complete interconnection schematic for interface signals. This should include all of the signal return reference lines which are necessary to carry the loop current for all signals that travel in the dielectric between transmission-line conductors. Good-quality capacitors must be located to provide high-frequency (sometimes referred to as rf for "radio frequency") decoupling at all signal transitions where the signal return reference changes from one conductor to another. Also shown must be all power supplies involved, their connections to logic devices and termination resistors, the power-supply return connections, and chassis ground connections.

All interconnections between signals and power supplies must be shown, especially termination resistors and all internal diodes, including diodes inherent in circuit elements, ESD (electrostatic discharge) protection diodes, and any reverse current-blocking diodes. All relationships between power-supply voltages, including the reference level(s), tracking, sequencing, and any interlocks or overvoltage shutdown protection, should be indicated. For example, it is recommended that a separate V_{TT} supply should be regulated as a fraction of the $V_{CC} - V_{EE}$ difference and should default to a high impedance or the V_{CC} level when OFF (especially when V_{CC} is not ground) to avoid overloading drivers if V_{TT} could become more than about 3 V below V_{CC}.

The ground return system and capacitors at signal transitions are an essential part of nearly all interconnection systems, even though they may be missing from a schematic of the signal lines only. Basic circuit theory tells us that all electric currents flow only in a complete circuit loop. Logic signals can neither create nor consume the electrons, so the return path is essential to allow the logic signals to travel in the forward direction. The high speed waveform travels as displacement current in the dielectric between the signal line and any return line. The return current tends to flow in the lowest-impedance path nearest to the signal line, but the high-frequency components may travel through the surrounding dielectric or flow in other signal lines as crosstalk or interference. If there is not an adequate decoupling capacitance at the transition from the signal return to the termination resistor, the end of the line appears open initially, instead of terminated. A line which appears open will cause signal reflections on the line unless *both* of the two terminals of a parallel termination resistor are adequately connected for high frequency to the corresponding signal and return sides of the transmission line.

Most conventional ECL systems connect V_{CC} to ground, so the driver and the transmission-line return have a common connection, usually in parallel with the chassis ground. The signals along the line are also referenced to ground, so the ground return must be a continuous, low-impedance path between all drivers and receivers in the ECL system. No interfering signals should share this ground system, unless the currents and impedance are limited so that the worst-case voltage gradient induced is less than the ECL noise margins.

It is obvious that both V_{CC} and V_{TT} cannot both be directly connected to ground. Some interconnection systems, especially for test, operate V_{CC} at +2 V so that V_{TT} can be at ground. This is convenient for test systems, so that the signals can be terminated by 50 Ω to the test system ground. It is important that the level that is dc connected to the signal return provides a continuous, low-impedance path close to the signal line and that any signal reference level that is a power-supply voltage must be adequately decoupled to the signal reference level at every point where the transient signal currents change reference levels. For example, if V_{CC} and the coax shield are the signal reference ground, there must be a low-impedance, high-frequency bypass capacitor between V_{TT} and the coax ground near the termination resistor that is connected from the signal to V_{TT}.

13.2.3 Positive ECL (PECL)

Smaller ECL systems may be required to share a common +5-V V_{CC} supply with standard TTL compatible systems. This case is commonly called positive ECL, or PECL. Such an interface is very vulnerable to the large TTL switching noise interfering with ECL operation, but it can work if carefully designed to maintain ECL signal integrity. In this case, it is essential that the common V_{CC} interconnecting the ECL systems must be low-impedance and relatively free of TTL power supply transients.

The ECL transmission-line signal return reference may be at ground to share a coax shield, chassis, or a ground plane but must be well decoupled to V_{CC} to avoid interference between the V_{CC} signal reference and the transmission-line ground reference. Differential interfaces may increase the V_{CC} shift allowable for proper operation, but any V_{CC} shift between ECL drivers and receivers must be limited to avoid heavy currents through internal clamp diodes. It is also recommended that the ECL signal area be physically partitioned so that TTL or other larger signals and their return currents do not flow near the ECL signals. When the V_{CC} is +5 V, V_{TT} is in the range of about +2.5 to a maximum of +3.0 V. Careful attention must be given that any combined regulation tolerance of V_{CC} and V_{TT} does not result in reduction of the minimum $V_{CC} - V_{TT}$ difference for proper operation or that the maximum difference (including fault conditions) does not overload drivers. Designing the V_{TT} supply to track the V_{CC} supply voltage is often recommended to minimize regulation tolerances. Of course, both V_{CC} and V_{TT} must be decoupled to the signal return at all signal reference transitions.

13.2.4 Split-termination load resistors

Smaller ECL systems may not have a sufficient number of single-ended interfaces to be terminated to V_{TT} to justify a separate V_{TT} power supply. In this case, a split termination between V_{CC} and V_{EE} can be used that is equivalent to a V_{TT} termination. The split termination at the far end of a line is illustrated in Fig. 13.3. If the minimum $V_{CC} - V_{EE}$ is 4.5 V, two resistors of $2 * R_L$ result in a termination impedance of R_L to a voltage of about 2.25 V below V_{CC}. If V_{EE} is more negative, the resistor on the V_{EE} end may be slightly higher to generate an equivalent V_{TT} recommended range of a minimum of 2.0 V to about 2.5 V. A larger (more negative relative to V_{CC}) V_{TT} equivalent termination voltage improves LOW state noise margin and may allow the resistance values to be larger to reduce power dissipation.

The split resistor termination is sometimes called a *Thevenin equivalent termination* or simply a *Thevenin termination.* The disadvantages of the split resistor termination configuration are that two parts are required per termination and that the bias current that always flows from V_{CC} to V_{EE} results in additional power dissipation in the termination resistors. However, the total *system* power consumption

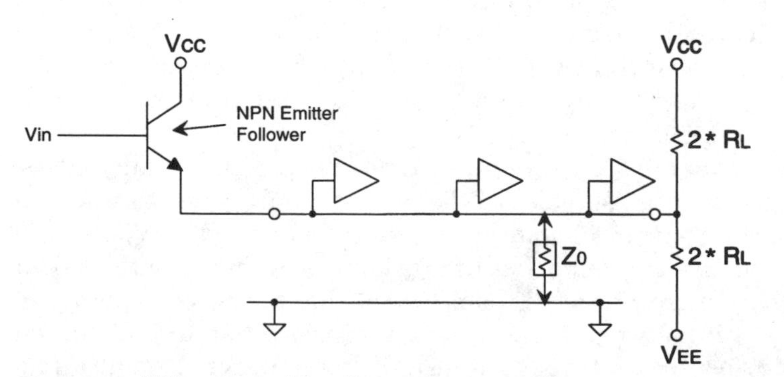

Figure 13.3 ECL output with split termination to V_{EE}.

is usually not significantly affected, because the inefficiency of a typical lightly loaded low-voltage power supply tends to offset the additional power dissipation in the split-termination resistors for small ECL systems. The primary difference is the location where the additional power is dissipated: either in resistors on the ECL signals for the split termination or in the separate V_{TT} power supply. Of course, the V_{CC} and V_{EE} voltages for split terminations must be either directly connected to the signal return or must be well decoupled to the signal return at the termination resistor as well as the location of all other signal reference transitions.

13.3 Multiple Drivers on ECL Party Lines

Either of the parallel terminated transmission-line cases illustrated above is usually suitable for multiple drivers on a single-ended party line bus. All inactive drivers are disabled to the LOW state, and only the active source is allowed to drive a HIGH onto the line. ECL transmission lines avoid bus contention due to multiple drivers because drivers are active only to the HIGH state. However, multiple drivers in different packages on an ECL line may require additional time for the current transients to settle when the currents switch or are shared between multiple drivers.[3]

An example of an ECL party line bus with terminations at two ends is illustrated in Fig. 13.4. The line impedance "seen" by a driver is $Z_0/2$, which doubles the load current to drive the same initial transition to travel in both directions on the line. Typically, the line is terminated at both ends, so the static load current on the driver is also doubled. The terminations at the ends should be nearly the line Z_0 to exactly eliminate reflections but may be higher to allow some overshoot to reduce rise time and compensate for some line losses. Each termination may be about 2.5 times Z_0 without exceeding the recom-

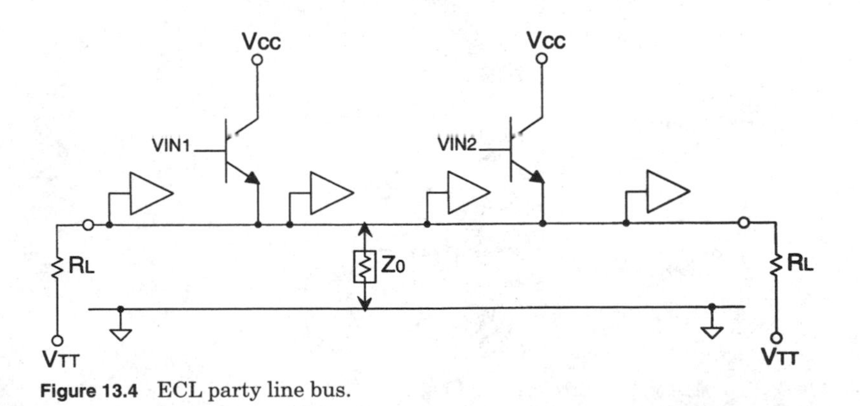

Figure 13.4 ECL party line bus.

mended overshoot limit of 35 percent.[3] ECL drivers specified to drive the middle of a line with two terminations are typically indicated by descriptions such as "25 Ω output drive."[8]

The inactive drivers in the LOW state are in a high-impedance state when the bus is in the HIGH state. However, if multiple drivers are at the same LOW level, they may share the LOW state current. Current sharing may cause a loss of the logic LOW level noise margin, especially if the pull-down resistor is to a V_{TT} of only 2.0 V below V_{CC}, or if there is any ground shift that results in a positive shift of the logic LOW level. Some ECL drivers are available with an output enable function. When the output enable is not active, the output of the disabled drivers will be more negative than a normal ECL LOW signal level. Although this control does not necessarily place the output into a high-impedance condition over a wide range of conditions (such as ground shifts between drivers or a partial power-supply loss), the disabled drivers will be at high output impedance if an active driver holds the bus at normal logic levels. Many of the parts with the output enable function also include the additional drive capability for 25-Ω outputs so they are suitable for multiple drivers on a party line bus. ECL devices with both features may be referred to as featuring "25 Ω cutoff" outputs.[9]

13.4 ECL Series Termination

Series termination can be used for ECL interfaces, as illustrated in Fig. 13.5. Series termination reduces the signal transient currents and eliminates the circulating currents that otherwise flow between the source and load for single-ended, parallel terminated interfaces. The series resistor also limits the currents which flow when driving the middle of a party line bus, or under partial power-supply loss when a powered output attempts to repower a lost voltage through

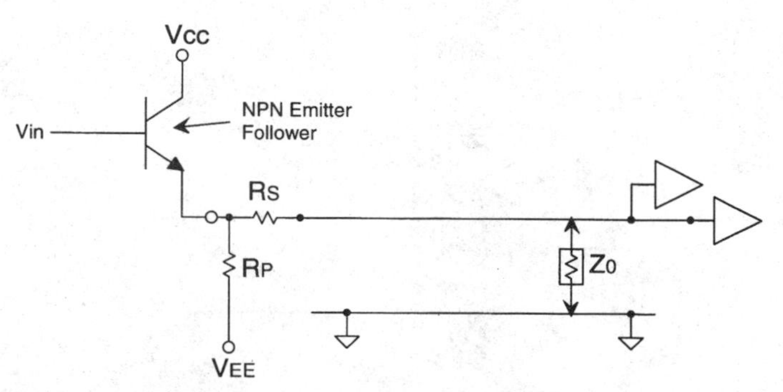

Figure 13.5 ECL output with source termination.

termination resistors and/or input clamp diodes. Series termination may eliminate the need for the V_{TT} supply but uses two components. The pull-down resistor to V_{EE} may dissipate more power than a parallel termination to V_{TT}. Therefore, the physical characteristics of ECL series termination are similar to the split termination.

The series resistor may be less than Z_0 to increase the initial signal level, which results in some overshoot and line ringing. This response is similar to a parallel termination load resistor of greater than the line impedance. The value of the pull-down resistor required depends on the value of Z_0, R_S, the initial signal level required, and the line response desired. The analysis also depends on the tolerances used for impedance values and power-supply regulation, as well as how well the signal has settled to a logic HIGH level before the negative transition.

The analysis of the relationship of the value of R_P and the signal levels can be determined from dc analysis of the circuit of Fig. 13.5. The signal return reference (shown with the ground symbols in the figure) is considered to be at the previous logic HIGH level. This concept is for *transient* analysis of the initial signal excursion; it does not represent the static currents which flow through the termination resistors only after the transmission-line transients have settled. The impedance for the initial signal excursion between the signal return reference and the signal line is Z_0.

A trial calculation can be made with these values and the series and pull-down resistors, assuming that the emitter follower is off, so the driver is open-circuit. The calculated open-circuit voltage at the R_S to R_P node is compared with the output logic LOW level of the driver. If the calculated open-circuit node voltage is more negative than the output LOW voltage, then the emitter follower is conducting some current into R_P and the output LOW level is substituted for the node voltage. Whichever node voltage is less negative is used to calculate the voltage across Z_0, which is the initial signal excursion that travels on the transmission line.

Some applications may require that the driver always conducts some current, even during transients. This assures that the response for HIGH-to-LOW transitions is nearly the same as for LOW-to-HIGH transitions. If the emitter follower must always conduct some current, then R_P must be low enough so that the calculated open-circuit voltage is always more negative than the minimum output LOW signal level.

The analyst and the system designer should be aware that ECL drivers typically have lower source impedance on the LOW-to-HIGH transition, because the emitter follower resistance is lower in the HIGH state. Therefore, ECL always tends to have the most ringing on the

LOW to HIGH transition, so the worst overshoot is above V_{OH}. Conversely, TTL typically has lower source impedance on the HIGH to LOW transition, so the worst overshoot is *below* V_{OL}. TTL is also more vulnerable to ringing on the LOW level because the threshold is closer to the LOW level, especially for some older TTL families, such as LS (low-power Schottky). Some of the faster advanced Schottky families have higher input thresholds and Schottky clamp diodes to control ringing, even though the published static noise margins are similar.[10,11]

13.4.1 Analysis of pull-down and series resistors

The analysis of the results of the pull-down and series resistors is illustrated by the following example.

Assume:

$$Z_0 = 100\ \Omega \qquad R_S = 50\ \Omega \qquad R_P = 500\ \Omega \Rightarrow R_{\text{sum}} = 650\ \Omega$$

$$V_{CC} = \text{GND} \qquad V_{EE} = -5.0\ \text{V}$$

No current is flowing in the line prior to the HIGH-to-LOW transition:

$$V_{OH}\,(\text{prev}) = -900\ \text{mV} \qquad V_{OL} = -1700\ \text{mV} \Rightarrow \Delta V = 800\ \text{mV}$$

Therefore,

$$I = \frac{(5.0 - 0.8)\ \text{V}}{650\ \Omega} = 6.3\ \text{mA}$$

$$V_{\text{node}} = V_{EE} + 6.3\ \text{mA} * 500\ \Omega = -1.85\ \text{V open circuit}$$

The calculated open-circuit V_{node} is lower than V_{OL}, so $V_{OL} = -1700$ mV is the appropriate node voltage and the output emitter follower is conducting additional current through R_P. The excursion at the node is the full 800-mV signal excursion, so the initial signal on the line for $Z_0 = 100$ and a 50-Ω series resistor is

$$V_{\text{init}} = \frac{800 * 100}{100 + 50} = 533\ \text{mV}$$

This initial signal excursion on the line is two-thirds of the 800-mV static level, or 33 percent short of the final static value with no-load termination. When V_{init} doubles at the open end, the signal level is about 1066 mV, or 33 percent *above* the final value (these are the result of $R_S = Z_0/2$). The resulting undershoot at the load end after the reflection from the driver returns to the load will be less than 100 mV, which is within typical ECL noise margins. The actual peak undershoot on typical lines is usually less than estimated by calculation for various reasons, including the following:

1. The signal does not quite double because the load is not perfectly open.
2. The driver has some internal impedance in series with R_S. Typical ECL internal impedance is about 7 Ω for outputs specified for 50-Ω loads, although the impedance is higher for very low currents.
3. Lumped capacitance at connection nodes and skin-effect losses on the line will increase the rise time and round the trailing edge of the signal waveform. This tends to reduce the peak amplitude of the ringing on the line.

Alternate calculations with other conditions may provide different results. Lower Z_0, higher R_P, and less negative V_{EE} result in less initial signal excursion if the calculated node voltage is less negative than V_{OL}. The analyst is cautioned that if the emitter follower is not conducting current because it is in cutoff, the source impedance is $(R_S + R_P)$, not just R_S. This may change the response to reflections or crosstalk traveling toward the source.

Analysis will also show that the transient response is altered if a load resistor to V_{TT} is substituted for R_P to V_{EE} at the source. A parallel termination resistor to V_{TT} must be located at the far end(s) of the line for proper transient response. When a parallel termination resistor is at the far end of the line, current flows in the line when the logic level is HIGH. The maximum HIGH-to-LOW excursion possible on the line occurs if the current into the line drops to zero. For example, consider the load-termination illustration of Fig. 13.2 (Fig. 13.3 is equivalent) with $V_{TT} = -2.5$ V, $R_L = 200$ Ω, $Z_0 = 100$ Ω, $V_{OH} = -900$ mV, and $V_{OL} = -1700$ mV (the same Z_0 and logic levels as above). The current flowing in the line for the HIGH state is (2.5 − 0.9)V/200 Ω = 8 mA. The maximum possible excursion is 8 mA $*$ Z_0 = 800 mV, which results in exactly the V_{OL} level of −1700 mV for this example. This means that for this case, even with a load-termination resistance of twice the Z_0 impedance, a full initial signal excursion propagates from the source. The maximum available excursion is determined by the current flowing in the line prior to the HIGH-to-LOW transition at the driver. The available excursion may be lower for other conditions, especially for less negative V_{TT}, lower Z_0, or higher R_L. The following conditions should be noted:

1. The load-termination resistor does not need to match Z_0 for adequate signal response. Even if the initial excursion on the line is

less than the full signal swing, an underdamped load termination will result in overshoot which usually exceeds the full signal excursion at the far end. However, the signal waveshape and ringing are not the same as calculated by linear analysis of a full initial signal excursion, and more settling time may be needed to approach static signal levels.

2. The transient response may be different for the LOW-to-HIGH transition than for the HIGH-to-LOW transition. The driver always generates a full excursion at the driver node for the LOW-to-HIGH transition. The series resistor is required to limit overshoot and ringing on the LOW-to-HIGH transition if the far end is not terminated.
3. The increase in both R_L and the magnitude of V_{TT} has resulted in a reduction in the static current drawn, compared to exact termination to −2.0 V. Operation remains proper, so power dissipation can be reduced and allowance included for impedance and regulation tolerances without performance degradation.

A load (or pull-down) resistor to V_{TT} at the source only may result in similar static conditions as if the same resistance is located at the end of the line, but the transient response of a long line may be affected. The static current in the line is zero for *both* logic states when there is no load termination at the end of the line. The only current that can be supplied for a HIGH-to-LOW transition is through the pull-down resistor.

For example, consider $Z_0 = 100\ \Omega$, $R_S = 50\ \Omega$, $R_P = 200\ \Omega$, and $V_{TT} = -2.5$ V (instead of V_{EE}), as was illustrated in Fig. 13.5. The initial HIGH-to-LOW excursion is only about 460 mV. This is just a little over one-half of a full signal excursion, which will double to about the static LOW level at the open receiver. There will be little ringing when this signal level returns to the source, because it is very near the static level. Therefore, this case responds similarly to a typical series-terminated case for the HIGH-to-LOW transition, but with less initial excursion than otherwise expected with $R_S = Z_0/2$. The initial signal excursion may be increased for lower R_P; or the initial signal excursion is reduced if V_{TT} is less negative.

For another example, $R_P = 100\ \Omega$, and $V_{TT} = -2.0$ V results in an initial HIGH-to-LOW signal excursion of about 440 mV. This is only slightly less signal than the previous case, but current draw is higher. The current drawn and the average $I^2 * R$ power in the pull-down resistor is summarized as follows for values of R_P:

	I_{HIGH}, mA	I_{LOW}, mA	P_{avg}, mW	V_{init}, mV
100 Ω to −2.0 V	11	3	6.5	440
200 Ω to −2.5 V	8	4	8	460
150 Ω to −2.5 V	11	5.3	11	533
500 Ω to −5.0 V	8.2	6.6	28	533

These results indicate that power dissipation in the resistor (also in the output emitter follower) is reduced when the pull-down at the source is to V_{TT} rather than V_{EE}. This may be preferred when a V_{TT} supply is available and the signal transient response characteristics are acceptable. For the signal levels used in this analysis, $R_P = Z_0 + R_S$ to −2.5 V results in a calculated open-circuit node voltage equal to V_{OL}. Therefore, V_{init} is equal for both positive and negative transitions. The more general form for the maximum value of R_P that results in V_{init} for negative transitions equal to V_{init} for positive transitions is the following: $R_P \leq (R_S + Z_0)(V_{OL} - V_{TT})/(V_{OH} - V_{OL})$.

The levels selected in the above analysis have $(V_{OL} - V_{TT}) = (V_{OH} - V_{OL})$, so $R_P = Z_0 + R_S$ is the maximum R_P that supplies just enough current in the LOW state so that the calculated node voltage just equals V_{OL}. However, it should be noted that when R_S is typically less than Z_0 (as in the above examples), R_P may be greater than this minimum value and still supply enough current for an initial HIGH-to-LOW signal excursion of at least one-half of the steady-state level. It may also be noted that if Z_0 is as low as 50 Ω and V_{TT} is as low as −2.0 V, it becomes more difficult to use series termination with the pull-down resistor to V_{TT} without very high currents in the HIGH state. Furthermore, driving the middle of a party line bus reduces the Z_0 seen by the driver by a factor of 2, and driving differential lines increases $(V_{OH} - V_{OL})$ by a factor of 2 because both lines switch. The pull-down resistor to V_{EE} may tend to be more practical for the heavy loading applications, even if the power dissipated in the resistor is greater.

13.5 Power-Supply Considerations

A typical single-ended ECL bus has common power supplies for all drivers and receivers. The signal return must be well connected to or decoupled to the signal return at any signal reference transitions. Any dc offset or noise between the V_{CC} of any driver and the V_{CC} of the receiver(s) reduces the noise margin of ECL signal levels.[12]

Many older ECL families had logic levels that varied with V_{EE}, but with less sensitivity than the 1:1 shift in logic levels with V_{CC}. ECL families with internal regulation for the reference levels and current sources provide signal levels which are insensitive to power-supply regulation over a wide range of V_{EE}. Typical ECL devices will operate over a V_{EE} range much greater than specified but at reduced speed capability. However, typical ECL drivers and receivers are not intended for "power down" capability, where a party line bus operates correctly even if some of the inactive devices are unpowered, including inserting a device onto an operating bus, called "live insertion." ECL inputs and/or outputs may have ESD diodes to V_{CC} and V_{EE} which would clamp bus signals to the minimum (least positive and least negative) power-supply voltage(s).

When a separate V_{TT} supply is used to provide pull-down or termination resistance, it should track the $V_{CC} - V_{EE}$ power supply. Particular attention must be made to layout and decoupling of this supply because it carries the high speed signal current transients. Additional precautions should be taken for especially critical systems (or the most critical portions of a larger system), such as very high speed, critical timing accuracy, or very long lines. All logic devices should be operated with both true and complement outputs loaded identically, whether actually used or not. This will help to reduce any changes in dc loading of the V_{TT} and V_{CC} supplies for different logic states. Parallel load terminations are recommended to avoid the change in load currents and possible standing waves due to reflections from unterminated loads.

If multiple-termination resistors are located in a common package, care must be taken to avoid crosstalk between signals. Multiple V_{TT} pins and internal high-frequency capacitance to V_{CC} are recommended for packages with multiple terminations. Component layout should be made to implement signal flow with the shortest line lengths practical. Differential lines or lines with critical timing skew requirements should be run in parallel with the same length, avoiding sharp corners, crossovers, and vias. There should be sufficient space between critical signals to avoid interference, even if a few devices available in a package are unused. Of course, an initial design should leave some spare space on board layouts, pins in connectors, and wires in cables, so that minor design tryouts or necessary modifications may be accomplished without compromising critical performance requirements.

A transmission line terminated at the source only may result in an initial signal excursion much less than the final static signal level. All of the loads for a series-terminated line may be located near the far end of the line to avoid an intermediate "step" in the signal level before the reflection arrives from the far end of the line. Loads near the source of a series-terminated line may have an extended propagation delay or may even be unstable during the delay until the reflection arrives from the end of the line. This was discussed and illustrated previously (Chap. 12, Fig. 12.6). ECL drivers may be connected in a loop with the series termination reduced to one-half of the value appropriate for one end of a line because the impedance looking into the middle of a line is $Z_0/2$. Signals everywhere around a loop-connected bus will recover from the intermediate "step" in the signal levels after one trip around the loop. The series resistance at the driver may be low enough that the initial signal excursion is sufficient to switch logic devices, but at reduced signal noise margin. This typically results in overshoot and ringing, which can be accommodated with additional settling time or Schottky diode clamps to limit overshoot.

Multiple ECL drivers with series terminations are often not practical on a party line bus because the active HIGH driver must drive all of the pull-down resistors on the line. This is unlike TTL and CMOS drivers in which the active transistors that provide both pull-up and pull-down current can be disabled. However, it may be practical to connect multiple ECL drivers to a single pull-down (this is called a "wired-OR" connection), or the number of pull-down resistors may be limited to the drive capability of an active driver. The use of drivers with capability to drive 25-Ω outputs increases the number of pull-down resistors than may be driven. Multiple drivers from a single package may be paralleled to increase drive capability or to integrate OR logic. An example of a single-ended bus with wired OR connection of multiple outputs to a single pull-down is illustrated in Fig. 13.6.

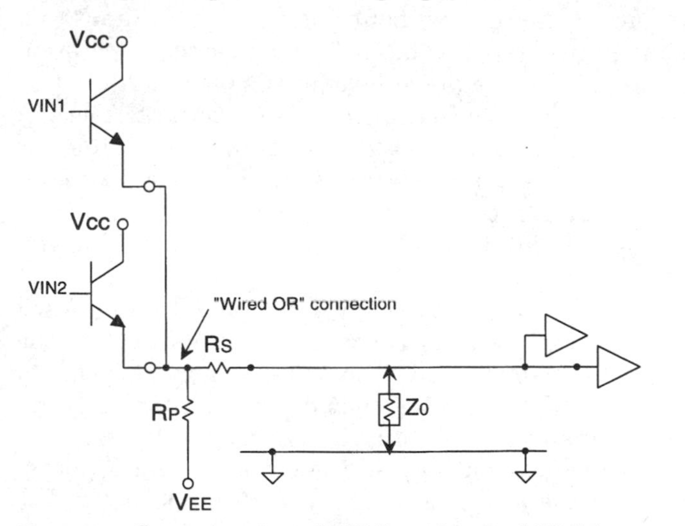

Figure 13.6 Source-terminated ECL line with wired OR drivers.

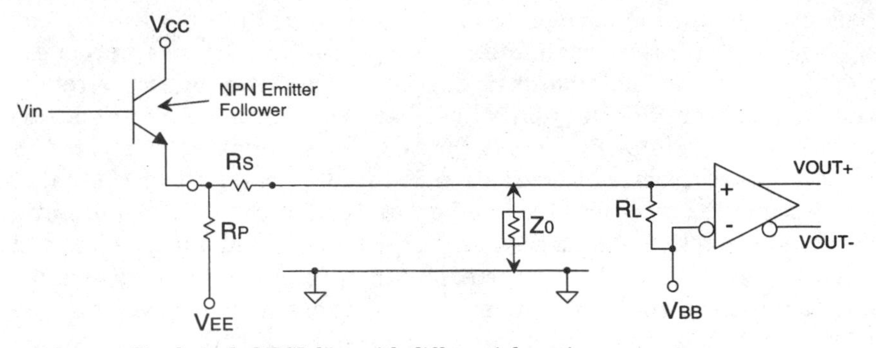

Figure 13.7 Single-ended ECL line with differential receiver.

The two output drivers may share current if both are in the LOW state, so this may shift the signal level slightly higher. The two output drivers should be located close together to limit the settling time required to switch the load current from one output driver to the other.

Single-ended lines may be terminated at both ends. Figure 13.7 illustrates a concept that provides the pull-down and series resistor at the source and a parallel termination to V_{BB} with a differential receiver at the load. V_{BB} is the ECL bias voltage that is very close to the midlevel between the ECL LOW and HIGH logic levels. Although the exact level may be slightly different for different ECL families, it is typically about V_{CC} − 1.3 V. Most of the ordinary ECL logic inputs use V_{BB} as an internal reference only, but typical differential receivers provide a V_{BB} output that can be used to receive single-ended signals.

The V_{BB} output pin can be used without buffering as an input to a high-impedance differential receiver input for single-ended operation without a load termination. This would be equivalent to no R_L in Fig. 13.7. It should be emphasized that unlike most ECL outputs, the V_{BB} output pin of a receiver is *not* intended to drive transmission lines or termination loads directly. V_{BB} must be buffered with one (or more) spare receivers with negative feedback and adequate pull-down resistance and decoupling capacitance to drive the transmission-line signals.

An example of using a differential receiver to generate a buffered V_{BB} is shown in Fig. 13.8. The negative feedback tends to keep the buffered output at the V_{BB} reference as the current changes owing to different line conditions. The high-frequency filter capacitor is recommended to control transients. The pull-down resistor R_B must be capable of sinking enough current to hold the output to the V_{BB} level when the signal on the line is V_{OH}. For example, $R_B \leq 300\ \Omega$ to a V_{TT}

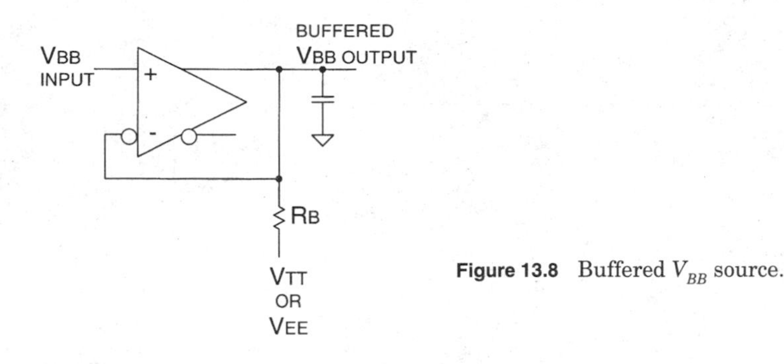

Figure 13.8 Buffered V_{BB} source.

of −2.5 V would be sufficient to sink 4 mA from a 100-Ω line at a V_{OH} of −0.9 V, V_{BB} of −1.3 V. A buffered V_{BB} reference may also be generated by a power-supply regulator that tracks the V_{BB} output of an ECL receiver.

The use of the differential receiver as was illustrated in Fig. 13.7 does improve noise margin for some types of interference that appear as common mode noise. If the interface cable is twisted pairs, most crosstalk noise will be common mode. However, offsets such as V_{CC} shifts with respect to V_{BB} directly interfere with single-ended signals. One of the alternatives would be to generate the V_{BB} reference level at the driver and transmit it as the reference to the differential receiver. Any ground (and V_{BB}) shift within the common mode rejection range of the receiver would not affect proper signal detection.

Termination to V_{BB} offers some advantages over either the source-only termination or the load-only termination of the previous figures. The double termination eliminates the dc bias current that flows in the ground loop for load-only termination to V_{TT} (or its split-termination equivalent). Of course, signal current flows in the return path. The load termination does control reflections, so longer lines can be operated at high frequencies without serious standing waves. The series resistor at the source limits the signal excursions and rise time, and dampens reflections and crosstalk. The double termination does attenuate both the signal and any ground shifts, so this configuration is a little more tolerant of V_{CC} differences between the source and the load. Termination at both ends means that the signal waveshape is less sensitive to the relationship of either termination resistor to the line impedance. For example, R_S may be less than Z_0, and R_L greater than Z_0 to reduce attenuation, but with very low signal ringing.

The single-ended transmission-line configuration referenced to V_{BB} is also suited for a party line bus.[13] The pull-down resistor and any series termination may be located at the drivers, subject to limita-

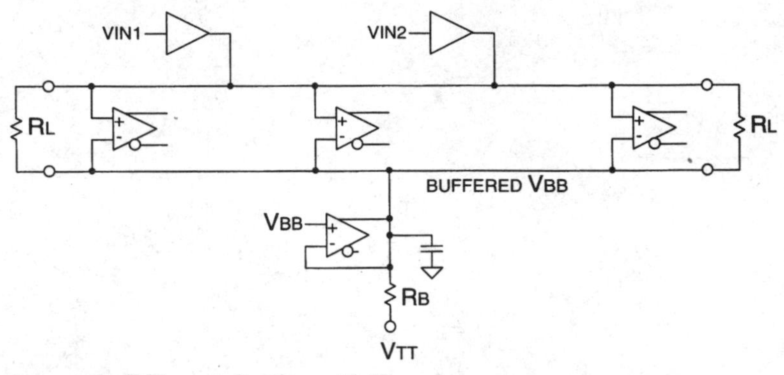

Figure 13.9 ECL party line bus with V_{BB} reference.

tions on multiple loading and distance between wired OR drivers for the frequency of operation.[14] An example of a single-ended party line bus referenced to V_{BB} with resistor terminations at two ends is illustrated in Fig. 13.9 This illustration does not separately show the details of the transistors and resistors that are internal to the driver symbols. The pull-down resistors may be located at the ends of the line if not incorporated with the drivers.

13.6 Schottky Diodes for ECL Load Terminations

A variation of single-ended, double-terminated line is illustrated in Fig. 13.10. The R_L resistor of the previous illustration is replaced with parallel Schottky diodes in reversed orientation. As for any line termination to V_{BB}, the V_{BB} connected to the line must be buffered, as illustrated previously. Although this diode configuration functions well as a termination, the circuit operation is quite different from a resistor.

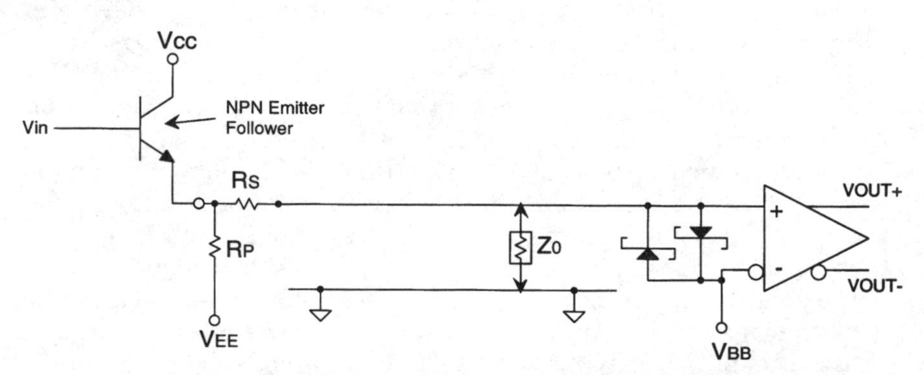

Figure 13.10 Single-ended ECL line with Schottky diode termination.

The diodes have little effect on signals that are near the V_{BB} reference level, so small signals are not attenuated until the signal level is well beyond the differential receiver thresholds. The diodes also operate as overshoot clamps, so there is essentially no overshoot beyond static signal levels and ringing on long lines is effectively suppressed. However, it should be emphasized that the primary operation of ECL terminations is different from overshoot clamp diodes for TTL or CMOS, which conduct no steady-state current. ECL terminations must absorb the line currents to eliminate reflections.

The characteristics of Schottky diodes result in more effective high-frequency ECL terminations than ordinary *p-n* junction diodes. The "threshold" of Schottky diodes is lower than that of *p-n* junctions and is well matched to the typical ECL signal excursions of about ± 400 mV with respect to V_{BB}. High speed signal Schottky diodes have a softer "knee," with a more lossy (resistive) characteristic at higher currents, and avoid the reverse recovery problems of ordinary *p-n* junction diodes. Figure 13.11 illustrates the forward current versus voltage for a typical Schottky diode suitable for ECL termination. Parallel diodes in reversed polarity will have the same characteristic for either polarity of signal. The load lines for resistors indicate that the forward characteristics of the Schottky diodes will match the current and voltage conditions over a range of at least 50 to 200 Ω at ECL levels of ± 400 to ± 500 mV. However, there is essentially zero current for signal levels less that ± 400 mV, so a Schottky terminated

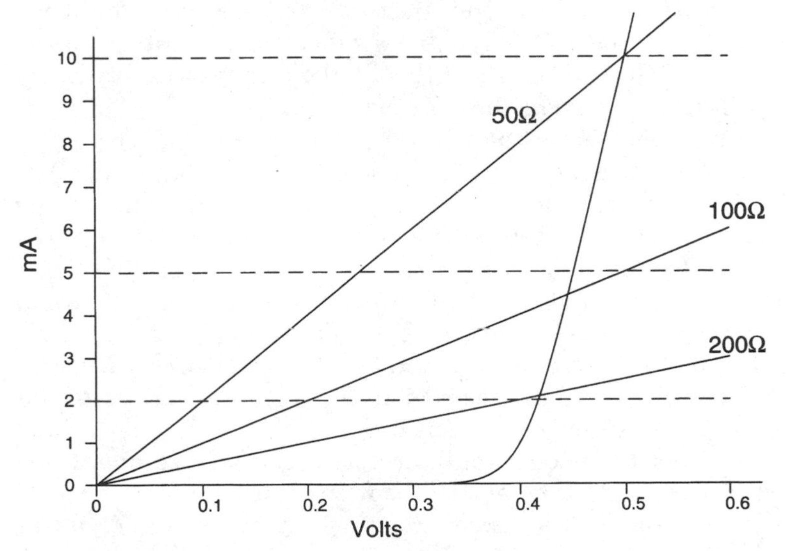

Figure 13.11 Schottky diode characteristics.

line appears open to small signals near V_{BB}. Because the Schottky diodes draw little current for signals less than full ECL levels, they are suited for the load terminations on a line with source termination. Schottky diodes may be placed at multiple locations on a transmission line (including every receiving input), because they do not attenuate signals until the levels are well beyond the minimums needed by differential receivers. As a general "rule of thumb," unless specified otherwise, $\pm$ 150 mV input is sufficient to switch the output of a single-stage differential receiver. The current drawn by Schottky diodes at a voltage this low is insignificant.

The nonlinear clamping characteristics of Schottky diodes as a load termination result in load signal voltage levels which tend to "adapt" to various transmission-line conditions, although there may be reflections of some of the line currents. Although these characteristics may seem intelligent or even "magical," they provide the desirable characteristics of digital processing, as well as artificial neural networks and biological neurons.[15] The characteristic of clamping the higher signal levels means that Schottky diodes will adapt to a wider range of series termination resistance, line impedance, losses on long lines, discontinuities, and ground offsets without attenuating small signals. The Schottky diodes will suppress the overshoot from larger signals but allow small signals near V_{BB} to increase to the Schottky diode threshold level. It should be noted that the Schottky diodes prevent overshoot and ringing at the load but do not necessarily prevent reflections on the line. If the initial signal excursion arriving at the Schottky diodes is lower that the diode threshold, the diodes tend to appear as an open load until the reflection increases the signal level to beyond about $\pm$ 400 mV. Any partial reflection will return to the source and establish static levels, although the steady-state static signal level will conduct current through the diodes.

Although the reflection establishes essentially static levels over a wide range of conditions, the analyst should be aware that any reflections may result in standing waves on high-frequency lines if the source launches a new signal excursion before the reflection returns to the source. This is characteristic of source-terminated lines. The reflection is about proportional to the difference between the initial signal excursion and the diode threshold. For example, if $R_S = Z_0/2$, the initial signal excursion is two-thirds of the open-circuit level, or about a 533-mV change from the previous level of either $\pm$ 400 mV for ECL. However, the signal excursion does not double at the far end because the Schottky diodes limit the total excursion to about 800 mV. Therefore, the reflection is only about one-third of the static level instead of two-thirds from an open circuit, which significantly reduces the standing wave effects on high-frequency lines.

The Schottky diodes referenced to the V_{BB} of the receiver as illustrated in Fig. 13.10 will tend to adapt to some ground shift between the driver and receiver because the diodes will attenuate only the level that is increased but not the level that is reduced by the ground shift. For example, even if the driver V_{CC} is shifted positive with respect to the receiver, the diodes will clamp the input HIGH level to about +450 mV above V_{BB} of the receiver. However, the input LOW level is not attenuated, so an initial negative excursion of only about 300 mV on the line is sufficient to achieve an input LOW level of 150 mV below V_{BB} $(+450 - 2 * 300)$ after the initial excursion doubles at the load. The analyst should be aware that offset currents will flow through the ground and V_{BB} supplies and that R_S, R_P, and V_{EE} or V_{TT} must be designed to be adequate to provide the negative excursion despite a ground shift.

The parallel Schottky diode termination also results in the suppression of small disturbances such as crosstalk, reflections from line discontinuities, or ringing. It should be noted that the small signal dynamic impedance of the Schottky diode characteristics in Fig. 13.11 is a low impedance throughout the region between 200 and 50 Ω total equivalent impedance with respect to zero. This means that when signal levels are in the region where the Schottky diode is conducting significant current, any disturbances that result in small changes in line current have very little effect on the signal voltage on the line or into the receiver. Therefore, Schottky diode termination can suppress small disturbances on the line without attenuating the intended signal. This is the "large signal capture, small signal suppression" nonlinear transfer characteristic of digital processing.[15]

These examples have indicated that parallel Schottky diodes can provide an effective alternative technique for terminating single-ended ECL lines. Of course, a pull-down load resistor is still required to provide the LOW state current. Although not essential for operation, series resistance at the source is definitely recommended, even if the source termination resistance selected is less than Z_0. The source resistance will limit the initial signal excursion, which limits transient currents and crosstalk. Source resistance will also limit the overshoot current that flows in the Schottky diodes. Source resistance should always be used on long lines that are subject to potential crosstalk and such discontinuities as connector pin fields, lumped loads, and stubs.[14] The source resistance tends to absorb backward crosstalk and reflections from discontinuities and crosstalk along the line.

13.6.1 Reflection diagrams for Schottky diode termination

The static current for Schottky diodes terminated to V_{BB} is very low, only about 1.0 mA for typical ECL levels.[13] This means that the parallel Schottky diodes operate primarily as an open load for steady-state static conditions. The dynamic operation of the parallel Schottky diodes is primarily as an overshoot clamp.

The low dynamic impedance of the diodes in the forward conduction region means that the diodes will accept a wide range of current at signal levels near the static V_{OH} and V_{OL} of typical ECL levels. This variable forward conduction characteristic results in stable voltage levels, but the currents may not stabilize until several reflection cycles. Since the dynamic impedance of the diodes in the forward conduction region is low, the combination with a low-impedance source results in an *overdamped* response to the recovery from overshoot. Although the voltage levels do not change significantly, the current slowly steps to lower and lower levels, approaching the steady-state static level. This is illustrated as a reflection diagram in Fig. 13.12, based on the diode characteristics from Fig. 13.11 and an output signal excursion of $V_{BB} \pm 400$ mV. Although the steady-state static conditions are not reached quickly, the overshoot currents and reflections do not result in the possibility for current growth for long lines. The overdamped response reduces current in a monotonic manner, without the ringing that could result in standing waves if the reflections on long lines are opposite phase with respect to the newer signal inputs. The reflections shown in Fig. 13.12 are all in the direction of reducing current, so all of the reflections are in the same direction as the next signal transition. It may be noted that

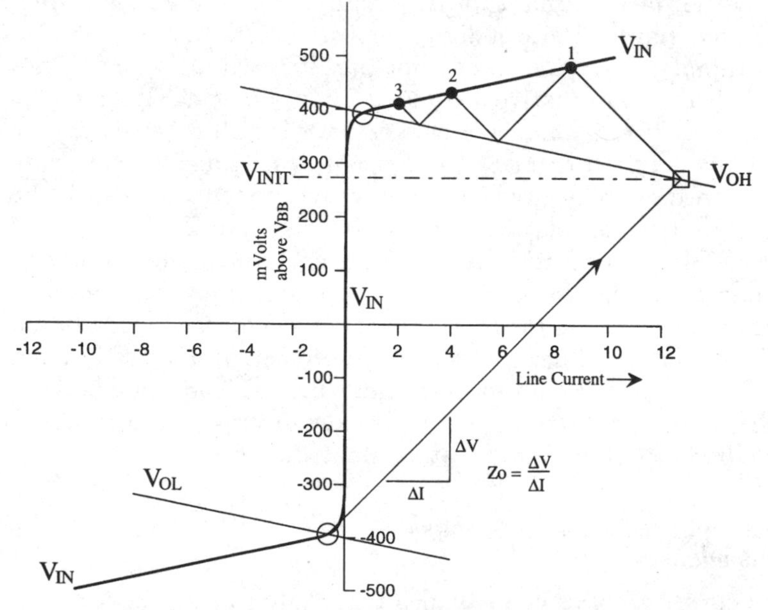

Figure 13.12 Reflection diagram for Schottky diode termination to V_{BB}.

the currents shown in this figure are the *line* currents and do not include the current in the pull-down resistor at the driver. This reflection diagram does indicate that the typical current that flows in the line during the initial signal transition may be about 14 mA. The pull-down resistor must provide at least this current, plus an allowance for tolerances and some bias current to maintain conduction in the output transistor for the LOW state output.

The source may be terminated to reduce the transient current and to reduce the number of reflection cycles to stabilize the overshoot current. Exact source termination would result in an initial signal excursion of one-half of the steady-state static level, which would double at the far end, as in conventional lines that are exactly terminated only at the source. However, exact termination at the source has the traditional disadvantages of an initial signal excursion that is in the threshold region: line losses reduce the signal excursion at the load and the rise time is slower than an underdamped response. Furthermore, the Schottky diodes provide suppression of small disturbances only when the diodes are in the forward conduction region. It is recommended that the source termination be underdamped for this case of Schottky diode termination so that the diode remains in forward conduction until reflections on the line have settled to lower currents.

Illustrated in Fig. 13.13 is a reflection diagram that shows a source termination of about $Z_0/2$. The maximum line current is reduced to

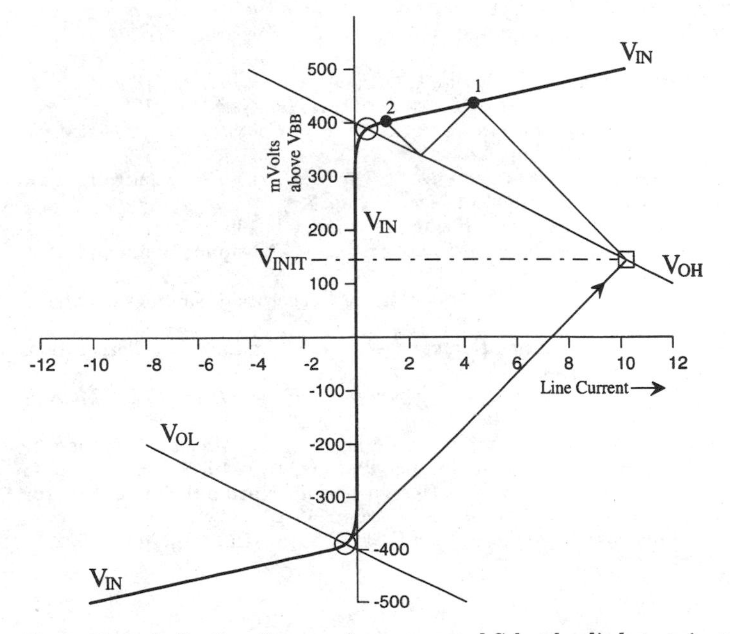

Figure 13.13 Reflection diagram for source and Schottky diode terminations.

about 10 mA, and the diode remains in the forward conduction region until after the second incident wave arrives at the load, when the current settles to nearly the steady-state static level. The initial voltage is above V_{BB}, so the initial signal should switch multiple loads distributed along the line. However, the noise margin is reduced, so a source termination value this high is not recommended for critical edge sensitive signals distributed to multiple loads that are far from the end of the line compared to the rise time.

13.6.2 Schottky diode termination summary

Schottky diode terminations tend to result in nearly ideal voltage waveforms at the end of a single-ended line. A pull-down load resistor is required on the line. The diode terminations do not prevent reflections on the line. The Schottky diodes draw transient current but little static current for terminations to V_{BB}, so they may be placed at multiple locations on a line. The Schottky diode terminations would not terminate undershoot disturbances on the line, but clamping the overshoot tends to prevent undershoot.

13.7 References

1. James E. Buchanan, *Signal and Power Integrity in Digital Systems: TTL, CMOS & BiCMOS,* McGraw-Hill, New York, 1996, p. 216.
2. *National Interface Databook,* National Semiconductor Corporation, Santa Clara, Calif., 1996 ed.
3. *MECL System Design Handbook,* Motorola, Inc., Phoenix, Ariz., 4th ed., 1988.
4. Hugh H. Skilling, *Electric Transmission Lines,* McGraw-Hill, New York, 1951.
5. Richard A. Matick, *Transmission Lines for Digital and Communication Networks,* McGraw-Hill, New York, 1969, chap. 4.4.
6. *F100K ECL 300 Series Databook and Design Guide,* National Semiconductor Corporation, Santa Clara, Calif., 1992 ed., sec. 5, chap. 7.
7. *Motorola MECL Data,* Motorola, Inc., Phoenix, Ariz., 5th ed., 1993, sec. 1.
8. *F100K ECL 300 Series Databook and Design Guide,* National Semiconductor Corporation, Santa Clara, Calif., 1992 ed.
9. *Synergy Product Data Book,* Synergy Semiconductor Corporation, Santa Clara, Calif., 1992.
10. *FAST Advanced Schottky TTL Logic Databook,* National Semiconductor Corporation, Santa Clara, Calif., 1988 ed., sec. 1.
11. James E. Buchanan, *Signal and Power Integrity in Digital Systems: TTL, CMOS & BiCMOS,* McGraw-Hill, New York, 1996, chap. 2.
12. *MECL System Design Handbook,* Motorola, Inc., Phoenix, Ariz., 4th ed., 1988, chap. 1.
13. *MECL System Design Handbook,* Motorola, Inc., Phoenix, Ariz., 4th ed., 1988, chap. 4.
14. *F100K ECL 300 Series Databook and Design Guide,* National Semiconductor Corporation, Santa Clara, Calif., 1992 ed., sec. 5, chap. 4.
15. C. H. Chen (ed.), *Computer Engineering Handbook,* McGraw-Hill, New York, 1992, p. 6.26.

Chapter

14

Differential Line Drivers and Terminations

This chapter considers the relationship of driver circuits for high speed signal transmission on longer lines with various line configurations and their terminations. This chapter describes differential lines for ECL, TTL, and CMOS technology. Differential lines allocate two conductors to each signal and receive the signal by detecting the signal difference between the two conductors. The load terminations of differential lines are typically connected between the two conductors to isolate signal currents from the power and ground distribution system. This chapter emphasizes that properly balanced source termination is important to control crosstalk interference, as well as to allow for tolerances in line impedance, load terminations, and lumped loading. Load termination is required to avoid pattern sensitivity for nonlinear sources and for high speed lines that may not settle to static levels between signal transitions. The use of Schottky diodes for ECL terminations is described and illustrated. Issues concerning multiple power supplies are discussed for ECL, TTL, and CMOS applications, including ECL powered from positive voltages and translators between ECL and CMOS/TTL.

14.1 Differential Lines

Differential lines are both similar and somewhat distinctive compared to single-ended lines. It has been emphasized previously that transmission-line propagation on wire lines requires two conductors to guide the signal which (hopefully) propagates at nearly the speed of light through the dielectric in the region between the conductors. Signals that are considered single-ended transmit the signal on one conductor and share the return line with multiple signals, as well as power distribution or other currents. Single-ended signal transmission is clearly the most popular and economical method for signal transmission over shorter distances. However, it requires a very low impedance, continuous ground return to control any interference with other currents that share the common return. Of course, it is to be understood that "ground" does not necessarily mean a connection to the earth. For example, the common "ground" for portable equipment, vehicles, or spacecraft is typically the conductive chassis or structure, which also connects to any cable shields.

Inductance is usually much more significant than signal line or ground return resistance for high-frequency impedance control. Single-ended signal transmission is usually not suited for long distances at high rates because the signal return system itself may have multiple signal transitions traveling on the common return. The common return system may include ground planes, chassis structures, and cable shields. The common return may have significantly different voltage levels at different locations if the distance between drivers and receivers is significant compared to the wavelength of the highest data and clock rates. Voltage differences due to combinations of high speed return currents flowing in the ground return may be referred to as standing waves.

Differential lines transmit the signal using at least two conductors, which is similar to single-ended signals. The distinction made here will be that both of the conductors are intended to carry only the current for an individual signal. Techniques are used to limit the other currents that can flow through either of the two conductors. This is normally achieved by the use of a differential receiver with a high input impedance for both of the inputs, with any terminations between the lines. The receiver will be powered by the local power and ground, but the signals are not terminated or decoupled to the local power or ground. Because interference results from two-way coupling, preventing other currents from flowing in the signal conductors also means that the signal currents are prevented from flowing in other paths. Differential signal transmission is effective in reducing crosstalk between signals and controlling interference to and from other systems.

The two conductors may have independent signals or may carry a single active signal and a reference to the receiver. A single signal transmitted with a reference does reduce the possible ground loop area and reduces the potential for other currents to return on the reference line.[1] If the two signals are complementary, the average signal level is a constant and the pair of conductors radiates less interference than a single signal and a reference. A differential line that carries complementary signals that sum to a constant and has the same impedance on both lines is called *balanced.* Conversely, an unbalanced line may have only one active signal, or the impedance may be significantly different on the two lines. For example, a TTL line with a diode to block V_{CC} charging may carry complementary signals, but the source impedance is unbalanced if the signal level results in a much higher source impedance for the HIGH state than for the LOW state.

Typical differential lines with only a single active signal use a signal that swings both above and below the reference line. For an example, a signal that switches between $\pm$ 5 V can be transmitted as an unbalanced differential line pair with a ground reference. The receiver detects the state of the signal with respect to the individual reference for that signal. The reference is connected to the receiver differential input, not the power-supply ground of the receiver. The power-supply ground of the receiver may be shifted compared to the reference, but the differential receiver detects only the signal difference, provided that the ground shift is within the common mode range limits of the receiver. The receiver detects the signal with respect to the *driver* reference (typically the driver ground), not the receiver reference. The receiver input is typically very high impedance, which essentially eliminates the possibility that currents from other loads could return on the reference line. The single active signal may be transmitted with respect to some other reference instead of ground. For example, a single active ECL signal could be transmitted with respect to the V_{BB} reference, as was illustrated in Fig. 13.9 for an ECL party line bus.

The most effective high speed digital data transmission lines over longer lengths are complementary, balanced differential lines. Both of the lines carry active signals, which switch simultaneously to opposite states. This means that the two signals sum to a constant, significantly reducing the coupling to and from other signals. Balanced lines also mean that the impedance and coupling is equal for both of the lines. This generally means that both of the complementary signals have the same source impedance and that the coupling to other lines is (nearly) the same to both of the lines. If balanced differential lines are terminated line-to-line, the line termination current flows only in the lines, not in the ground return.

Differential pairs may be located between two ground planes in printed circuits (including flex cables) to minimize the coupling with other signals in construction called stripline. Typical balanced differential lines in wire cables are twisted pairs to further reduce the interference both to and from other signals, including ground and/or shield noise. The twisted pairs may be individually shielded to minimize the coupling with other signals. Both the signal voltages and line currents sum to zero, so there is no change in the currents flowing in the ground return as the signal changes state. Currents that do flow in the ground reference return do not flow in the signal lines, so these currents have essentially no effect on the differential signal into the receiver. Ground and shield noise that does exist is coupled as common mode noise to both conductors nearly equally, so it is rejected by the differential receiver. Balanced termination impedance prevents conversion of common mode noise to differential interference . However, it should be understood that although there is no *signal* current flowing in the ground reference for a perfectly balanced signal pair, a ground reference is still required to carry whatever unbalance currents do occur for real systems, to carry the dc offset currents which flow because most receivers have a finite input bias voltage and current, to control the common mode voltage within the capability of the receiver, and to provide the ground system for cable shielding.

14.1.1 Comparison to single-ended lines

Single-ended transmission lines are characterized by information transmitted with respect to a reference that is common to multiple signals. Multiple data lines of a parallel bus transmitted above a common backplane (or mother plate or flex cable) ground are typical examples of single-ended signals. The common ground may be carried through connectors on multiple pins distributed throughout the connector pin field to continue the common reference for the multiple signals. The common reference may be shared with any other load currents that can return through the same path.

Coaxial cables (including ordinary oscilloscope probes) are also examples of single-ended signals. For usual applications, the coaxial shield carries no intentional signals except the return current for the signal on the internal conductor. The coaxial shield is usually connected to ground at both ends for high-frequency applications. The shield (like the ground reference plane for flex cables) is in parallel with other grounds. The system designer should be aware that single-ended ground references and cable shields may be return paths for currents from other system components. Therefore, alternate paths for large return currents are needed to avoid signal interference or the possibility of damage.[1] Of course, the safety ground is always required to prevent damage due to excessive voltages, including ESD damage if the signal reference may be disconnected when cables are removed. Connectors should be designed to discharge any static charge and to connect to the chassis ground before any signals can make contact.

The receiver for single-ended signals typically detects the input signals with respect to the common reference. Therefore, the signal reference at the receiver must be maintained at very close to the same level as at the driver to allow reliable signal detection. Examples of single-ended receivers include the following:

- TTL inputs which have an input threshold of about 1.5 V above ground, depending on the logic family.[2]
- CMOS inputs which have an input threshold that is proportional to the power-supply voltage. Typical full CMOS inputs have a threshold about midway between the V_{CC} power supply and ground, when the complementary input transistors conduct with about equal voltage across either transistor. However, some devices, especially those described as TTL compatible inputs, may have a larger pull-down transistor to lower the threshold closer to ground. However, the threshold is still proportional to V_{CC}, so the threshold increases at higher voltages. The input threshold may exceed the TTL maximum threshold of 2.0 V when V_{CC} exceeds 4.5 V.[3]
- ECL inputs which have an input threshold of about 1.3 V below the V_{CC} power supply rail. V_{CC} is ground for conventional ECL powered by a negative power supply V_{EE}. The ECL power supplies may be shifted positive, so that V_{CC} is positive (usually about +5 V shared with CMOS/TTL), and V_{EE} is at ground.

14.2 Differential ECL Interfaces

Most ECL devices are capable of providing true and complement differential outputs that are suited for driving balanced differential transmission lines. Differential lines are capable of eliminating the dc bias and signal currents that flow in the ground return loop of single-ended interfaces. Balanced differential lines generate less crosstalk and tolerate more common mode interference and ground offset than single-ended lines. Differential ECL receivers suitable for long differential lines are typically specified for $\pm$ 1.0 V of input voltage with respect to V_{BB}. This common mode range may be adequate for interface lines within a system that can maintain a low-impedance V_{CC} (ground) reference between the drivers and receivers.

14.2.1 Common mode signal range

Performance specifications should be carefully evaluated for parts being considered, as the performance as well as the input signal range may be specified differently for different parts. For example, a *Synergy Product Data Book*[4] specifies an xxE116 receiver common range as V_{CC} – (0.6 to 2.0 V) and notes that this means that the most positive of the differential inputs must be within this range and the differential signal within ± 1 V. The xxE416 receiver with two stages of internal gain (and a dc input sensitivity of ± 50 mV) is specified at a common mode range of V_{CC} – (0.0 to 1.5 V), with the same note as for the xxE116. However, the xxS314 receiver (S and E designate different ECL families) is characterized at a permissible common mode range of V_{BB} ± 1.0 V without an explanatory note. If this specification is interpreted in accordance with the notes for the xxE116 and xxE416, and $V_{BB} = V_{CC} - 1.3$ V, this results in a signal range for the most positive input of V_{CC} – (0.3 to 2.3 V).

The National Semiconductor *F100K ECL 300 Series Data Book and Design Guide*[5] describes the xx314 receiver as providing common mode rejection of ± 1.0 V, but the dc electrical characteristics specify a common mode voltage as V_{CC} – (0.5 to 2.0 V), without further definition. The xx316 differential receiver and line driver with 25 Ω cutoff outputs also specifies common mode voltage as V_{CC} – (0.5 to 2.0 V).

The user of any differential receiver is cautioned that descriptions of common mode performance do not necessarily mean that the receiver will operate properly in the presence of a ground shift of the claimed common mode in the presence of full-amplitude signals, with possible overshoot for underdamped lines. Performance is uncertain unless there is a definite specification and/or test for the specific condition. Since most receivers are actually sensitive to input signal voltage range, a common low-impedance V_{CC} reference between ECL drivers and receivers will tend to keep signals within the specified range. Reducing the differential signal excursions also reduces signal range, leaving more tolerance for ground shifts and common mode interference. Double terminations at both source and load reduce signal amplitude and overshoot at the input to the receiver. Schottky diodes for load termination effectively limit maximum signal amplitude without attenuating small signals. Users are also cautioned that unless a receiver is specified for input loading at signal levels more than a diode drop beyond the power rails or with power off, an active driver may be overloaded if connected to an unpowered receiver.

14.2.2 ECL signal levels

When the common mode range of the receiver is adequate, there are several alternative connections for driving and receiving differential ECL signals. The typical ECL signal swing for a single-ended line is about $\pm$ 400 mV with respect to V_{BB}. The typical differential signal swing is double that, or about $\pm$ 800 mV because both signal lines are active. Although both signals are active, transient analysis of a differential line can usually be considered as if it were a single line with double the signal amplitude compared to a single-ended line with only one line active. The differential signal can be conceptually considered as $\pm$ 800 mV with respect to an arbitrary reference such as ground, or it may be considered at 0 and 1600 mV, also with respect to an arbitrary reference, although the individual electrical signals are typical ECL signal levels.

Differential signals may be observed on an oscilloscope by combining the complementary signals in an "algebraic add" or "$A + B$" mode, with one of the channels inverted. If one channel is not inverted, the sum is double the common mode, which should be a constant for perfectly balanced complementary signals. The sum mode will display any complementary skew or other unbalance when the logic states switch. A typical switching unbalance is that the rise and fall times are not quite equal, so the complementary signals do not quite cancel during signal switching transitions.

14.3 Differential ECL Terminations

Differential ECL lines may be connected simply as a duplication of the single-ended ECL lines illustrated previously. One of the high-performance differential interface connections is simply to send each of the two signal lines on separate coax lines. In this case, the total differential line impedance is exactly double the Z_0 of each coax. The coax shield carries the signal return currents for each of the two signals, but the two opposite polarity return signals approximately cancel dynamic currents. A more common (and economical) interconnection for differential interfaces is some form of shielded pairs, in either twisted pairs, flex cables, or flat ribbon cable form (Fig. 8.11). Differential pairs in printed circuits are run as adjacent signal lines with a line-to-line spacing appropriate for the desired Z_0, and either sufficient spacing or shielding from adjacent signals to avoid differential mode interference. Differential pairs run as adjacent signals with controlled impedance between them will nearly cancel dynamic current transients and interference.

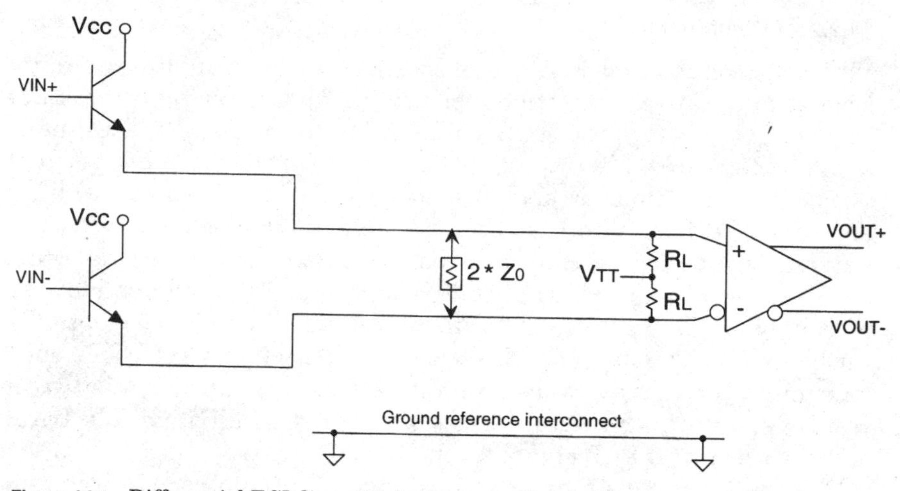

Figure 14.1 Differential ECL line with load termination to V_{TT}.

Differential lines may be terminated as if the complementary signals were two separate lines. Differential ECL signals run in separate coaxial lines may operate well if the lengths of coax are equal, but the individual return currents flow in each of the coax shields. An example of a differential ECL line terminated only with two load resistors to V_{TT} at the far end is illustrated in Fig. 14.1. The lines are terminated to a common reference, so dc currents do flow in the signal lines, and the signal returns are not isolated from other currents. Load and bias currents flow from the driver to the termination resistors, but the currents for the two signal switching signals sum to the same value. Therefore, there is no significant net dynamic signal current change between the source and load. As with the single-ended lines terminated at the load only, sufficient current must flow in the load resistor in the HIGH state to avoid cutoff of the output emitter follower in the driver on the falling edge transition.

When differential lines are terminated at the load only with no pull-down at the source, each line is terminated with half of the total appropriate load resistance to V_{TT} or the split-termination equivalent. Load termination will result in dc bias currents flowing in the ground return loop between the driver and the load(s). Although no significant dynamic signal-switching currents travel in the ground reference, a low-impedance return is required for the dc bias currents and to accurately control the relationship of V_{TT} at the load and V_{CC} at the source. Any shift of V_{TT} more negative increases the line currents, which increases power dissipated in the driver. Any shift of V_{TT} more positive may result in cutoff of the output in the LOW state.

Any crosstalk or line reflections of either differential mode or common mode traveling toward the load end are terminated at the load if the sum of the two load resistors is well matched to the differential line Z_0. However, any crosstalk or line reflections of either differential or common mode traveling toward the source end are reflected at the source because the driver is low-impedance if both outputs carry current. However, if the load resistors are too large or if V_{TT} is shifted too far positive, the LOW side may be in cutoff. If one side is much higher impedance than the other, then common mode disturbances traveling toward the source are converted to differential mode reflections because the cutoff side is high-impedance. Therefore, load termination only is not usually practical and is generally not recommended for long lines unless the transmission-line conditions and the ground reference are very well controlled.

14.3.1 Series terminations at the source

Source terminations are commonly used for differential lines because the dc load and bias currents only flow at the source. The load end may be left unterminated to minimize static power consumption if the line is short enough to avoid standing waves on the line due to reflections. The differential equivalent of two individual lines with the pull-down load resistor and the series termination at the source only is illustrated in Fig. 14.2. The characteristics of source termination only are that the initial signal excursion is smaller, which causes less crosstalk to other circuits, and that crosstalk, reflections, or other disturbances traveling toward the source are terminated at the driver.

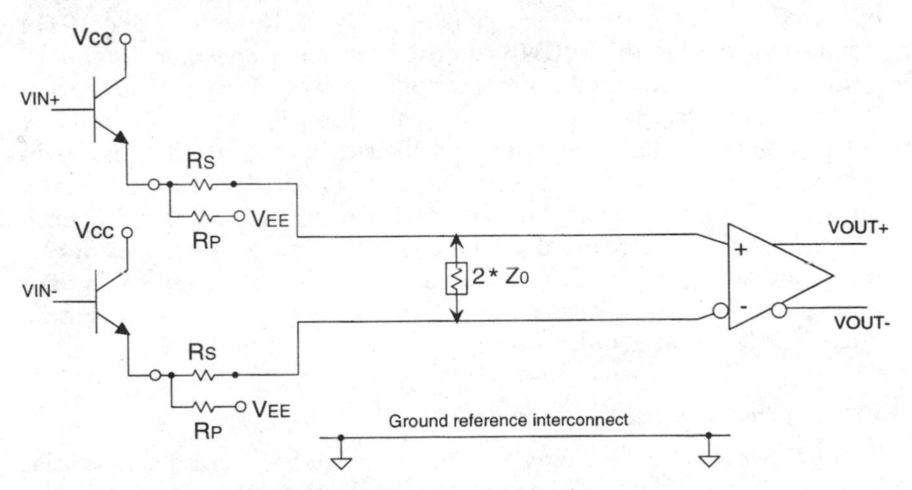

Figure 14.2 Differential ECL line with source termination.

The pull-down resistor at the source may be somewhat larger because the initial signal level required is less, and the signal level will increase at the far end. Both signals and any disturbances traveling toward the open load are nearly doubled at the far end.

One of the issues that should be considered for differential lines terminated at the load or open at the load is the result of disconnecting the interconnection when the receiver is powered. A differential line terminated at the load will tend to result in equal voltages on the two inputs to the receiver when the source is disconnected. If the differential line is terminated line-to-line, the inputs will tend to be at the same bias level as for open inputs. The results for open inputs tend to depend on the input circuitry of the receiver. Some receivers are protected against oscillation or other upset when inputs are equal levels or open by positive feedback called *hysteresis*. Receivers with internal hysteresis tend to hold the previous state after input signals are lost, or to power up to a preferred (but maybe undefined) state on power up with unconnected inputs. Typical ECL receivers do not include hysteresis.

Some differential ECL receivers include an internal offset bias so that equal inputs will result in a defined state; others have an internal bias so that open inputs will result in a defined state. Other differential ECL receivers require special attention to the results of open inputs because open inputs on one receiver may upset the current source to other receivers in the same package. For example, it is recommended that any unused inputs of the xx115 (quad) and xx116 (triple) receivers should be tied to the V_{BB} pin to avoid upsetting the current source bias network for all devices in the same package.[6] However, if the inputs are disconnected or otherwise unintentionally open, all devices in the same package may malfunction. The 10114 (triple), as well as the xxE116 (quint) inputs are specified to go to a defined state if both inputs are left open. However, inputs held exactly equal may result in outputs at undefined levels about midway between logic levels.[4,6] Allowing a receiver to operate in this high-gain region may result in amplification of noise or oscillation due to unintended feedback. Some of the other differential ECL receivers specify that internal bias circuits define what output state results from certain conditions on the differential inputs. For example, the xx314 (quint) receivers are biased so that the output goes to a defined state if the inputs are at equal voltages between the power rails.[5]

14.3.2 Schottky diode load termination

Parallel Schottky diodes may be placed line-to-line across differential lines for a nonlinear load termination, as illustrated in Fig. 14.3. The

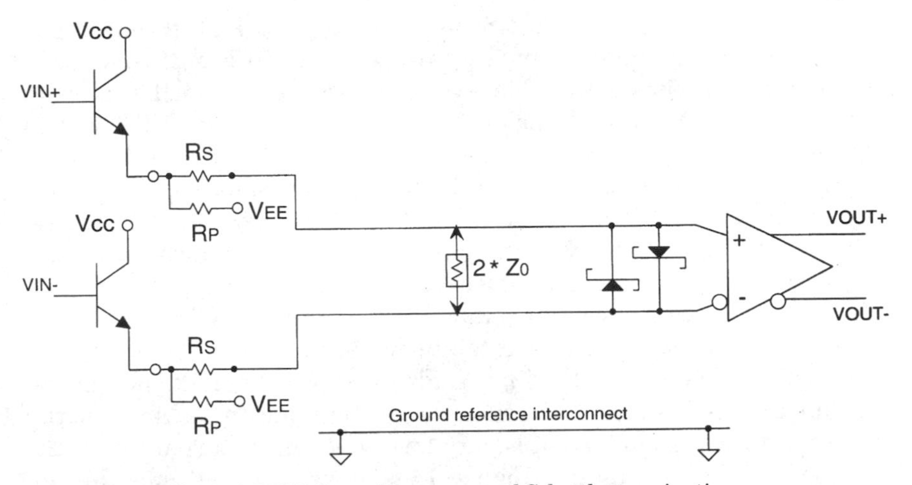

Figure 14.3 Differential ECL line with source and Schottky terminations.

parallel diodes provide a high impedance to differential signal levels that are less than the threshold at which the Schottky diodes begin to conduct. The operation of a differential transmission line in this low-current region can be considered as similar to an open line. Signal transitions in the low differential signal range would be nearly doubled, as if the line were open. This characteristic will result in a faster transition through the threshold region than if the line were terminated with a parallel resistor at the load.

When the differential signal level is large enough to begin forward conduction in the higher differential signal range, the current increases rapidly with little increase in differential voltage. The ECL differential voltage is typically about ± 800 mV, which is approximately double the typical forward voltage of a Schottky diode. The diode forward current is determined by the series resistance, which can be located at both the source and in series with the diodes at the load. The resulting differential voltage at the receiver will be slightly greater than for lines terminated with Schottky diodes to V_{BB} if the diodes terminating a differential line are conducting more current.

The variable conductance characteristic of the Schottky diode means that the diode current determines the equivalent termination impedance of the diode. This means that the source termination determines the equivalent termination impedance of the diodes at the load. The load will be exactly terminated with no reflections of the driven signal if the diodes are operated at a voltage-to-current ratio that exactly matches the transmission-line impedance Z_0. The source will be exactly terminated if the Schottky diode forward voltage that matches the line Z_0 is exactly one-half of the ECL differential voltage.

The forward voltage of typical Schottky diodes for ECL terminations is approximately one-half of the typical ECL differential voltage. Therefore, both ends of the differential line can be operated at close to exactly terminated conditions with minimal reflections of the driven signal.

There are several alternatives to consider in selecting the series resistance for an ECL differential line terminated by Schottky diodes. Lower series resistance will increase the currents, which may compensate for high-frequency losses but increase power consumption. Higher series resistance will reduce currents but requires reflection at the load to reach steady-state static signal levels. The Schottky diodes at the load end of the line result in nearly constant voltage levels but do not prevent reflections. The dynamic impedance of the diodes in forward conduction is very low, so small disturbances on the line will not be terminated. Figure 14.4 illustrates that one alternative is to add resistance in series with the Schottky diodes at the load. The dynamic impedance of the load end of the line can be well terminated, but the signal input to the receiver remains very nearly constant when the signal is sufficient so that the diodes are in forward conduction. This configuration provides an additional degree of freedom in limiting static power consumption, compensating for high-frequency losses, and providing for termination of small disturbances at both ends of the line. The voltage across the resistors is the difference between the line voltage and the receiver input. Small line disturbances may increase or reduce the line voltage but will have almost no effect on the receiver input signal.

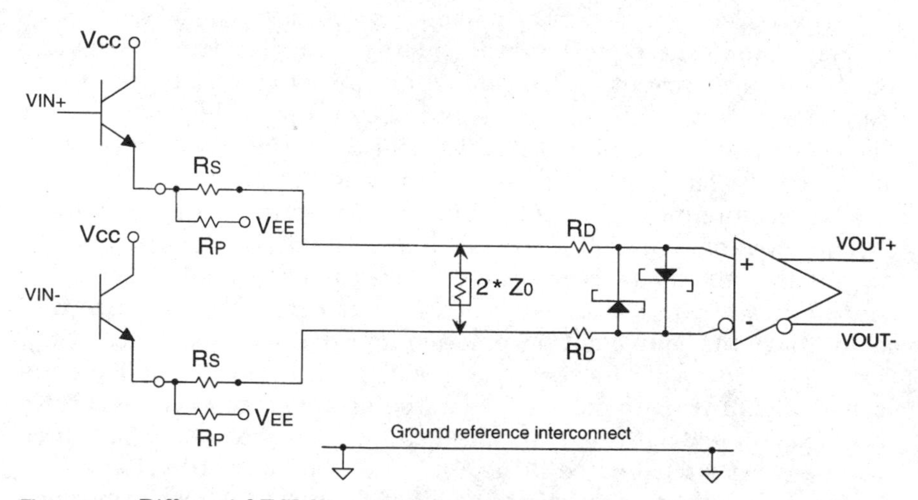

Figure 14.4 Differential ECL line with source and diode series resistors.

14.4 Differential Reflection Diagrams

The Schottky diode terminations are nonlinear, so reflection diagrams are appropriate for illustrating how a differential line responds to various termination resistance values. The input and output lines are symmetrical for differential lines, so only the rising edge transitions are illustrated. Figure 14.5 is a reflection diagram that illustrates an example of source termination that is exactly matched to Z_0, with Schottky diode forward voltage slightly greater than one-half of the differential ECL voltage. The steady-state static operating points where the two output characteristic lines cross the Schottky diode input line are indicated by an open circle. The diagram dimensions assume that the differential ECL signal voltage is $\pm$ 800 mV. The voltage and current scale is chosen to illustrate a differential Z_0 of 100 Ω at a unity slope (45° angle) for convenience in drawing the dynamic transmission-line impedance load lines.

The initial voltage excursion on the line is almost equal to the final diode voltage for this example. There is a very small reflection as the voltage is increased (and current reduced) slightly at the load. The reflection is exactly terminated at the source, so there are no further reflections or ringing. An operating point at slightly higher current would be required to reach the diode voltage and current that exactly match Z_0 for this example. Reducing the source resistance so that the

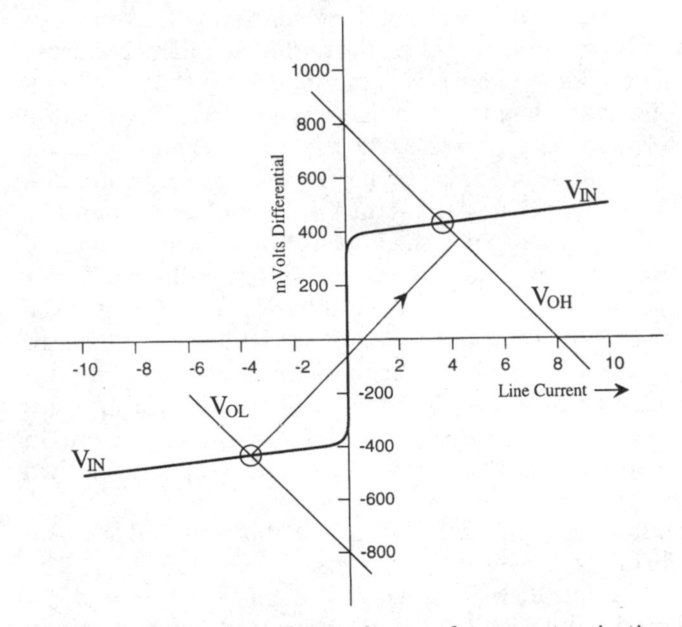

Figure 14.5 Rising-edge reflection diagram for source termination $= Z_0$.

initial voltage excursion matches the steady-state static operating level would eliminate the reflection at the load. Reducing the source resistance further would increase the initial voltage excursion beyond the steady-state static operating level, which may compensate for losses on high speed lines.

It should be noted the line current flows in a loop for differential lines that are terminated line-to-line. The differential V vs. I curves are mirror image identical about the zero voltage and current axis because the line current flows through one HIGH output and one LOW output. The differential output impedance of the source is the total of both of the internal impedance and the external series termination resistors. This is different than for single-ended lines, where the current for the HIGH output flows through a different path than for the LOW output. Because the differential V vs. I curves are mirror images, the two static operating points lie on a line which passes through the zero voltage and current origin. The slope of this line determines whether the initial voltage excursion is greater or less than the final static level. Lower static line currents result in a slope that is more vertical than Z_0, so the initial voltage excursion is less than the final static level. Conversely, higher static line currents result in a slope that is more horizontal than Z_0 and a greater initial voltage excursion than the final static level.

The only distinction between V_{OH} and V_{OL} of a differential line is the polarity of the current. The traditional convention is that positive current flows out of the true terminal and into the complement terminal of the driver when the true output is more positive than the complement output. The same magnitude of negative current flows out of the complement terminal and into the true terminal when the driver is in the opposite state. It should also be noted that the reflection diagrams for differential ECL lines illustrate only the current in the lines. The V_{OH} and V_{OL} lines are not the output lines of the individual states of a single ECL output. The current load of the pull-down resistor is in addition to the line current. If the pull-down current for the LOW output is not greater than the line current, then the transistor is in cutoff and the output impedance includes the pull-down resistor in series with the other impedance in the loop. The linear output lines in the reflection diagram assume that there is adequate current in the pull-down resistor to maintain the LOW transistor in forward conduction throughout the range of line currents indicated.

The variable conductance characteristics of Schottky diodes mean that the line performs well over a wide tolerance of conditions. The diodes will suppress small disturbances on the line with only slight changes in the signal voltage. The Schottky forward voltage is well above the threshold of differential receivers, so input signal voltages

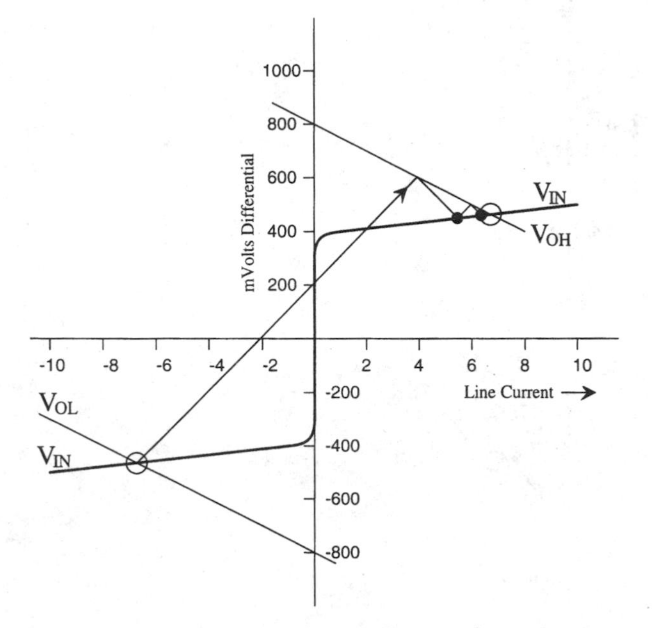

Figure 14.6 Rising-edge reflection diagram for source termination = $Z_0/2$.

near the Schottky forward voltage will result in stable output levels from the receiver. The Schottky diode clamping also limits the input signal levels, so the receiver can tolerate more common mode signal shift within the common mode signal range.

Figure 14.6 is a reflection diagram that illustrates an example of a lower source termination of $Z_0/2$. It is usually recommended for most lines that the source should be terminated with series resistance of less than Z_0. A good compromise for source termination for many lines is typically about $Z_0/2$. The static operating point is at higher current than for an exact termination at the source. Therefore, the initial signal excursion on the line is *greater* than the final voltage. Reflections result in an initial voltage at the load that is lower than the initial line voltage but then steps toward higher diode current at a slight increase in voltage. Both the source impedance and the dynamic (small-signal) impedance of the Schottky diodes in this region are less than the line Z_0, so the reflection steps are monotonic toward higher current and voltage at the load. There is very little change in the signal voltage at the receiver input at the end of the line.

A significant result of the source impedance lower than Z_0 is that there are reflections on the line, and the operating currents are higher. The higher initial signal excursion on the line may cause more crosstalk and other disturbances, but more signal amplitude may

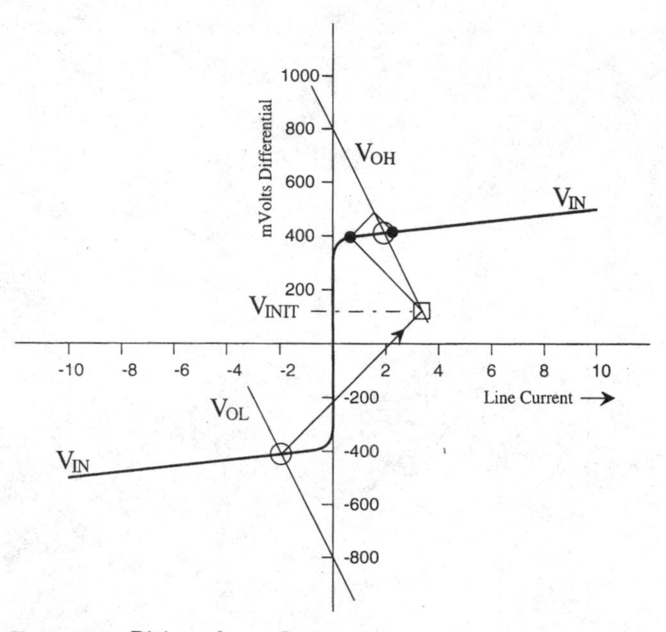

Figure 14.7 Rising-edge reflection diagram for source termination $= 2 * Z_0$.

compensate for severe losses on very long lines. Of course, higher pull-down currents are required to provide the higher line currents with the lower source impedance.

Figure 14.7 is a reflection diagram that illustrates an example of source termination that is twice the line Z_0. The steady-state operating conditions, compared to nearly matched to Z_0, are much lower line currents through the diodes but only slightly lower signal voltage at the receiver input. The initial signal excursion on the line is less than the final signal level but well above the zero voltage level. The reflection at the load increases the signal level at the input to the receiver, into the forward voltage conduction range of the Schottky diode. The second reflection increases the load current to just slightly above the steady-state static level. The source impedance is higher than Z_0, and the Schottky diode dynamic impedance in this region is lower than Z_0, so the slight reflections result in ringing in this region. Signal voltage at the input to the receiver is only slightly affected by the reflections that adjust the diode forward current to steady-state static level. All signal levels at the far end of the line are well beyond the receiver threshold, so there is no concern for reduced noise margin and no need for any additional time for signals to settle to valid levels. Any disturbances on the line that are small enough to stay within the forward conduction region of the diode will have no significant effect on the signal voltage at the input to the receiver.

The above examples of source termination resistance indicate that the Schottky diode termination of differential lines results in nearly ideal signal levels at the receiver over a wide range of conditions, with essentially no overshoot or undershoot under any reasonable condition. The Schottky diodes clamp the differential signal to nearly the single-ended ECL voltage excursion, which is well beyond the receiver threshold for excellent noise margin performance. The reduced signal excursion compared to an open line at the load also reduces signal transition times and reduces the dynamic power dissipated in lumped capacitance.

14.4.1 Three cases of source termination

The three cases illustrated in the reflection diagrams for differential ECL terminated with Schottky diodes may be summarized as follows:

- Exact source termination nearly eliminates reflections of the driven signal. The transient line current and the steady-state static diode forward current is about 4 mA for a 100-Ω differential line if the diode forward voltage is about 400 mV.
- The static diode forward current may be reduced by increasing the source termination resistance above Z_0, but a transient line current of almost 4 mA is still required to achieve full signal levels after the current is reduced at the far end of the line. Higher source resistance reduces the transient signal energy that travels on the line and reduces the static power dissipation but has almost no effect on the signal level at the receiver if losses are ignored.
- Reducing the source termination resistance below Z_0 increases the transient signal energy that travels on the line and power dissipation, with essentially no increase in signal levels at the far end. The higher initial signal level on the line may be useful to compensate for heavy losses on very long lines.

Signal performance at the far end with Schottky diode termination at the load end is essentially unaffected by operating conditions, so the choice of source termination resistance depends on the relative importance of the other issues. Source termination resistance less than Z_0 is recommended when it is important to increase the initial signal excursion to compensate for heavy losses. Nearly exact matching of the total source impedance to the line Z_0 results in the most nearly ideal performance, with no reflections of signals or crosstalk at the source end. Increased source termination resistance reduces power dissipation with lower signal excursions on the line but has no significant change in performance at the far end when losses are ignored.

14.4.2 Both source and load resistance with Schottky diodes

The previous reflection diagrams for series resistance only at the source illustrated that the low dynamic impedance of the Schottky diodes in forward conduction maintains a nearly constant differential signal voltage at the receiver when the diodes are in forward conduction. However, this means that small line disturbances traveling toward the receiver will be reflected with a change in current to maintain nearly the same voltage across the Schottky diodes. A reflection diagram with series resistance at both the source and the load with Schottky diodes is illustrated in Fig. 14.8. This reflection diagram illustrates a total series impedance of Z_0 at both the source and in series with the Schottky diodes at the load. The characteristic of the far end of the line, including the series resistors, is labeled V_{LINE}. The input to the receiver remains as the V_{IN} operating characteristic, but this is no longer the total load impedance seen at the end of the line. The difference between the V_{IN} and V_{LINE} characteristics is the drop across the R_{IN} series resistors at the load. Total impedance at

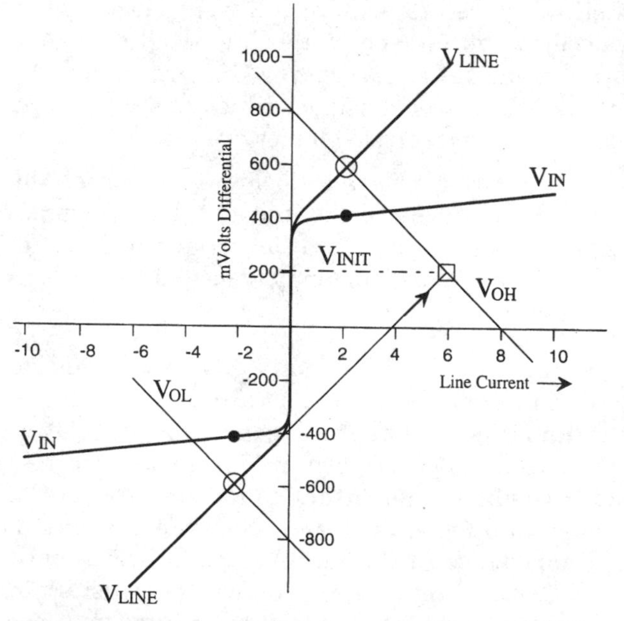

Figure 14.8 Reflection diagram for both source and diode terminations = Z_0.

both the source and load equal to Z_0 means that sufficient resistance has been added in series with both ends so that the slope of both the output and input characteristics is the same as the dynamic impedance of the transmission line. The actual resistance added at each end is a little less than Z_0 to allow for the internal impedance of the driver transistors at the source and the Schottky diodes at the load. One-half of the resistance is added in each line to maintain balance. The voltage and current scales for these reflection diagrams have been chosen for convenience so that a dynamic impedance of 100 Ω is drawn at a slope of unity.

The total loop resistance in the line for this case is $2 * Z_0$, so the steady-state static current on the line is the same as for the $2 * Z_0$ resistance at the source only. However, the line response is different because there is resistance at both the source and load ends of the line. The steady-state static voltage on the line is greater than the receiver input voltage. The conditions illustrated result in a steady-state static current of about ± 2 mA for $Z_0 = 100\ \Omega$. This means that there is about a 200-mV drop across the source resistors and about ± 600 mV across the line. The source impedance is equal to Z_0, so the initial signal excursion into the line at the source is one-half of the ± 800 mV (total of 1600 mV) no-load differential voltage, or about an 800-mV excursion one way. This initial excursion results in an initial voltage signal of about 200 mV, which is indicated by the square at the intersection of the Z_0 load line and the V_{OH} output characteristic. The initial signal level on the line is greater than for $2 * Z_0$ resistance at the source only. When the initial signal arrives at the load, the voltage increases (and current decreases) to the steady-state static level. The steady-state static conditions on the line are indicated by the large open circles at the intersection of the V_{LINE} input characteristic and the V_{OH} and V_{OL} output characteristics. The steady-state static receiver input voltage is indicated by the solid dot, which is located on the V_{IN} Schottky diode characteristic at the same current as the line, ignoring any current into the receiver itself. The line is at static equilibrium conditions when the reflection returns to the source, ignoring any losses on the line. The dynamic signal levels at the load will not be as large as illustrated by the reflection diagrams if the line losses are significant.

A desirable characteristic of dynamic impedance exactly matching Z_0 at both ends is that small differential disturbances on the line are exactly terminated at both ends. Exact termination does prevent reflections of small disturbances, but it also means that any line disturbances traveling toward the load will change the load voltage to absorb the signal energy without reflection. However, the diodes on the receiver input mean that although the line voltage responds to

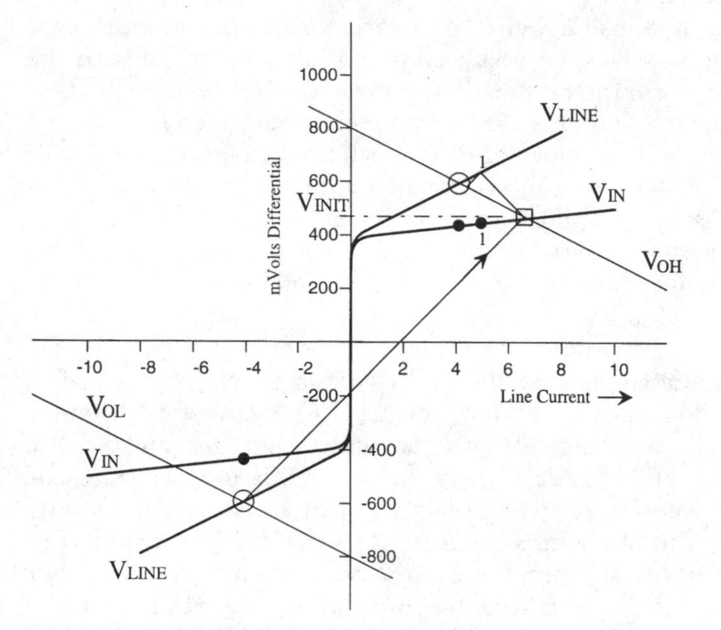

Figure 14.9 Reflection diagram for both source and diode terminations = $Z_0/2$.

line disturbances, the diodes provide almost constant input signal voltage to the receiver. The nonlinear load means that there may be a reflection of the large signal transition, although there is no ringing. It should also be noted that line-to-line differential terminations do not terminate common mode disturbances on the line, although they are ignored by the receiver if the signal levels remain within the input signal common mode range.

The resistance values for the terminations in series with the source and the diodes may be different from Z_0 to trade off signal levels and line currents. Figure 14.9 illustrates both the source and diode impedance at $Z_0/2$. The steady-state current is higher, but the 600-mV line voltage is the same as for both the source and diode impedance at Z_0. In general, all cases of equal source and diode impedance result in a line voltage at the average of the diode voltage and the no-load output voltage. The ratio of diode impedance to source impedance determines how far the line voltage rises above the diode voltage after the reflection when the signal reaches the diodes.

The initial signal excursion on the line is significantly increased for $Z_0/2$ at both ends, to about 400 mV as indicated by the square for the rising edge on the reflection diagram. This level nearly coincides with the V_{IN} characteristic of the Schottky diodes for this case, but this is just coincidence (although generally desirable) for the conditions for this case. The lower source and diode impedance may be considered a more aggressive configuration because the initial signal excursion is very nearly the full steady-state signal levels, and the steady-state currents are higher. The higher initial signal excursion also means that the reflection that returns to the source is significantly reduced.

There is a small reflection and slight overshoot at the far end of the line, ignoring any losses. The first incident wave signal level at the far end of the line and at the input to the receiver are indicated by the number 1 in the reflection diagram. The dot on the receiver input V_{IN} characteristic for the steady state static level is only slightly lower than the level for the first incident wave. The current decreases monotonically toward the steady-state static level because the dynamic impedance of both the source and load is less than Z_0.

Figure 14.10 illustrates the midpoint waveforms on the line for the two cases of both source and diode impedance equal to Z_0 (*a*) and $Z_0/2$ (*b*). The input to the receiver is the minimum of the line signal level or the diode voltage of approximately 400 mV. The $Z_0/2$ case has an initial signal excursion of at least the diode voltage, so the signal rises to about the diode voltage at all points along the line as the first incident wave travels to the far end. The line voltage at the far end rises above the diode voltage with a very slight overshoot, ignoring losses. High speed lines with significant skin-effect losses would have less overshoot and lower signal levels on the line than illustrated by the reflection diagrams. The case of Z_0 termination at both ends has an initial step at the midpoint, with no overshoot on the line.

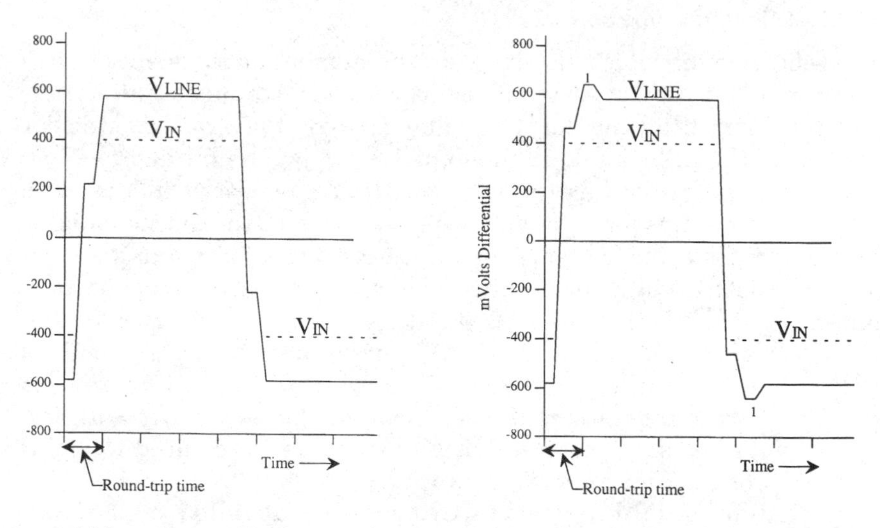

Figure 14.10 Midpoint waveforms comparing Z_0 and $Z_0/2$ source and diode terminations.

The steady-state line current is high enough that even significant line disturbances will have very little effect on the receiver input signal. Line losses may significantly attenuate the leading edge of the initial signal excursion without much reduction in the receiver input signal level. For example, the termination with $Z_0/2$ at both ends of the line can tolerate an initial signal excursion at the far end of as low as only 70 percent of the initial signal launched by the source with very little change in the signal level at the receiver input. The termination with Z_0 at both ends tolerates as low as 80 percent of the initial signal launched by the source. However, any significant loss of the signal launched with $2 * Z_0$ at the source only will reduce the signal excursion at the far end. Lower source resistance may be appropriate to compensate for line losses because the initial signal excursion on the line is greater.

The combination of source and diode resistance of about $Z_0/2$ may be suitable for multiple terminations on lines with multiple receivers along the line because the initial signal excursion on the line is very nearly equal to the diode voltage. Differential disturbances on the line traveling to either end are reasonably terminated because both the source and load dynamic impedance are close to the line impedance. The resistance in series with the Schottky diodes absorbs any changes in the signal level on the line, so that the input signal level to the receiver is essentially unaffected by disturbances or losses on the line.

14.4.3 Differential line currents

The reflection diagrams also indicate the maximum transient or static current that the pull-down resistor at the source must sink to provide the differential current to the line. A current of 4 mA is sufficient to support the full signal excursion of 400 mV at the far end for $Z_0 = 100\ \Omega$ and no resistance in series with the Schottky diodes. The reflection diagrams for no resistance in series with the diodes indicate that differential line currents do not exceed 4 mA for a source termination resistance of Z_0 or greater. This current can be supplied by a pull-down resistance of about 200 Ω to a V_{TT} of −2.5 V, or 800 Ω to −5 V to maintain a V_{OL} of −1.7 V. Of course, the actual design values must include an allowance for tolerances. Additional current is required for the higher signal levels for resistance in series with the Schottky diodes, so a lower pull-down resistance is required to maintain a V_{OL} of −1.7 V at the driver output.

If the differential line current exceeds the capability of the pull-down to keep the LOW output at V_{OL}, then the LOW output impedance becomes $R_S + R_P$. Figure 14.11 illustrates the change in the differential V_{OH} and V_{OL} operating characteristics when the LOW output

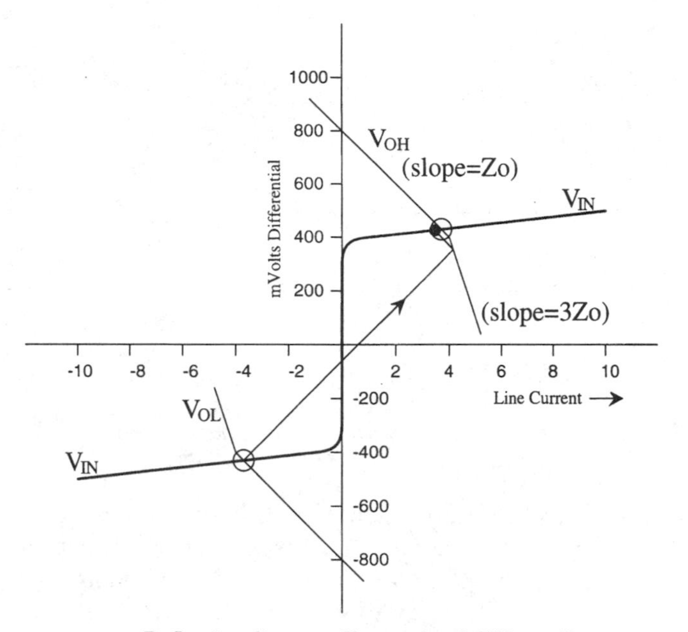

Figure 14.11 Reflection diagram illustrating LOW cutoff.

transistor goes into cutoff at 4 mA of line current. This figure illustrates the sum of the series resistance equal to $Z_0 = 100\ \Omega$ and the pull-down resistor of 200 Ω to $V_{TT} = -2.5$ V. Therefore, the source impedance changes from Z_0 below 4 mA line current to $3 * Z_0$ above 4 mA of line current because the LOW output transistor goes into cutoff. This change in source impedance has essentially no impact on the line response and the input signal levels into the receiver. However, it is not recommended that the driver be allowed to remain near cutoff for steady-state operating conditions because the driver impedance is unbalanced. If the driver output impedance is not balanced, common mode crosstalk traveling toward the driver will be converted to differential mode crosstalk and reflected to the receiver as a differential disturbance.

When there is no resistance in series with Schottky diodes at the load, it is recommended that the sum of R_S in the true and complement lines should be high enough so that the initial signal excursion that arrives at the load is a little *less* than the static operating point of the diodes, to minimize both reflections and static currents. This generally means that the forward diode should conduct current at an operating point that results in a ratio of voltage to current that is a little *higher* than the total differential line impedance. Since the dynamic slope of the diode forward conduction characteristic is low-impedance, this means that the steady-state static current is less

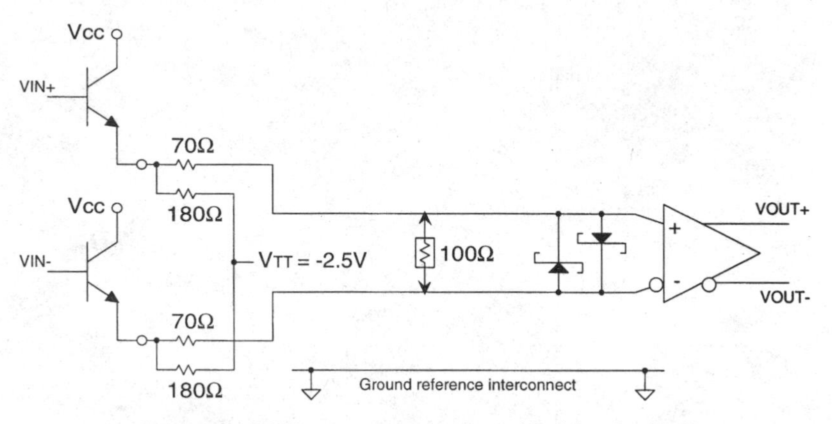

Figure 14.12 Example of ECL source termination with Schottky diodes.

than for an exact match to the line impedance, but the steady-state signal voltage is only slightly lower. The line response is similar to a slightly underdamped double-terminated line, but the variable conductance characteristic of the Schottky diodes prevents ringing. The source impedance is also a little *higher* than Z_0. The signal level at the receiver is about one-half of the unloaded differential swing, or about exactly equal to the single-ended ECL signal level. The Schottky diode termination maintains nearly the same differential signal level over a wide range of conditions.

Figure 14.12 illustrates an example of pull-down to $V_{TT} = -2.5$ V and source termination resistance values for a 100-Ω differential line terminated with parallel Schottky diodes. The series resistance ratio of 1.4 provides line performance about midway between the exact match and the $2 * Z_0$ illustrations, with good source impedance matching so that either differential or common mode reflections and crosstalk traveling toward the source are well terminated. The actual series resistance added to the output would be less to allow for the impedance of the output transistor, which is typically about 5 to 10 Ω for ECL drivers. Differential disturbances traveling toward the load are reflected with a current reversal, but very little change in voltage across the Schottky diode inputs to the receiver. Line-to-line terminations do not provide common mode termination, so common mode voltage disturbances traveling toward the load are nearly doubled. The ground reference interconnect does not carry signal current but is required to carry any offset currents and maintain the receiver input signal within the common mode signal range of the receiver.

Figure 14.13 is a reflection diagram illustrating the source impedance of $1.4 * Z_0$ with Schottky diode terminations at the load. The steady-state static current is less than for the source impedance equal

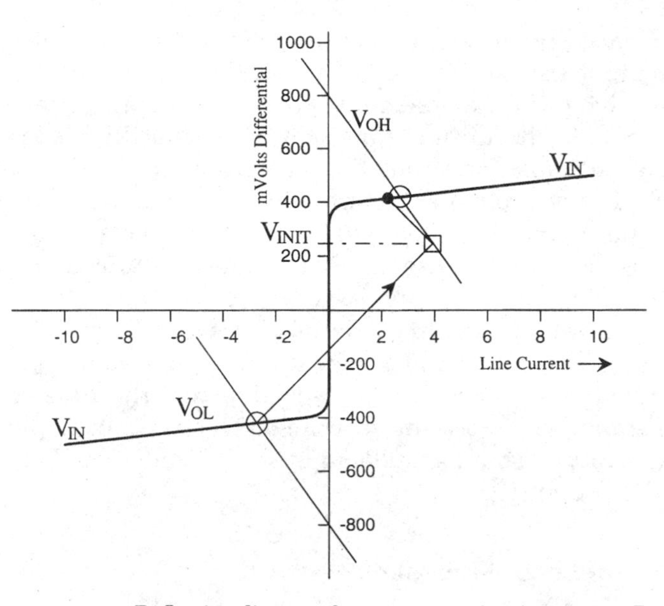

Figure 14.13 Reflection diagram for source termination = $1.4 * Z_0$.

Z_0 case, and the initial signal excursion on the line is a little higher than for the source impedance equal $2 * Z_0$ case. Higher initial signal excursion on the line means that there can be more loss of initial signal amplitude while achieving forward conduction on the Schottky diodes so that the receiver input signal is unaffected.

The parallel Schottky diodes tend to "adapt" to the input signal level. When the input signal level is lower than the diode threshold, the diodes are high-impedance and the positive reflection increases the signal level. When the signal excursion is beyond the diode threshold, the current increases rapidly to clamp the signal level. Once the signal level is sufficient so that the diode is conducting current, the dynamic impedance to further signal excursion or to any smaller disturbances is very low. Therefore, the signal level is stabilized above the diode threshold. Any smaller disturbances such as overshoot, ringing, reflections from line discontinuities, or crosstalk have little effect on the differential signal amplitude.

A differential line terminated with Schottky diodes at the load should be terminated with series impedance at the source to operate the diodes well into the forward conduction region for both dynamic and steady-state static conditions. The typical ECL differential signal swing is significantly greater than the Schottky diode threshold, so the source impedance determines the steady-state static current in the diodes as well as the reflections that adjust the initial signal level to the static signal levels of voltage and current at the load.

When high-frequency line attenuation on long lines is of more concern than minimizing current, the sum of R_S at the source may be reduced to increase the initial excursion into the line. If high-frequency skin effect attenuates the initial transition, the Schottky diodes will allow the signal to transition through the receiver threshold as if the line were open but will clamp the subsequent line charging to limit the static signal excursion. This will reduce the effects of line losses on pattern sensitivity because the line achieves static conditions more rapidly than if the lossy line is allowed to continue slowly charging to higher levels during periods of inactivity. The diode clamping eliminates overshoot and undershoot effects and maintains nearly constant signal levels to reduce the variability in the time at which new signal transitions cross the threshold. This is an example of the "large signal capture, small signal suppression" nonlinear characteristic of digital processing.[7]

14.5 Parallel Resistor Load Termination

Figure 14.14 illustrates a differential ECL line with a traditional line-to-line resistor termination at the load. When differential lines are terminated line-to-line at the load, the transient and static signal currents flow in the signal wires. The ground current is insignificant if the receiver input impedance is high. Many ECL differential receivers even eliminate the input pull-down resistor that stabilizes most open inputs at a logic LOW.

When the common mode noise and possible ground shifts are of more concern than line losses, a resistor divider at the load may be used to attenuate the common mode voltages. Of course, the differential signal is also attenuated, so the attenuation must be limited to

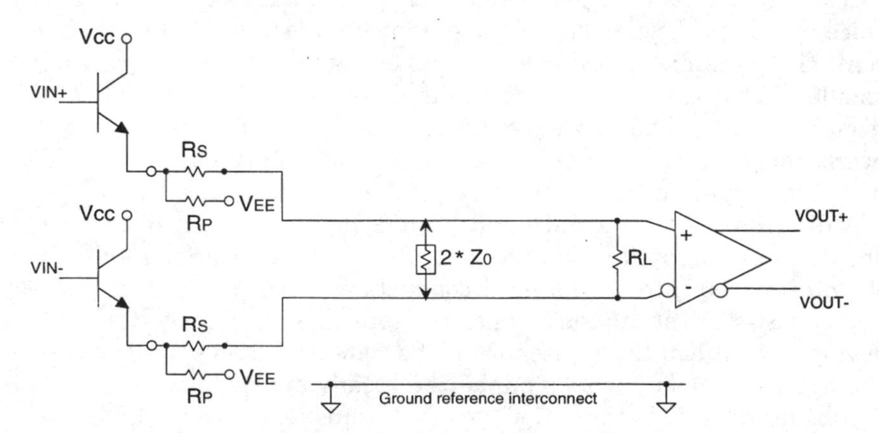

Figure 14.14 Double-terminated differential ECL line with resistor terminations.

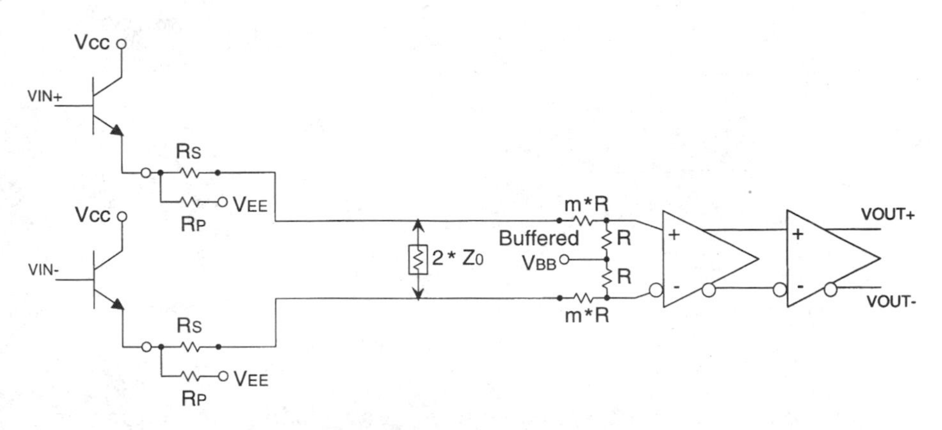

Figure 14.15 Differential ECL line with attenuation to V_{BB} reference.

provide at least the minimum signal required by the receiver. The common mode attenuation must be with respect to some reference at the receiver. The attenuation reference should be reasonably centered in the input signal range of the differential receiver to achieve maximum common mode range with minimum signal attenuation.

An example of an ECL line with an attenuator referenced to V_{BB} at the receiver is illustrated in Fig. 14.15. The total termination load resistance across the differential line is the sum of the series and parallel resistors, or $2(m + 1)R$ as shown. The load resistance is center tapped to the receiver reference to shift input signals toward the reference. The attenuation ratio at the receiver input is determined by the ratio of the series resistance to the parallel resistance. When the series resistance is m times the parallel resistance, the attenuation ratio is $1/(m + 1)$ at the receiver. The attenuation means that the common mode and the differential signal at the receiver input is reduced by a factor of $1/(m + 1)$ relative to the levels on the line before the attenuator. Of course, the total attenuation with respect to the driver open-circuit voltage includes the series resistor at the source and any line losses at the frequency of concern. Common mode voltage on the line includes both ground shifts and crosstalk or other interference which shifts the signal level on both of the differential lines.

V_{BB} is shown as the attenuation reference because it is the center (average) of the ECL signal HIGH and LOW signal levels. Therefore, there is no current in the ground return loop from the driver through the receiver V_{BB} supply when there is no ground shift. However, current will flow through this return loop if there is a ground shift caused by external sources. Differential lines are typically operated with a low-impedance ground connection between both ends of the line, including cable shields to provide a ground current return path.

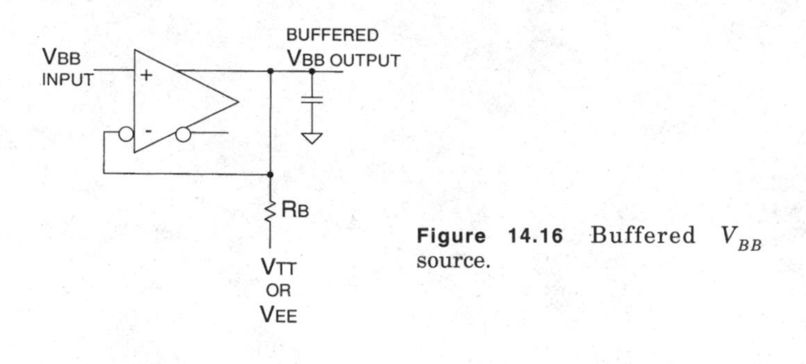

Figure 14.16 Buffered V_{BB} source.

Higher currents and transients from heavy loads such as motors, solenoids, and other power devices must be isolated from low-level signal returns to prevent malfunction or damage to sensitive signals.

The V_{BB} reference output pin of an ECL receiver is not intended to drive transmission-line currents, so it is important that the V_{BB} reference to the transmission line is buffered to provide a low impedance for the attenuation. Figure 14.16 is an example of the use of a differential receiver to provide a buffered V_{BB} reference source. The buffered output also has a high-frequency bypass capacitor to provide low output impedance to high-frequency line disturbances.

V_{BB} is an appropriate reference for the common mode attenuation if the receiver input signal range for proper operation is reasonably centered about V_{BB}. However, if the appropriate range is not centered about V_{BB}, then the input common mode range may be greater if a different reference is provided. Some ECL receivers without internal level shifting may function properly for much more negative shifts than positive shifts of the logic levels. The common mode range of these may be improved by a more negative attenuation reference.

Figure 14.15 also includes two stages of differential receiver to increase the differential gain. The additional receiver gain is not necessary if the signal input to the first stage is always large enough to assure full output levels. However, higher gain may be desirable if the attenuation ratio is high or if there are significant losses on the line at the frequency of concern. Furthermore, the gain of typical ECL receivers drops with increased frequency, so additional gain will increase the usable upper-frequency range. Receivers incorporating two stages of internal gain and a specified minimum input signal requirement in a single package, such as the xxE416, are preferred if available.[4] Two separate stages of individual receivers would require the additional load resistors on each output and result in the uncertainty of the unspecified minimum input signal required to overcome any differential offset in the first stage.

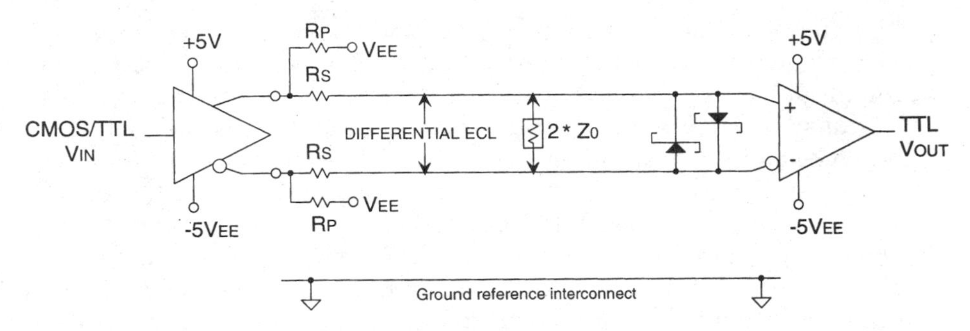

Figure 14.17 Differential ECL line for CMOS/TTL interface.

14.6 ECL Translators to CMOS/TTL

Translators between differential ECL and CMOS or TTL are available to implement differential ECL transmission lines with direct translation to and/or from TTL at either end of the line. Dual-voltage translators for interfacing with conventional ECL powered by −5 V are typically available for common ECL logic families. Both the ECL and TTL signals are referenced to the common ground. Illustrated in Fig. 14.17 is an example of a differential ECL interface between TTL at both ends. The differential ECL transmission line provides high-frequency, low-level, low-noise interconnections with low timing skews. Although this figure illustrates TTL interfaces at both ends, the signals at either end may be purely ECL. The figure illustrates the double-terminated ECL transmission line with both series terminations at the source and parallel Schottky diodes at the load. Schottky diodes are recommended to provide noise immunity in the presence of CMOS/TTL switching noise. However, source termination with an open load or a line-to-line parallel resistor load termination are alternate options.

Source termination only, with no load termination, is often appropriate for these lines because the TTL at one or both ends tends to limit the data rate to well below the maximum ECL capability. If the line has only a single load and is short enough that the reflections from the open receiver return to the source before another transition is launched, signal levels will stabilize to nearly static levels before each transition. This configuration requires the ground reference interconnection which is commonly integrated with cable shields and chassis connections.

14.6.1 Power-supply considerations

The differential ECL transmission line with V_{CC} at ground generally tolerates differences in the voltages, including loss of any power-supply voltage, or the sequence in which power is applied or removed. Since standard negative ECL signals are referenced to the common ground, differences between the power-supply levels at the ends of the transmission line are of little consequence as long as the common ground remains connected between them. The principal limitation is that receiver inputs are clamped to V_{CC} and V_{EE}, so unpowered receivers may prevent signals from exceeding ground by more than about a diode drop in either direction. ECL outputs do not have an active pull-down transistor, so the outputs are not overloaded by a clamp to ground from unpowered inputs. If there are multiple receivers on a party line, any unpowered receiver may result in malfunction of the entire bus. Therefore, an ECL party line bus is not suited for "live insertion" unless the receiver is fully powered before the bus signals are connected. All receivers must be powered for proper operation of the bus, and the common ground must remain connected.

Although an unpowered receiver clamps inputs near ground, this does not overload active ECL driver transistors which only source current to the HIGH state. The pull-down resistors may draw current through an unpowered receiver clamp diode, but this should be within ratings and is further limited by series resistors for source termination. Typical TTL translators have no sequencing requirements for the +5-V TTL supplies, so loss of +5 V for the TTL interface at either end is of no concern except that normal operation requires appropriate power. Therefore, typical ECL interfaces tolerate different −5-V power supplies at either end. Typical dual-supply ECL interfaces with translators to positive logic at either or both ends tolerate different power supplies at either end, or power-supply faults. The power supplies do not require power-supply sequencing controls. The principal requirement is for a common ground reference that keeps the receiver signals within the common mode range, as well as providing a low-impedance path for returning any unbalanced offset currents. The ground reference system is usually integrated with the shielding system for EMI and ESD protection.

14.6.2 V_{TT} termination supply considerations

ECL interfaces that use a V_{TT} termination supply do need to ensure that it does not become excessively negative with respect to V_{CC}. This is typically ensured by regulating V_{TT} as a proportion of $V_{CC} - V_{EE}$ and referenced to V_{CC}. If the termination supply fails, it defaults to V_{CC} = ground. Since unpowered drivers result in both true and complement outputs at a logic HIGH, the total current loading to the termination supply may increase beyond the normal operating conditions. Therefore, if a termination supply may remain ON when the source of signals is unpowered, the maximum loading to all signals at HIGH levels should be considered. The failure of the −5-V supply to the driver does not overload the driver; it does result in all outputs at a logic HIGH. This increases the total current to the V_{TT} supply. However, if the V_{TT} supply fails to V_{EE} instead of to V_{CC}, this may seriously overload drivers terminated to V_{TT}.

Caution: Failure of V_{TT} to excessively negative levels may overload the drivers.

One of the disadvantages often considered for the use of an ECL interface between larger TTL systems is that the standard configuration requires both TTL supply voltages and power-supply voltages for the ECL drivers, terminations, and receivers. The separate V_{TT} termination supply may not be cost-effective for small systems or just a few interface lines, so it may be eliminated. ECL signals may be parallel terminated with a split termination to ground and −5 V_{EE}, and/or series termination with the pull-down resistor to −5 V. The split termination to ground and −5 V_{EE} automatically defaults to ground when the −5 V_{EE} fails to ground. Line-to-line terminations (resistor or Schottky diodes) for differential lines do not require a separate termination supply at the receiver.

14.6.3 Distributing a single −5-V ECL supply

Long line interfaces between units are often powered separately at the two ends. Providing an ECL −5-V supply at both ends of the line may be considered a significant impact, especially for only a few interface lines between ends that are primarily CMOS/TTL. A single −5-V ECL supply can provide current to both ends of the interface if additional lines in the interface cable are used to provide −5-V current to the other end. Figure 14.18 illustrates sending the −5 V_{EE} supply voltage for the receiver through the interface cable. The −5-V supply should usually be located at the end with the most drivers, since drivers and the pull-down resistors draw more current than receivers. Power lines in the interface should be provided as twisted pairs or equivalent forms of wire pairs to minimize the current loop area and reduce interference to or from other ground return currents. Pull-downs and series termination at the source, and line-to-line termination (resistor or Schottky diodes) at the load are recommended to limit signal reflections and signal return currents in the ground reference interconnect.

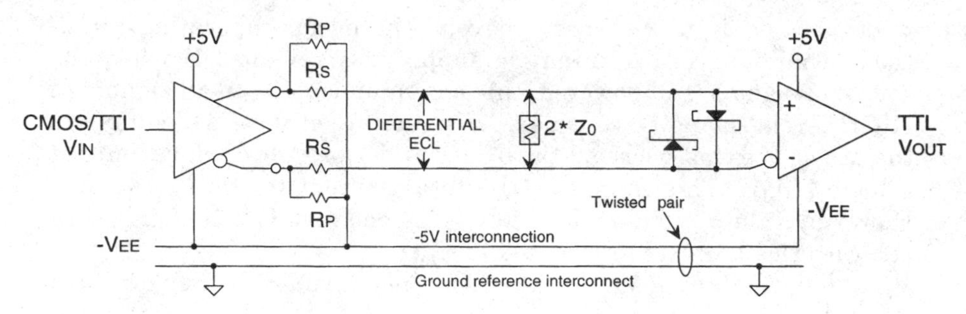

Figure 14.18 Differential ECL line with $-V_{EE}$ interconnection to receiver.

14.6.4 Single supply PECL and CMOS/TTL

ECL can be operated on a single +5-V power supply, shared with CMOS or TTL. In this application, the ECL V_{CC} is the same +5-V supply as for the CMOS/TTL, and the ECL V_{EE} is ground. In this case, the ECL is called positive ECL, or PECL (some consider that PECL stands for pseudo ECL). Of course, ground does not necessarily mean a connection to the earth in the case of such applications as portable equipment, vehicles, and spacecraft. Ground generally means the common reference of the equipment chassis and the cable shields. One of the important issues in ECL and TTL interfaces is that ground is the level to which the power-supply voltages default when unpowered. Also significant is the driver output level, which is lower-impedance; TTL outputs are lowest-impedance in the LOW state, which is ground when normally powered from a positive power supply. ECL outputs are low-impedance in the HIGH state, which is ground when normally powered from a negative supply. Therefore, both TTL and normal ECL powered by a negative supply tolerate loads to ground. ECL outputs from a negative supply are not overloaded by loads or termination voltages that fault to ground, or even short circuits to ground. However, there is serious risk of excessive currents if an ECL driver powered by a *positive* supply is loaded to ground.

Single supply translators can be used as the drivers and/or the receivers between CMOS or TTL and differential ECL transmission lines operated on the same +5-V power supply.[5] A differential PECL transmission line with CMOS/TTL interfaces is illustrated in Fig. 14.19. The figure illustrates a double-terminated PECL interface with parallel Schottky diodes as the load termination. Schottky diodes are recommended to stabilize the PECL signal levels in the presence of CMOS/TTL switching noise or other line disturbances. Schottky diodes also allow the use of series termination resistors that are larg-

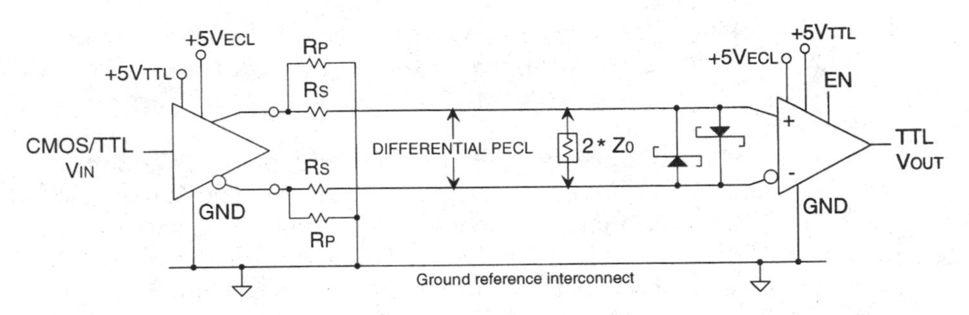

Figure 14.19 Differential PECL line for single-supply CMOS/TTL interface.

er than Z_0. The larger series resistors will tend to limit return currents that may flow if the +5 V_{ECL} at the driver is more positive than the +5 V_{ECL} at the receiver. However, output series resistors appropriate for line termination are not necessarily sufficient for protection during power-supply faults.

Figure 14.20 illustrates typical ESD clamp diodes that may be present at inputs and outputs of typical high speed semiconductor devices, including ECL, CMOS, and the various TTL families. Not all devices have all of the diodes illustrated for ESD protection, but some of the interfaces may have other semiconductor devices that tend to conduct if the signal lines are driven beyond the power rails. Vendors may change the internal structure of their parts, or alternate suppliers may have different structures if there is no specification that limits currents when signals are driven beyond the power rails. Some parts specifications may define the presence of input clamp diodes with an input clamp voltage specification. However, the device does not usually include internal current limiting to protect the diodes from excess currents. Typical devices with clamp diodes usually specify an absolute maximum rating for the maximum external input and output current and/or voltage that is required to protect the clamp

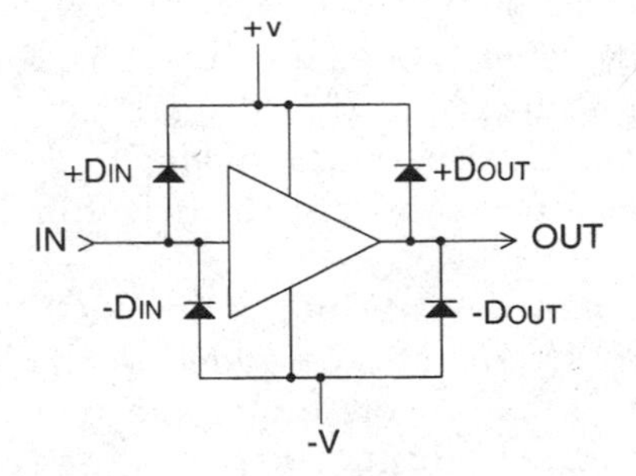

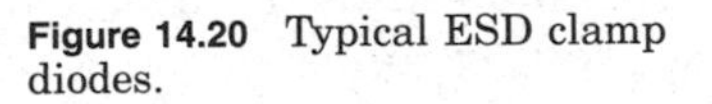
Figure 14.20 Typical ESD clamp diodes.

diodes. Unless input and output devices incorporate and specify internal current limiting, the designer must add external protection at interfaces which may otherwise result in conditions which exceed safe limits.

+5-V power is supplied to both the TTL and ECL power pins of each translator. The power pins are separate in the translator package to limit the interference between the low-level PECL signals and high-level CMOS/TTL switching noise. The layout of the parts and power planes should separate the PECL power and ground from high-level switching noise, but the +5-V power must be at nearly the same dc level at all times. Receiver power must be applied at all times when drivers are powered to prevent overloading of the drivers by the clamp diodes of unpowered receivers.

A PECL interconnection system does have unique characteristics and limitations which must be considered. One of the most important is that the signal ground reference interconnect, as was shown in the previous figures for traditional ECL, becomes a +5-V reference interconnect for PECL. It is recommended that the +5 V for the PECL be physically separated (i.e., separate voltage planes) from the +5 V for the TTL to avoid TTL switching noise interference on the PECL signals. However, the +5-V supplies must be electrically common to avoid damage to the parts. This normally means that they should be actually wired to the same power source, or at least slaved together so that the voltage ramps up and down together. Damage may result if the TTL and PECL voltages are significantly different on either translator or if the receiver +5-V power defaults toward ground while the PECL line driver is powered.[5]

Caution: Damage may result if the PECL driver V_{CC} power supply is more positive than the receiver V_{CC} power supply, unless the receiver is specified for inputs more positive than V_{CC}.

Although series terminations at the source will tend to limit the current on a receiver power-supply fault, the termination resistance alone is not sufficient to limit excess currents to safe levels under most conditions. A method to avoid damage is to provide the power to the receivers from the driver +5-V power supply with power wires through the interconnection, as also suggested above for conventional ECL. Other methods include interlocks that assure that drivers cannot be powered unless the receiver power is valid, and a diode OR to power the PECL receiver from either the +5-V supply at the driver or the +5 V at the receiver.

Figure 14.21 illustrates the input clamp diodes internal to the PECL receiver, as well as input series resistors that may be added to the receiver input to limit input currents to safe levels under receiver power-supply fault conditions. Although the input resistors may

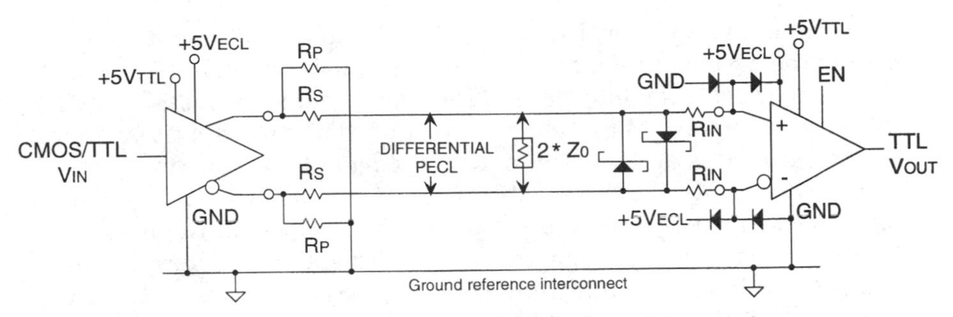

Figure 14.21 Differential PECL line showing ESD diodes and input resistors.

slightly impact maximum speed performance, this is usually not significant compared to the speed capability of the typical CMOS/TTL interfaces. These diodes will conduct current if the input signal levels exceed the power rails of the receiver. Therefore, the diodes to the receiver $+5\ V_{ECL}$ will conduct current if the receiver power supply is lower than the signals from the driver. A minimum of about 100 Ω is required in *each* ECL line for a +5-V power supply to limit the output current to 50 mA, which is a typical absolute maximum current rating for ECL driver outputs. Conservative design practice should limit the maximum current to much less than the absolute maximum rating.

14.6.5 Multiple power-supply limitations

The usual motivation for using a PECL interface is to share a common power supply. When this means that both ends of the PECL interface are powered from the *same* identical power bus, this avoids the multiple power-supply limitations. The problem of multiple positive power supplies is common to PECL, as well as most CMOS and TTL interfaces if the inputs have clamp diodes to the positive supply. This means that even if a PECL receiver input is protected by power provided from the driver, the TTL output of the PECL receiver must be protected against overload by TTL loads on a different supply that may fail to ground. This problem must be considered also for interfaces between TTL powered by +5 V and CMOS powered by +3 V.

An obvious solution for devices without power-down specifications is to provide the power to all of the PECL and TTL devices from a common +5-V supply, or to slave all supplies together with interlocks to prevent any significant difference between them. For larger systems and interfaces where a common supply is not practical, the use of conventional ECL with a separate −5-V power supply may be the most cost-effective and reliable. A single −5-V supply can power the ECL at both ends and can be ON or default to ground in any combination with the +5-V TTL supplies at either end of the interconnection.

The ground and power distribution system is usually a critical element of the design of high speed interconnection systems. It has been emphasized that a common ground reference interconnection is essential for single-ended signals that travel on the ground system. The common ground is an essential part of the signal interface for typical ECL, CMOS, and TTL interconnections. In the absence of specifications that define the performance of interface devices in the presence of power and/or ground faults, the system designer must provide common power and ground to both drivers and receivers, as well as common power to the other signal interfaces of the drivers and receivers.

The system designer is warned to be aware of the power and ground distribution requirements for high speed interfaces. The system designer may be tempted to simplify the power-supply design and distribution problems by dividing a larger power distribution system into smaller portions that are independently powered with distributed supplies with independent regulation. However, the signal interfaces at the power-supply boundaries may result in problems that are more difficult to solve than providing a comprehensive power and ground distribution system.

The issue of multiple power supplies must be considered at all interfaces where drivers and/or receivers are powered from a power distribution system that can have significantly different voltages, including failure of any supply, or "live insertion" of temporarily unpowered drivers and/or receivers onto a powered interface. The typical problem is that powered drivers may be overloaded by either receivers or other drivers on a party line bus. Inserting an unpowered board into a powered system may result in high inrush currents that may overstress components. Devices with special features to tolerate power-down conditions may not be designed for compatibility with adequate line terminations for high speed operation. A typical problem for differential drivers is that the source impedance must be reasonably balanced to prevent common-mode-to-differential-mode conversion. Designing systems to simultaneously meet requirements such as high speed performance, ESD protection, multiple power supplies, power faults, live insertion, ground shifts, EMI shielding, and reduced power consumption can be challenging.

Party line drivers suitable for multiple power supplies or live insertion should be designed and specified to limit the maximum loading that may occur when the part is unpowered and the output is driven to active signal levels. For example, a National Semiconductor DS26C31M differential driver lists as a feature that outputs won't load the line when $V_{CC} = 0$ V.[8] The specifications for this part limit I_{OFF} to $\pm$ 100 μA when V_{OUT} is 0 to +6 V and $V_{CC} = 0$ V. It should be noted that this is different from the I_{OZ} output leakage current when the output of a powered driver is disabled. For another example, a similar National Semiconductor DS96F174M differential driver lists as features that it meets EIA-485 standards and an extended output voltage range of −7.0 to +12 V. The specifications for this part limit I_{OFF} to ± 50 μA when V_{OUT} is −7.0 to +12 V and V_{CC} = 0 V. This performance not only allows for other drivers on a party line bus to be powered and remain active when this part is unpowered but also allows for up to approximately 7 V of ground shift between active drivers and powered drivers that are disabled.

A ground shift between drivers of 7 V is equivalent to an output voltage range of about −7 to +12 V for the nominal +5-V power supply. The description of "common mode voltage range" (or similar descriptions) for line drivers or receivers should include both a ground shift and the maximum signal excursion. This definition of common mode range is the limit for total signal range, not just the average level of the input signal(s).

14.7 TTL Differential Drivers

This section briefly describes and illustrates various TTL output circuits that may be used to drive complementary signals on differential transmission lines. Although TTL devices are not considered high speed on the same scale as ECL, it has been a very popular technology, and a large proportion of system interfaces operate with TTL or TTL-like circuits. Because TTL outputs have significant limitations in a transmission-line environment, this section attempts to describe some of these limitations. The large signal swing is more conventional in traditional designs than the lower ECL levels. TTL circuits have some characteristics that allow operation at reasonable data rates, so many designers could ignore the need for providing a transmission-line environment with such components as solid ground planes for signal returns and line terminations to control ringing. TTL traditionally operates on a single positive 5-V supply and generates differential output signal levels of about 2 or 3 V difference between the HIGH and LOW states, depending on the line loading and the drive current capability and series resistance at the source. Differential lines with no dc load may overshoot to almost double the initial signal excursion and tend to hang at the higher level because TTL drivers become high-impedance to signal levels above the current cutoff level, but below the power supply.

The level of integration of the number of transistors in early generation devices was so low that complex functions required that the signals propagate through a large number of individual devices before a useful result was realized. The slow propagation delay of the individual devices and the large number of devices in series meant that the clock speeds and data rates were primarily limited by the propagation delays of the logic devices. Designers could reasonably estimate the speed performance of a complex TTL system by the sum of the published propagation delays of the parts in the longest delay path. If typical parts performed significantly better than the published delays, the designer may be lucky enough that the first system built may work at the design speed after the functional logic was verified to perform the intended function. It was easy for digital designers to concentrate only on logic design until signal and power integrity problems forced them to consider electrical design[1] and "analog effects."

There are several contributors to the need for designers of TTL systems to consider the electrical design, including some of the following. More complex functions are integrated within the silicon chip, where much higher data rates can be achieved in a limited environment. As device propagation delay was reduced with advanced technology, the transmission-line delays became a more significant proportion of total propagation delay. Higher pin count packages resulted in more signal pins for a limited number of power and ground pins. More accurate testing allowed device vendors to specify the speed closer to the limit of the device capability. Large production runs allowed device vendors to separate parts into speed grades, significantly increasing the chance that the failure limit of parts is very close to the specification. The speed of TTL parts tends to be tested under nearly ideal conditions (not worst case) to simplify the test and improve repeatability (and may conveniently increase the advertised performance compared to competitors).

14.7.1 Beware of F_{max}

The speed that is specified for most TTL parts is maximum propagation delay and a maximum register clock rate, often called F_{max}. Unfortunately, maximum propagation delay and maximum clock rate specifications for individual parts under ideal test conditions are very optimistic and only vaguely related to the much more conservative clock and data rates that can be achieved for reliable operation of a network of actual parts in producible, maintainable systems. Propagation delay is usually tested at a low data rate (typically 1 MHz), so the thermal effects due to switching current transients, line loading, and forward conduction recovery are not included. Propagation delay and F_{max} numbers are primarily useful as a qualitative measure of relative performance for comparing different devices and families when tested under similar conditions.[3]

It might seem intuitively obvious that a device should be able to propagate a pulse width that is as short as the maximum propagation delay. This might be reasonable if the propagation delay were from the beginning (e.g., 10 percent) level of the input to the output near the final (e.g., 90 percent), with reasonable input rise (and fall) times, and a transmission line load that remains for the duration of the round-trip delay of a reasonable transmission-line length. This would mean that the output transition is nearly complete before the next input transition begins to reverse the state of the output. If this were appropriate, the designer would simply double the minimum pulse width value for the minimum period of a 50 percent duty cycle square wave, then use the inverse of this period as the maximum frequency of a square wave that may propagate through the device. The prudent designer would adjust this for duty cycle variations (for example, 80 percent adjustment for 40 percent minimum duty cycle), and derate for design margin. The result would be a maximum frequency of about 30 to 40 percent of the inverse of the maximum propagation delay.

However, devices have been observed to fail to propagate signals at lower frequency than 30 percent of the inverse of propagation delay. For example, a 54S140 in a clock driver application did not operate much over 20 MHz, although the maximum propagation delay was less than the specified 10 ns,[9] and 26LS31 line drivers with a propagation delay specification of 15 ns[8] have failed to operate beyond 10 MHz under actual loaded conditions.

Part of the problem is that propagation delay is measured at the 50 percent crossing points, which is reasonable for propagation delay through cascaded stages, not from 10 percent of input to 90 percent of output. The individual device is tested with a reduced amplitude input, a dc load (typically 500 Ω) to ground to reduce the output swing and ringing, and a lumped capacitance load. The measured propagation delay tends to be dominated by rise time (especially for the low-to-high output t_{PLH}), not intrinsic internal pipeline delay or storage time. Therefore, the propagation delay includes only the first 50 percent of the output rise time and should be approximately doubled for estimating frequency if the rise time were linear. Furthermore, the test input is typically an unreasonably fast rise time waveform, so about another 50 percent of the output rise time should be included to allow for a realistic input rise time if the input is from a similar device in actual operation. Therefore, the propagation delay should be about tripled to account for the additional rise time for realistic inputs and for the output to approach final value. This means that the 30 to 40 percent of the inverse of the maximum propagation delay defined above should be reduced by another factor of about 3, for a net of about 10 percent of the inverse of the maximum propagation delay. This is still not considered very conservative if it does not include reasonably worst case input signal and output loading conditions.

Rule of thumb for estimating a preliminary relationship of F_{max} and propagation delay to the maximum operating frequency capability of TTL: 10 percent of F_{max} or 10 percent of the inverse of the maximum propagation delay.

For example, an individual device with a propagation delay of 10 ns may be suitable for operation in a practical system application at less than 10 MHz, or a register with an F_{max} of 60 MHz may be suitable for system operation at less than 6 MHz. These are rough estimates, with actual performance limits depending on the specific application. Most advanced Schottky TTL devices do not achieve propagation delays of much less than about 10 ns,[3] so operation of TTL systems of any significant physical size is realistically limited to about 10 MHz. The worst-case timing examples in *Signal and Power Integrity in Digital Systems*[1] show that a basic register-to-register transfer with buffers between boards on a common backplane is limited to a maximum of about 20 MHz for the FAST family. Differential transmission-line interfaces between separate chassis would reduce the maximum frequency capability, owing to the higher device propagation delays for the differential drivers and receivers, and the rapid rise in driver power dissipation above about 10 MHz.[8] More complex processing logic or clock distribution timing skew between the registers would further reduce the maximum frequency capability from these examples.

What is really needed is a specification or method for determining the maximum frequency at which a worst-case input signal results in a full amplitude output into a worst-case output load. This means that the device must provide a large signal gain greater than unity, which is required to restore the signal to proper levels at each device output, despite signal distortion and line losses at the input.

14.7.2 Schottky TTL outputs

Figure 14.22 is an example of a typical Schottky TTL output stage. The HIGH and LOW state outputs have very different transmission-line drive characteristics, so it is not well suited as a balanced transmission-line driver. The output impedance is very high for outputs above about 3.5 V, so a dc load resistor to ground is required to keep the HIGH output in forward conduction. The R_{HIGH} resistor in series with the HIGH output provides a reasonable source termination and limits currents when both the HIGH and LOW transistors are conducting during switching transients. The high-frequency failure is that the LOW transistor remains in conduction longer at high temperature, and the low-to-high rise time becomes slower, so that the output remains near the LOW output level. Although the device conducts heavy internal current during high-frequency failure, the driver

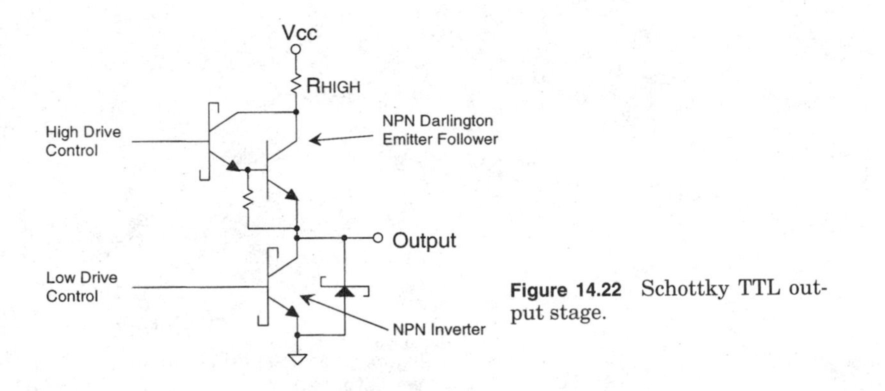

Figure 14.22 Schottky TTL output stage.

is not immediately destroyed and usually recovers at reduced frequency and temperature.

The LOW output transistor has no current limiting and may result in a very fast fall time on the falling edge. The total falling voltage excursion may be large if the high level was allowed to hang near the +5-V level. Negative-going overshoot may be tolerable if all loads have a shunt Schottky diode to ground as shown on the output. The Schottky clamp from the base to collector of the LOW output transistor reduces the LOW drive when the output overshoots below ground and usually prevents conversion of the overshoot to undershoot.

A series resistor is typically required for high speed TTL outputs when waveshape control is required for line drivers. Figure 14.23 illustrates a driver output with resistance in series with the LOW output. Most TTL devices do not integrate this into the device, so an external series resistor performs the equivalent function. Although

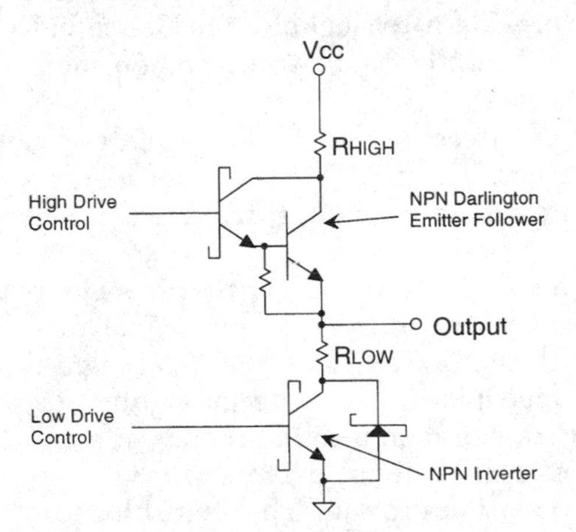

Figure 14.23 Schottky driver output stage.

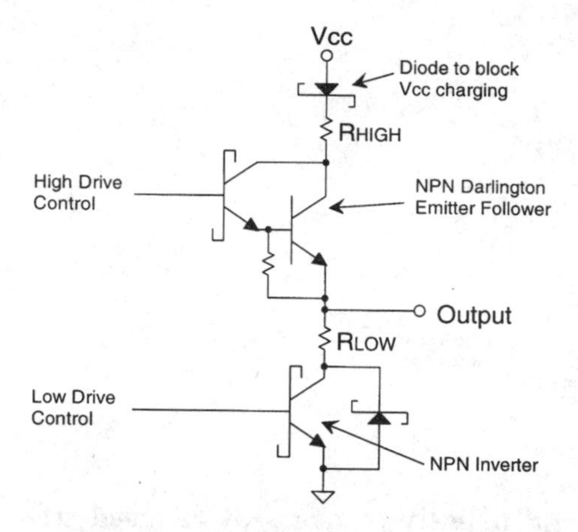

Figure 14.24 Schottky bus driver output stage.

this also adds a little more resistance to the HIGH output, this has little effect on operation.

The typical TTL outputs illustrated in the previous figures result in a high impedance in the HIGH state when the output signal level is high enough to cut off the HIGH output emitter follower. However, the transistors will begin to conduct if the output is driven significantly above the power supply. Active HIGH output signals would tend to attempt to charge the V_{CC} line of an unpowered driver of a party line bus with multiple drivers. Figure 14.24 illustrates a bus driver output with a Schottky diode in series with the HIGH output to block V_{CC} charging when a driver is unpowered. The series diode does reduce the output HIGH level, but the operation in a transmission-line environment is not significantly altered. As for other typical TTL outputs, a dc load to ground is required to maintain the HIGH output in conduction in the presence of overshoot on the line, and series resistance is required to provide a reasonable impedance balance between the HIGH and LOW states to control common mode conversion to differential mode noise.

Figure 14.25 is an example of an alternate line driver configuration for limiting the HIGH state current. The HIGH resistor is in series with the emitter follower output, and provides negative feedback to limit the output current.

Figure 14.26 is an example of a party line bus driver that has blocking diodes in series with both the HIGH and LOW output transistors. The additional diode in series with the LOW output allows for significant ground shifts between the multiple drivers on a party line bus. Although most interface signals are based on a solid ground interconnection, a ground shift between conventional drivers would draw ground current from the most positive ground. The dual blocking diodes illustrated in this figure mean that the HIGH and LOW levels

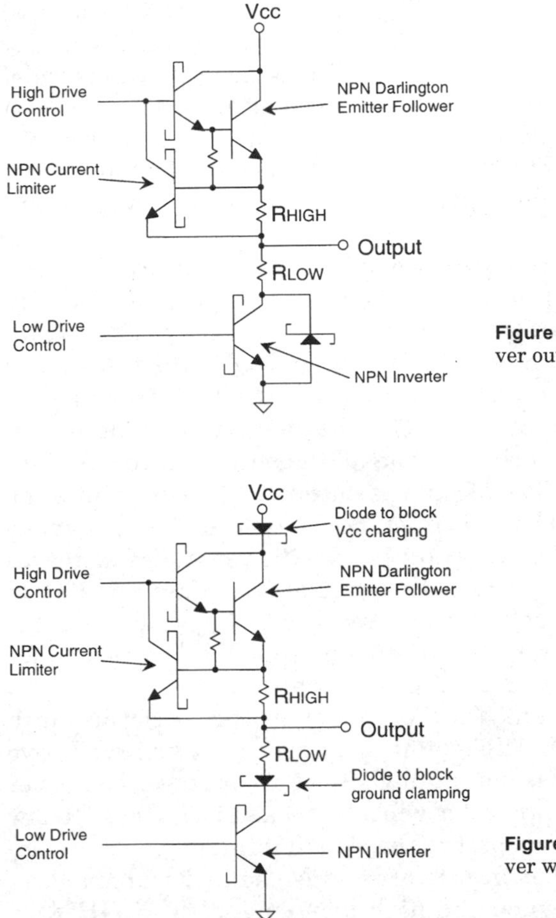

Figure 14.25 Schottky line driver output stage.

Figure 14.26 Schottky bus driver with blocking diodes.

of a party line bus are set by the active driver. Even though the drivers tolerate a range of ground shifts between units on the party line bus, the ground shifts must be limited to within the capability of the blocking diodes of the drivers and the common mode signal range of the receivers, as well as to control ESD and provide high-frequency shielding grounds. Furthermore, a dc load must be maintained in both directions so that the diodes of the active driver remain in conduction to maintain balance of the driver output impedance.

14.8 CMOS Differential Drivers

This section briefly describes and illustrates various CMOS output circuits that may be used to drive complementary signals on differential transmission lines. CMOS continues to be an emerging technology for transmission-line drivers. CMOS offers the capability to drive signals to near both HIGH and LOW power-supply rails without minority carrier recovery time issues. CMOS signals may be operated at lower power-supply levels and can be much better balanced in both states than TTL. CMOS transistors tend to conduct equally well for either direction of current for output signal levels either inside or beyond the power-supply rails, so CMOS drivers can effectively terminate reflected overshoot.

CMOS transistors are primarily a majority carrier device that acts much like a resistor when turned ON by a voltage applied to the gate input. Therefore, an output transistor must be physically large to conduct high transmission-line currents with low ON resistance. The CMOS output transistor requires a rather large voltage excursion on the gate to switch the output stage. This means that large intermediate buffers are required to drive the gate of the output stage, and that Miller effect negative feedback from the output to the gate limits the output switching speed. The ON resistance of typical CMOS transistors has a positive temperature coefficient, so the resistance increases at higher temperature. This means that overload conditions on CMOS driver outputs tend to be self-limiting; because as higher output current increases the temperature of the transistor, the ON resistance increases to limit the increase in output current.

Figure 14.27 illustrates an NMOS output stage that performs much like a TTL output stage. This totem pole configuration with two NMOS output transistors is not a full CMOS output stage. The lower transistor is an *N*-channel inverter which is turned ON when the low drive signal is HIGH with respect to the source terminal which is connected to ground. The upper transistor is an *N*-channel source follower which is turned ON when the high drive signal is HIGH with respect to the output. The operation of the source follower is analogous to the Darlington emitter follower pull-up of the TTL output, so the output impedance is high for overshoot to signal levels above the minimum conduction level of the source follower. The output impedance is also high for signal levels above the V_{CC} power supply, so this

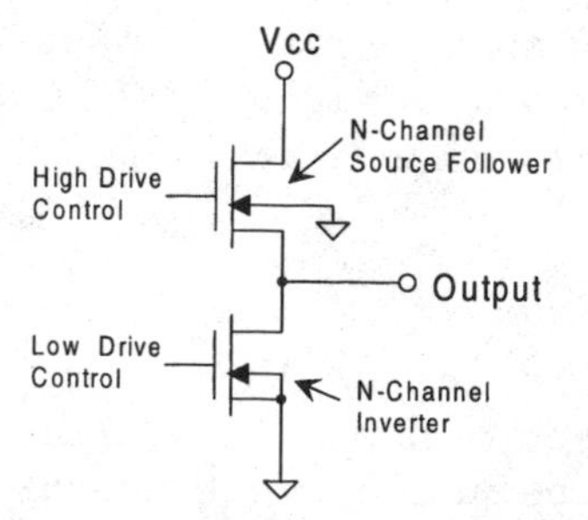

Figure 14.27 TTL-compatible NMOS output stage.

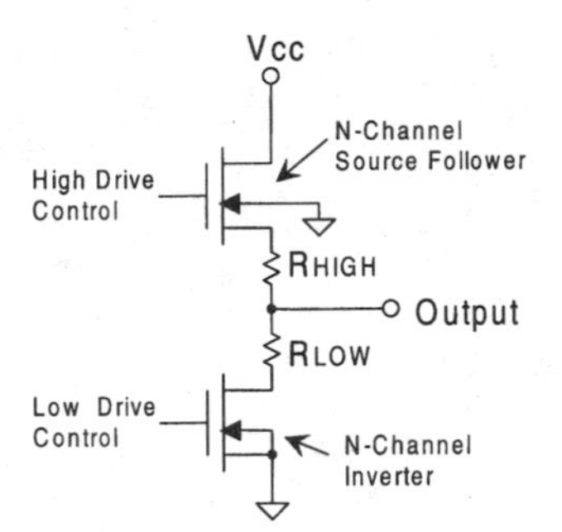

Figure 14.28 TTL-compatible NMOS output with resistors.

output configuration also tolerates differences in power-supply voltages and power off of multiple drivers on a party line bus.

The body diode of the NMOS transistors is indicated by the arrow on the transistor symbol. The body diodes of the NMOS transistors would conduct if the output (or the V_{CC} supply) were driven more *negative* than the ground to which the body is shown connected. This does not affect normal operation in MOS interconnection systems with good ground connections because a LOW output transition is driven by an NMOS inverter which remains conducting for output signal levels below ground.

Figure 14.28 illustrates series resistors internal to the TTL compatible NMOS output. The LOW resistor significantly reduces ground bounce and negative overshoot and ringing, as well as eliminating the need for external series resistors.[10] In actual practice, the HIGH resistor may represent only the dynamic impedance of the source follower in the normal operating region, not a physically separate resistance. The source impedance for the positive excursion is not critical because the source impedance to overshoot is high and will not convert the overshoot to ringing. However, a dc load is required to maintain the HIGH output in conduction in the presence of overshoot on the line, as for the TTL outputs. If the differential signal is allowed to hang at a level that is large enough so that the HIGH output is high-impedance, common mode noise on the line will be converted to differential mode noise and reflected toward the receiver(s). High input impedance receivers and line-to-line differential terminations only reflect common mode noise, so the driver must terminate common mode noise with balanced impedance to dampen common mode noise and avoid conversion to differential noise.

Figure 14.29 is a basic CMOS inverter that drives the output to either the V_{CC} power rail or to ground with a low ON resistance. The complementary inverter means that CMOS can provide a much more balanced impedance for the HIGH and LOW states, and the signal excursion can be nearly the power-supply level. For example, a CMOS

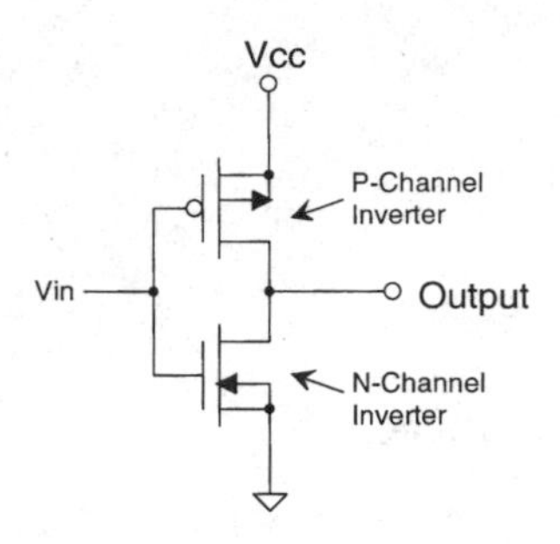

Figure 14.29 Rail-to-rail CMOS output stage.

driver powered from +3.3 V can provide a higher differential signal to a dc load than TTL powered from +5 V because there are no current-limiting resistor or base-to-emitter drops in the HIGH state and no Schottky diode clamp to limit the LOW output voltage. Control logic can disable the gate drive to both output transistors.

Full rail-to-rail CMOS outputs are not well suited for driving most long differential lines directly because the low ON resistance launches a full-amplitude signal into the line and reflects with voltage reversal any crosstalk or signal reflections, resulting in ringing on the line. If the ringing has not subsided before the next driven transition is driven on the line, there may be standing waves due to the impedance mismatch. Although an exactly matched load termination at the far end could eliminate reflections in theory, a practical line has tolerances on the line impedance and load termination, discontinuities at connectors, lumped loading, and crosstalk.

Figure 14.30 illustrates the CMOS output with internal resistors to provide source termination of the driver output. Although a single resistor in series with the output would provide similar line termination characteristics, the internal termination is illustrated as a separate resistor in series with each transistor. Including separate resistors offers the ability to tailor the resistor value to compensate for different transistor ON resistance, tends to limit the simultaneous flow-through current in both transistors during signal transitions,

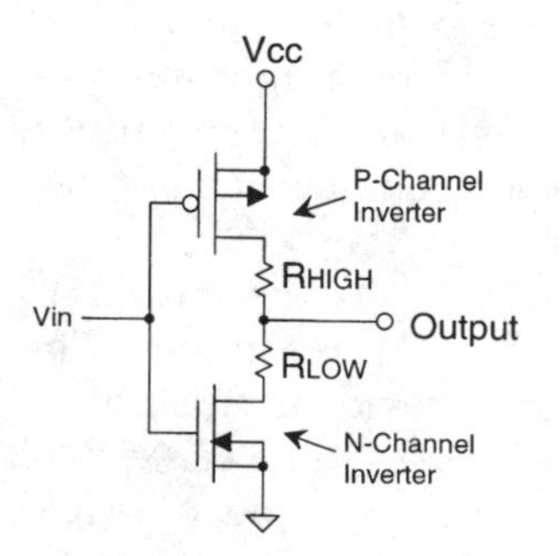

Figure 14.30 CMOS output with resistors.

and reduces the average power dissipation per resistor. The resistors should be dielectrically isolated so that any signal excursions beyond the power rails could not result in forward-biased isolation junctions. The value of the internal resistance is not particularly critical, as has been shown in previous chapters. The industry practice of about 25 Ω for the internal resistor versions of TTL-compatible NMOS outputs is also appropriate for most applications of rail-to-rail CMOS outputs.

The body diode of the PMOS inverter is indicated by an arrow connected to the V_{CC} supply as shown in the previous figures because the body (sometimes referred to as the well) of the *P*-channel enhancement mode transistor is *N*-doped material. The body is connected to the positive supply to avoid forward bias of the *n*-*p* junction under normal operation. However, the junction would become forward-biased if the output were driven above the V_{CC} power supply. This usually does not occur for differential lines with only one source because the PMOS inverter conducts both forward and reverse current to keep the HIGH output near the power rail. However, if there are multiple drivers on different power supplies, an active HIGH signal from one source may charge the V_{CC} of an unpowered source through the body diode of the PMOS inverter. Furthermore, when the PMOS gate of an unpowered inverter is at ground, an output more than the PMOS threshold above the gate will begin to turn the unpowered PMOS transistor ON. Therefore, the PMOS transistor has two paths for HIGH outputs to charge V_{CC}, both the body diode and the drain to source. Both paths should be blocked to eliminate charging currents into V_{CC} by a HIGH signal level on the output of an unpowered driver.

Figure 14.31 is an example of a CMOS driver with a Schottky diode in series with the HIGH output to block V_{CC} charging in applications with multiple drivers on different power supplies, but a common ground. This is essentially the same driver with Schottky diodes described in Chaps. 10 and 11 for single-ended applications. Although a single-ended driver on a long line may be easier to control, using differential CMOS drivers and differential receivers on a party line bus will improve noise margins and maintain signal integrity on low-voltage systems. A differential line can be terminated line-to-line, eliminating ground loop currents and the need for a low-voltage V_{TT} termination supply. Differential receivers eliminate the need for a precision receiver reference and reduce sensitivity to power-supply variations and ground noise. Differential drivers nearly double the total differential signal available at the receiver for the same power-supply voltage, and the two signal lines nearly double the line impedance compared to single-ended signals referenced to ground. Of course, differential lines still need to satisfy the termination and load-

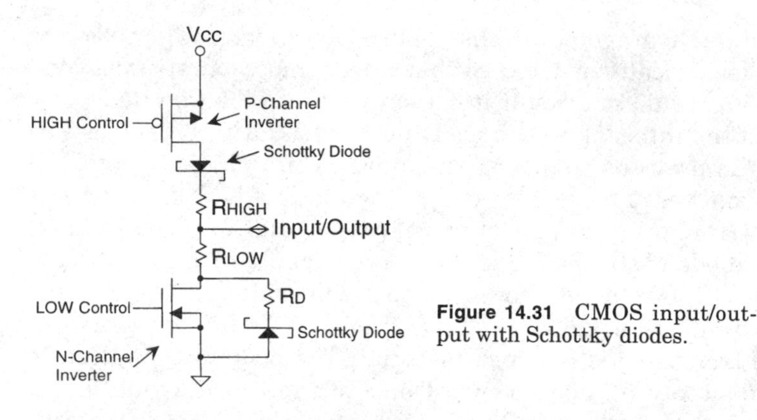

Figure 14.31 CMOS input/output with Schottky diodes.

ing requirements for high speed operation. High speed lines without sufficient settling time available for series termination require load terminations at the physical ends of the line to limit reflections. Furthermore, drivers with series diodes should have load terminations to maintain conduction in the diodes to provide balanced source termination at the active driver.

Figure 14.32 is a differential reflection diagram for the CMOS driver with the Schottky diode in series with the HIGH output. The values used for the diagram are that the R_{HIGH} and R_{LOW} are each equal to $Z_0/2$. The differential impedance of a single line is considered as $2 * Z_0$, so the line impedance seen by a driver at a center point is Z_0. Therefore, the differential dynamic source impedance for this exam-

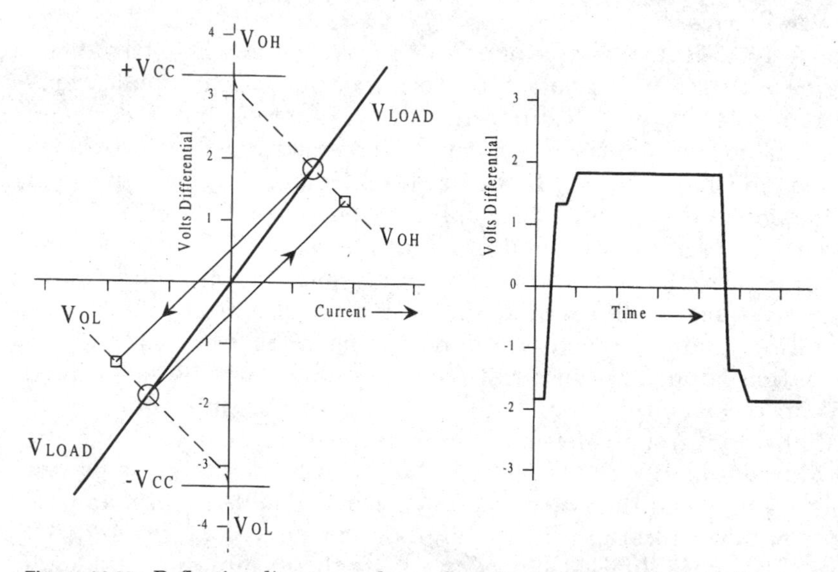

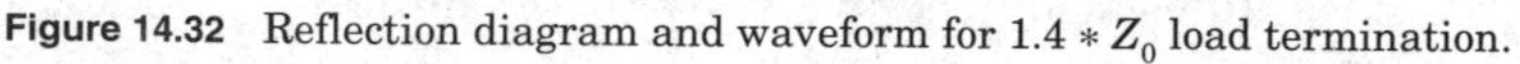

Figure 14.32 Reflection diagram and waveform for $1.4 * Z_0$ load termination.

ple is equal to the impedance of two lines in parallel when driving a center point of a line, so the source is exactly matched.

The load termination at each end is considered to be 1.4 times the differential impedance of a single line for this example, so the parallel combination of the load termination at both ends is $1.4 * Z_0$. The load termination of greater than Z_0 reduces the static power consumption compared to an exact match but results in a small reflection. The reflection diagram is drawn to illustrate the case of the terminated driver located midway between the terminations at the ends, so the signal arrives at the two ends simultaneously, and the reflection returns to the active driver simultaneously. When the active driver is not located exactly in the center of the line, any reflections from the ends arrive at different times. When a reflection arrives from only one side, an exactly terminated source reduces the reflection by 50 percent, as described in Chap. 12.

The voltage scale of the reflection diagram is labeled for a 3.3-V power supply, but it could be for a lower voltage, until limited by the driver capability, the noise margins required, and the receiver sensitivity. It should be noted that the voltage scale is differential; the actual voltage swing on each signal line is part of the difference between the power-supply voltage and ground. For R_{HIGH} and R_{LOW} each at $Z_0/2$ (including the internal dynamic impedance of the transistors and the Schottky diode), but the load termination at each end reduced to match the differential line impedance of $2 * Z_0$, the differential signal swing is reduced to $\pm V_{CC}/2$, less the forward drop of the Schottky diode. Lower source impedance and/or higher load termination increases the steady-state static signal levels but results in overshoot and some ringing. Minor overshoot is usually desirable to reduce rise time and to compensate for skin-effect losses on the line.

Figure 14.33 is an example of a basic balanced differential CMOS interface for a single driver, so there are no partial power-down issues for multiple drivers. The line is inactive if power is lost to the driver. Multiple receivers are acceptable if they do not load the line if unpowered, or if all receivers are always powered together. Both signal lines are terminated through the source resistors and body diodes if the signal lines are driven either below ground or above the unpowered V_{CC} supply of the driver. Positive and negative disturbances on the line are terminated at the source to V_{CC} and ground, respectively, for normal operation. As has been emphasized, source termination resistors are important to terminate both differential and common mode noise and reflections, and to limit common mode to differential mode conversion due to unbalanced source impedance. The load termination is important to limit standing waves on high speed long lines and to compensate for inexact termination at the source.

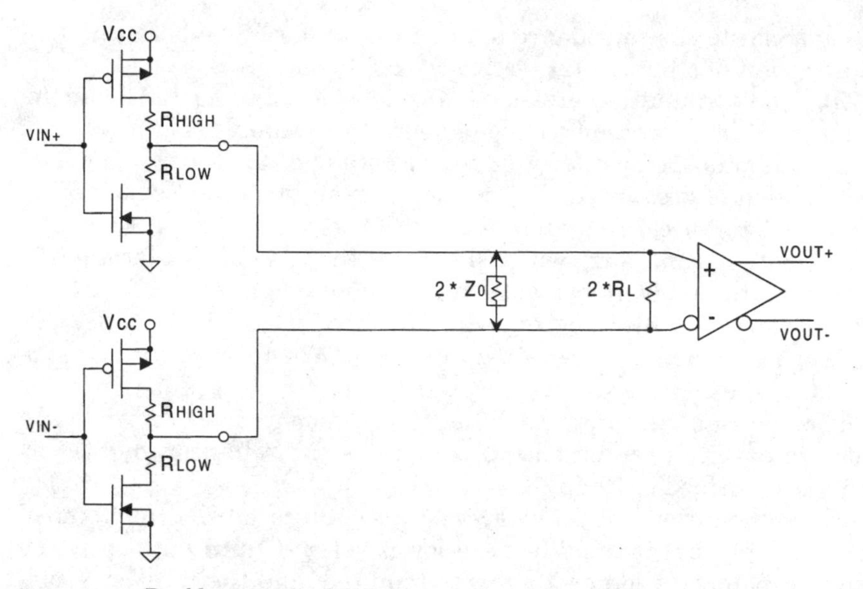

Figure 14.33 Double-terminated differential CMOS line.

This example illustrates only a single receiver and load termination, but multiple receivers are acceptable, and the termination can be divided at locations clustered near the far end of the line. Multiple high input impedance receivers may be placed along the line if the stub lengths off the main line are minimized and the load termination near the far end of the line is reasonably close to the loaded dynamic impedance of the line, shown as $2 * Z_0$. As for typical resistor terminations, lower source resistance and higher load termination will increase the steady-state static signal levels. The overshoot of underdamped lines results in faster initial rise time and tends to compensate for high frequency skin effects and other losses on longer lines. Although the interface as shown does not directly depend on a ground interconnection for operation, a ground reference to control the differential line impedance and for shielding is typically required. A ground path is also required to keep the signal levels within the common mode range of the receiver and to provide a return path for any unbalanced or common mode currents.

14.9 Summary of Some Types of Line Interfaces

This section is a short list of some basic types of typical line interfaces with a brief description of their characteristics. Refer to the *National*

Interface Databook[8] or the published standards for more detailed information.

MIL-STD 188C is a low-level specification for a point-to-point (one driver, one receiver) unbalanced (single-ended) polar interface. The single active signal levels are about ± 6 V with respect to ground, with an output slew rate specified at 10 percent of the bit rate, usually controlled by an external capacitor on the driver output.

RS-232 is a popular Recommended Standard (RS) of the Electronic Industry Association (EIA) and the Telecommunications Industry Association (TIA) for a point-to-point unbalanced polar interface, similar to MIL-STD 188C. Signal levels are also about ± 6 V, the slew rate is a maximum of 30 mV per ns, with a maximum data rate of 20 kilobits per second. Receiver threshold is ± 3 V, or about one-half of the signal swing.

EIA/TIA 423 is a hybrid standard for a modified RS-232 unbalanced polar driver and RS-422 differential receivers, which isolate the signal reference ground of the driver from the power-supply ground of the receiver. The minimum driver signal is lower, so ± 5-V power supplies may be used, and multiple receivers are allowed. Maximum data rate is 100 kilobits per second for up to 40 ft of cable length.

EIA/TIA 422 is a popular standard for a balanced differential interface operating at a data rate of up to 10 megabits per second with a single, optional line-to-line termination at the far end of the line. Multiple receiver operation with a single power supply and up to ± 7 V of common mode shift on receiver inputs is allowed. Receiver input threshold is ± 200 mV. Most drivers for 422 applications include features to permit multiple driver party line operation and partial power-down operation, but with a common ground between all drivers.

EIA/TIA 485 may be considered as an upward compatible upgrade to RS-422 that specifies party line operation of multiple drivers with significant (± 7 V) ground shifts between up to 32 transceivers with a line-to-line termination of about 120 Ω (± 10 percent) at both ends of the line. The loaded driver differential output may be as low as ± 1.5 V, but most other signal characteristics are similar to RS-422. The intended transmission line is twisted pair terminated line-to-line at both ends. The RS-485 driver and receiver standard is the basis for multipoint interface data links, including the balanced differential interface class for the Small Computer Systems Interface (SCSI) standard for interfaces between personal computers and peripherals such as printers and disk drives, at rates up to 4 megabits per second.

EIA/TIA 644 is a new low-voltage differential signaling (LVDS) technology based on current source (actually current limited in the HIGH state) drivers instead of conventional voltage source drivers.

The architecture can be considered somewhat analogous to TTL, but with a very high resistance in series with the HIGH state. The current sources provide inherent limiting of short circuit and switching currents, which permits higher data rates than TTL or CMOS before the power-supply current rises rapidly. The typical LOW level output is about +1.0 V; the HIGH output is current limited at only about 3.3 mA, so the signal excursion driving one end of a 100-Ω differential transmission line is about ± 330 mV. This signal excursion is less than differential ECL (or BTL), and the driver requires no pull-down resistors and no V_{TT} termination power supply. The initial excursion on the line is proportional to the dynamic impedance of the line. Because the differential source impedance is much higher than the line impedance, the load termination impedance value is quite critical. The load termination impedance should be slightly *lower* than the impedance of the line for an underdamped response with a little overshoot so that the initial signal excursion crosses the threshold sooner and to compensate for rise time degradation and line loading.

The current source in the HIGH level output means that the source impedance is not balanced, so common mode noise traveling toward the driver is reflected and may be converted to differential noise. A typical line-to-line termination at the end of the line does not terminate common mode noise either. However, the center of the line-to-line load termination may be ac terminated to dampen common mode noise.[8] This requires additional components but does not increase power dissipation. It is quite important that the far end of the line is well matched to the loaded impedance of the line to avoid significant reflections because the driver source impedance is not well balanced. Lumped loading and stub lengths should be minimized to avoid reflections to the unbalanced source.

IEEE 488 is a popular standard used for laboratory and test equipment data and control interfaces. It is a complete bus standard, defining signal protocol as well as electrical and mechanical specifications. The signal levels are essentially TTL compatible, with an 8-bit parallel data interface for up to 15 devices on the bus.

IEEE 1194.1/BTL (*backplane transceiver logic*) is a class of multipoint (party line) single-ended bus transceivers intended to provide lower signal swings and higher speed than TTL-based interface standards. BTL is the technology for the IEEE 896.2 Futurebus+ standard. BTL uses a Schottky diode in series with the LOW output, which isolates the inactive drivers from line capacitance and ground charging but increases the LOW level above ground. Typical output logic LOW is about 1 V with a 33-Ω termination to 2.1 V at both ends of the bus.

PI-Bus (parallel interface bus) was developed in the 1980s for the military very high speed integrated circuits (VHSIC) program. It is mandated for many military airframes and may be applied to commercial avionics. Open collector LOW outputs are driven through a Schottky diode to isolate inactive drivers. The bus is terminated with 30 to 40 Ω to 2.0 V at both ends of the bus, so the PI-bus interface is similar to BTL. A separate high-level control is provided to limit the output HIGH signal levels to the interface controller if it is powered with a lower power-supply voltage.

Futurebus+ and GTL+ represent two of the more popular of the proliferating list of low-voltage-swing interfaces with an objective of achieving ECL-like signal performance in low power supply CMOS technology.[11,12] These interfaces can be considered analogous to upside-down PECL (positive ECL), with an open-drain NMOS transistor driving the LOW signal toward ground through a Schottky diode for isolation. The HIGH level is provided by a pull-up resistor to a V_{TT} termination supply at each end of the bus. The V_{TT} termination supply for GTL (gunning transceiver logic) moved from +1.25 up to +1.5 V to improve noise margin. The V_{TT} termination supply for Futurebus+ (IEEE Standard 896.2) is +2.1 V.

9614-9615 differential driver-receiver pair were introduced about 1970 by Fairchild Semiconductor as a single 5-V power supply balanced driver with low, symmetrical output impedance to drive long terminated lines, with a receiver differential input threshold of ± 500 mV over a ± 15 V of receiver input common mode range. They are replaced by 55114-55115 National Semiconductor part numbers. The 7830-7820 driver-receiver pair is functionally similar.

26(x)31-26(x)32 differential driver-receiver pair were originally introduced in low-power Schottky (x = LS) technology to meet the RS 422 standards. Later versions were introduced in FAST (x = F) advanced Schottky and CMOS (x = C) technology. The receiver differential input threshold is ± 200 mV over a ± 7-V input common mode range. The 26LS33 alternate receiver has a differential input threshold of ± 500 mV over a ± 15-V input common mode range, which matches the 9615/55115 receiver.

14.10 References

1. James E. Buchanan, *Signal and Power Integrity in Digital Systems: TTL, CMOS and BiCMOS,* McGraw-Hill, New York, 1996, chaps. 6, 11.
2. *FAST Advanced Schottky TTL Logic Databook,* National Semiconductor Corporation, Santa Clara, Calif., 1990 ed., sec. 1.
3. James E. Buchanan, *Signal and Power Integrity in Digital Systems: TTL, CMOS and BiCMOS,* McGraw-Hill, New York, 1996, chap. 2.

4. *Synergy Product Data Book,* Synergy Semiconductor Corporation, Santa Clara, Calif., 1992.
5. *F100K ECL 300 Series Databook and Design Guide,* National Semiconductor Corporation, Santa Clara, Calif., 1992 ed.
6. *Motorola MECL Data,* Motorola, Inc., Phoenix, Ariz., 5th ed., 1993.
7. C. H. Chen, ed., *Computer Engineering Handbook,* McGraw-Hill, New York, 1992, chap. 6.7.
8. *National Interface Databook,* National Semiconductor Corporation, Santa Clara, Calif., 1996 ed.
9. Randall D. Lewis and Kenneth W. Awkward, "Method for Determining a Maximum Operating Frequency for TTL Gates," *1990 Proceedings Annual Reliability and Maintainability Symposium,* IEEE, 1990, p. 372.
10. David Wyland, "Resistor Output Logic Gives High Speed with Low Noise," AN-07, *Quality Semiconductor Databook,* Quality Semiconductor, Inc., Santa Clara, Calif., 1991, p. 6-61.
11. Greg Edlund, "Noise Budgets Help Maintain Signal Integrity in Low-Voltage Systems," *EDN,* July 18, 1996, p. 111.
12. Samuel H. Duncan and Robert V. White, "Designing 2.1 V Futurebus+ Termination System Requires System-Engineering Approach," *EDN,* May 26, 1994, p. 117.

Appendix

Composites and Their Role in Electronic Packaging

Composites are well known for their use in structural applications such as reinforced concrete and control surfaces on aircraft, and in early electrical applications such as glass-reinforced phenolic cases for housing small transformers. A printed wiring board is a form of composite and is an early example of composite usage in electronics. In a broad sense, a composite is a combination of materials which has properties that none of the constituent materials have by themselves. Materials are selected to be combined with each other to take advantage of one or more characteristic properties that the partner material possesses. The properties can be mechanical, or such as tensile strength or thermal conductivity, or electrical, such as high resistivity or low dielectric constant. A material can be made less brittle by combining it with a softer one. A material's tensile strength can be improved greatly by combining it with a reinforcement which binds itself well to the weaker material, thereby imparting some of its strength to the combination.

Composites today are classed into three groups according to the matrix material or main body of the combination. These groups are polymer matrix (PMC), metal matrix (MMC), and ceramic matrix (CMC). Polymer matrix composites are by far more numerous in variety or types than either metal matrix or ceramic matrix materials, but in spite of this, significant contributions have been made to the advancement of electronic packaging by metal matrix and ceramic matrix technologies. This section describes some of the developments in all three types of composites with special emphasis on the materials and technologies that are available for the packaging of electronics.

Polymers have found extensive use in all aspects of electronics packaging, and the employment of polymer matrix composites is a natural evolution to be expected as polymer technology continues to advance. Since printed wiring board technology was well entrenched as an interconnect medium for a large variety of electronic components, it was also used for supporting and interconnecting leadless ceramic chip carriers (LCCCs) shortly after their introduction. Ceramic substrates with leads attached for connection to sockets or, more commonly, printed wiring boards (PWBs) were used to a minor extent. The initial leadless chip carriers were small (0.250 × 0.250 in) and had metallization tracks underneath the carriers which were intended to be soldered to the supporting substrate. This substrate could be ceramic or a PWB. The ceramic substrate (which could carry a small number of LCCCs) was itself a high-temperature cofired multilayer circuit and could be used to provide some of the interconnection required for the LCCCs. As long as the LCCCs remained small the popularity of the ceramic substrates was very low. However, as the LCCCs grew in size and pinouts with the complexity of the devices they carried, the practice of soldering them to PWBs began to experience difficulties. Larger LCCCs grew by 0.05 in on a side for every additional four input-output connections that the packaged chip contained. At 40 leads per carrier a critical juncture was reached. The package had grown to 0.550 in per side and the expansion difference between the chip carrier and the PWB beneath it which contained the solder pads began to cause cracking of the solder joints. This was not immediately noticed because it took many hundreds of thermal cycles between −55 and +125°C to cause cracks that were observable to the unaided eye. After several hundred of these thermal cycles, expansion difference-induced stress lines began to appear in the solder joints. After further thermal cycling, these stress lines developed into actual cracks in the solder.

Chips continued to grow in complexity and size and so did the LCCCs. The number of thermal cycles required to cause solder joint cracks went down, causing the problem to surface more rapidly. Consequently the awareness of this expansion difference problem grew quickly, particularly among military equipment suppliers and their customers. Surface mount technology (which is the name that evolved to describe the practice of using leadless devices mounted onto PWBs) began to be used more extensively, and this caused both users and furnishers of PWBs to seek materials for constructing the PWBs which could have lower coefficients of expansion than standard epoxy-glass and polyimide. Such board materials would provide a better match to the ceramic packages which contained integrated circuit chips. Another approach was to restrain the expansion of the PWB with a core material. These cores consisted of several combinations of metals whose coefficients of thermal expansion were intended to match those of LCCCs, which have a coefficient of thermal expansion of between 6 and 10 ppm/°C.

A large variety of PWB materials were formulated (and continue to be formulated) to match the needs of surface-mounted components. Both the matrix materials and the reinforcements underwent change and development. A material with a high T_g (glass transition temperature) has additional advantages, such as resistance to chemical etchants used in the processing of PWBs and their assembly, and can withstand repeated component assembly and disassembly with very little damage to the bond between the copper cladding and the laminate material.

Aramid reinforcements had held the coefficients of thermal expansion of polymers to their design values but at the expense of cracking of the surface matrix material (epoxy, polyimide). The cracking was limited to the surface or buttercoat of the PWB but nevertheless presented a serious drawback to the full acceptance of this promising material. Consequently considerable research was expended on lowering the coefficient of thermal expansion (CTE) of the resins or matrix materials. Constraining the expansion of the PWB in the plane of the board simply transfers the expansion to the out-of-plane direction *z*. This aggravated expansion can be detrimental to the copper plating on the walls of the plated through holes because of severe stressing during thermal cycling in service. Reinforcements other than woven fabric may be used for the electrical properties they impart. Microspheres of glass or plastics and powdered fillers are sometimes used to lower the dielectric constant. The CTE of this kind of PWB is isotropic because the fillers are not tied together.

A.1 Solid-Core PWB Expansion Restraint

Solid cores have been laminated to PWBs to serve as expansion limiters. Printed wiring boards may be bonded to one or to both sides of the restraining material, thus the name *core*. The out-of-plane expansion of the PWB is increased with their use just as it is when a woven fabric is the board reinforcement. The core is usually designed to serve as a heat sink as well, but the cost and weight penalties their use extracts limits their use. For a time the copper-Invar-copper was used extensively for PWB assemblies which contained large LCCCs of 48 or more pinouts. This combination exploited the high thermal conductivity of copper and the very low CTE of Invar to its advantage. MMCs for the most part are not used simply as reinforcements and heat sinks but rather as complex shapes and special microwave packages.

A.2 Graphite-Epoxy Heat Sinks

The thermal conductivity of graphite has long attracted interest in its use for managing the problem of heat removal from components which generate considerable heat as they operate. Just how to utilize carbon efficiently in electronic assemblies remained a major impediment until work began on graphite-epoxy composites. This combination has had a long history of applications in replacing or reinforcing metals in the aerospace industry. As such it was developed, matured, and became well quantified. Its use in combination with other materials such as aluminum and titanium was well established. This considerable experience provided a reasonable foundation for graphite-epoxy to serve as a principal mechanism for heat removal from electronic assemblies. Some research was funded by the U.S. government in support of the Standard Electronic Module activity at the Naval Weapons Support Center in Crane, Ind. Additional work has been done by industry to supplement the navy studies. Much of this work has been focused on using graphite-epoxy composites as heat sinks, thermal paths, and module frames. In certain constructions the module frame serves as the heat-removal mechanism because the frame is clamped to rails which are the principal mechanical support as well as the heat-conduction medium.

Heat can be conducted along the carbon fibers very efficiently. The difficulty in using this composite lies in transferring heat to the fibers through the epoxy matrix (epoxy has a very poor thermal conductivity) and out of the fibers to the interface (chassis). The issue of heat transfer to the fibers perpendicular to their length is one of minimizing the amount of epoxy resin between the fibers and the heat source, such as a high power package. If the module frame is made from a lay-up of graphite-epoxy, the layers are typically placed one on another at 90 or 45° so that the strength is more or less uniform in all directions in the plane of the frame. Placing the first several and last several layers in the direction of the heat exchangers assures that the heat will flow more toward the exchangers rather than at right angles. The relatively poor thermal conductivity of the composite at right angles (from fiber to fiber) helps to assure that the heat is conducted away from the source to the exchanger efficiently. Removal of this heat is accomplished in the standard electronic module by the wedge retainer clips which clamp the frame firmly to the exchanger rib. Heat is then transferred from the uppermost fibers to the rib. An improvement to this approach is described in U.S. Patents 4849858, 4867235, and 5002715. In those patents the inventors depict a method of bending the fibers near the heat exchangers in order to have the ends of the fibers make contact with the exchangers. Since most of the heat is conducted along the fibers rather than at right angles to them, removing the heat from the ends rather than at right angles to the ends results in a greater flux from the fibers to the heat exchangers.

A.3 Metal Matrix Composites

High strength, high thermal conductivity, low density, and good toughness are all available at low risk and low cost. This sounds hard to believe, but for the most part, these features which designers seek in the materials they specify for electronic packaging are possible to realize through metal matrix composites. While no one material has all the characteristics listed, it is possible to pick and choose how much of each feature one can obtain in a given material. This happens when one or more materials and reinforcements are combined. For example, a metal with high thermal conductivity can be combined with a reinforcement which has a high elastic modulus and low density to achieve a lightweight, high strength, high thermal conductivity composite. In practical terms, combinations like these exist. They are combinations of copper and graphite, aluminum and graphite, or aluminum and silicon carbide. These composites are being manufactured and marketed as electronic packaging materials.

To see why MMCs are so attractive as a design consideration, consider the properties of materials which match the CTEs of silicon devices. The CTE of aluminum is 22×10^{-6}/K and silicon is 3.8×10^{-6}/K. If one must mount the silicon on aluminum and expect the bond between the two to survive thermal cycling, it is necessary to use a number of different materials between the silicon and aluminum which have CTEs somewhere between those of aluminum and those of silicon. This is not a desirable solution. Another approach would be to modify the CTE of the aluminum with a material which would constrain the aluminum so that the combined CTE would more closely match that of silicon. One could use carbon fibers or particles of silicon carbide (SiC) for this purpose. The choice of which to use could be governed somewhat by factors other than the resultant CTE, such as cost. While commercially available carbon fibers have a Young's modulus of 130×10^6 lb/in^2 (this is about 12 times that of aluminum) and a thermal conductivity of 600 W/m-K ($1\frac{1}{2}$ times that of copper and 3 times that of aluminum), the cost of carbon fibers is many times higher than the cost of high-purity silicon carbide particles. Carbon fibers are sold for over \$100 per pound while SiC particles are being sold for between \$10 and \$25 per pound. This in itself accounts for the popularity of SiC as a MMC reinforcement over carbon fibers. Other features to consider are (1) carbon fibers are generally used in long lengths and laid in the matrix where they are placed, and (2) the thermal conductivity of the fibers is much higher along the major axis than it is perpendicular to that axis. This causes the composite to have orthotropic physical properties. Particles of SiC are used to reinforce aluminum and to endow it with isotropic properties. By varying the volume percentage of the SiC in the matrix it is possible to tailor the composite's CTE from roughly 10×10^{-6} to over 20×10^{-6} per °C.

Metal matrix composites have been used in a variety of applications for electronics. The principal examples are for substrate supports or carriers and hermetic packages. An emerging application is the standard electronic module heat sink. The main reasons for using MMCs as substrate carriers are (1) tailored matching to the CTEs of substrates which carry devices such as gallium arsenide and silicon and (2) improved thermal conductivity over the composite matrix material. Although these features are addressed to some extent by other technologies (PMCs), metal matrix composites are also required to become hermetic housings, and this brings several additional requirements into the picture. These are low cost, isotropy of physical properties such as expansion characteristics and thermal conductivity, and isolation from the surrounding environment through hermetic sealability. Two material combinations have extensive usage, and the experience gained with their use has been valuable in influencing further development. These are aluminum/silicon carbide and aluminum/graphite. These composites have helped to spur further research resulting in carbon- or graphite-reinforced copper, which is a newer MMC with promise of outstanding thermal conductivity and very low in-plane CTE. This CTE is a close match to silicon and makes this composite attractive as a candidate for packages containing high-power-dissipating silicon devices if the in-plane CTE is the same in all directions. Uniform in-plane CTE can be achieved by constructing the composite so that the reinforcing continuous fiber tows are placed at 0° and the next layer at 90°, the next at 0°, and so on. If the layers are all placed at 0°, the in-plane CTE will be quite low in the fiber length direction but much higher (nearly that of copper) in the transverse direction. In order to take full advantage of the benefits and properties the reinforcements can bring, it is important to use them in an orientation that results in essentially uniform behavior unless anisotropy is desired for the application.

Composite properties vary as the amount of reinforcement in them is increased or decreased, whether that reinforcement be in the form of short filaments or fibers, continuous fibers, or particles. More materials are used as reinforcements than listed earlier. Some of these additional reinforcements are boron, aluminum oxide (alumina), and refractory metals. Bonding between the matrix material and the reinforcement is the key to performance of the MMC. This performance depends on the bond integrity which in some materials combinations is enhanced by interface coatings to decrease reactivity between the matrix and reinforcement. Treatments of the reinforcement are sometimes used to decrease lubricity and prevent premature pull-out of the fiber from the matrix.

An example of a reactive combination is the combination of graphite fibers and aluminum. This composite has enjoyed success in structural applications, but its use in electronics as a hybrid package material is nonexistent. Its low density (weight) and high thermal properties render it a candidate for heat sinks in electronic assemblies. To avoid the reaction between aluminum and graphite at liquid aluminum temperatures, vacuum deposition of the matrix onto the graphite fibers is used. Alumina fibers are another material used to reinforce aluminum. To strengthen the bond between matrix and reinforcement in this case, small amounts of lithium are added to the melt.

Glossary*

Accelerated Aging The rapidly induced deterioration of a material, system, or device in a short time by increasing voltage, temperature, humidity, or a like factor above normal anticipated levels. The results yield predicted equipment life under normal operating conditions.

Accelerated Stress Test A test performed at a higher stress level than in normal operation and at a shorter time in order to produce a failure. Alternatively, a test in which conditions are intensified to allow data to be obtained in a shorter time.

Acceptance Tests Various tests necessarily performed to determine the acceptability of a device or assembly agreed to by the buyer and supplier of the product.

AC Termination A transient load termination technique with a capacitor in series with a parallel load termination. The voltage on the capacitor depends on the previous signal history and the time constant of the termination resistor and capacitor.

Active Components Electronic devices that act on a signal to amplify it, rectify it, and so forth.

Active Devices Transistors, diodes, or integrated circuits in monolithic form or in hybrid form.

Adhesive A material used for attachment of two surfaces. Attachment is accomplished by the attraction between molecules of the mating surfaces being attached or bonded together.

Aluminum A lightweight metal with good corrosion resistance and very good electrical and thermal conductivity, used in semiconductors to provide thin film interconnections between devices on a chip or integrated circuit. Much of electronic equipment housings (chassis) are composed of this metal in bulk form.

Aluminum Nitride A high thermal conductivity ceramic used in electronic packaging. Its symbol is AlN.

Ambient Temperature The temperature of the surrounding cooling medium, such as gas or liquid, that is in contact with the heated parts of an apparatus.

---

*For further information, refer to Charles A. Harper and Martin B. Miller, *Electronic Packaging, Microelectronics, and Interconnection Dictionary,* McGraw-Hill, 1993; James E. Buchanan, *Signal and Power Integrity in Digital Systems,* McGraw-Hill, 1996; and Charles A. Harper (ed.), *Electronic Packaging and Interconnection Handbook,* McGraw-Hill, 1997.

American Wire Gage (AWG) A measure of wire diameter; larger gage numbers represent smaller wire sizes. AWG 30 is about 10 mils in diameter and has about 0.1 Ω resistance per foot of copper wire. A change of 10 gage numbers changes cross-section area and resistance by a ratio of about 10.

Analog Pertaining to representation by continuously variable physical quantities the values of which are displayed by a needle or other indicator.

Analog Circuit A circuit that provides a continuous relationship between input and output rather than a discontinuous or switching condition.

Analog Integrated Circuit A linear integrated circuit intended to be used so that the output is a continuous mathematical function of the input.

Analysis of Binary Combinations (ABC) A structured factorial test method for evaluating the response of a function by performing analysis at combinations of two levels of the variables that may affect the function. The number of variables that can be analyzed is linearly proportional to the number of runs (tests) performed. The number of runs is selected as some power of 2, such as 8, 16, 32, etc.

AND Gate A logic circuit whose output is a 1-state only when every input is also in the 1-state.

Anisotropic Description of a material whose electrical, thermal, or optical properties are different in different directions through the material. (*See also Isotropic.*)

Asynchronous Inputs Signals with no fixed relationship to the system clock.

Asynchronous Operation The condition where one operation triggers another, irrespective of the timing signal from the system clock.

Attenuation The reduction of signal energy due to losses such as resistance or radiation.

Balanced Differential Communication Transmission of digital data on a pair of signal lines where each signal is the inverse of the other.

Ball Bond A bond made with gold wire which has been flame cut to melt the end of the wire into a ball which is then pressed down on a bonding pad on a component.

Beam Lead Chip A chip device that has its electrical terminations formed as short ribbons cantilevered over the edges of the chip.

Beam Lead Device A device that has beam leads as its interconnection mechanism.

Beryllia Beryllium oxide, a ceramic with high thermal conductivity used in high power applications. Its symbol is BeO.

Blind Via A via that extends from one outside layer to one or more internal layers but not through a multilayer board or multilayer ceramic substrate.

Bonding Wire Fine gauge wire (between 0.0007 and 0.025 in diameter) used to interconnect chips and other very small components to each other and/or to substrate and package terminations.

Bridging, Electrical The existence of a conductive path between conductors, usually unintended and representative of a defect.

Built-In Self-Test (BIST) The integration of additional circuits to perform tests intended to detect improper system performance. High speed BIST typically includes such functions as a controller interface, pattern generators to stimulate the system, and signature registers to compress the results.

Bump A small metal lump formed on a chip bonding pad or on interconnecting ribbon used to enhance interconnection.

Buried Layer An inner conductor in a multilayer circuit board, also a conductor beneath a semiconductor region and within the chip.

Buried Vias Vertical connections between adjacent internal layers of a printed circuit that do not extend to the outside layers. These smaller connections cause less line loading than through-hole vias. (*See also Blind Via.*)

Burn-In The storage of components at elevated temperatures for a specified time to cause failure of marginal components.

Cable Several conductors separated from each other by a dielectric, sometimes in the form of a central conductor and outer shell conductor.

Capacitance The property of metal conductors and dielectrics that allows storage of electricity when a potential difference is applied to the conductors that are isolated from each other.

Capacitive Coupling The electrical interaction of two or more circuits through the capacitance between them.

Center-Fed Routing Launching a signal in two directions into the center of a line. Also called *center-driven,* or *bidirectional,* lines.

Characteristic Impedance The impedance that a transmission line presents to an electromagnetic wave.

Circuit Board A sheet of metal cladding on an insulator on which the cladding has been formed into conductors to form a circuit pattern.

Circuit Element A basic component of a circuit.

Clock The source of a periodic signal used to synchronize systems that depend of synchronicity.

Clock Frequency The clock repetition rate.

Coating A thin layer of material that may be conductive or insulative applied over components or circuit base material for improvement of electrical properties or protection against contaminants.

Coaxial Cable A cable consisting of a central conductor over which is formed a dielectric that is in turn covered by a metal jacket which shields the inner conductor from external radiation and prevents radiation from the inner conductor from exiting the cable.

Cofiring A process in which multiple layers of conductors and dielectrics are fired or baked together to form a multilayer circuit.

Cognitive Bias The human tendency to underestimate or overlook subtle differences from typical or expected results. The electrical analog is *small signal suppression.*

Common-Mode Conversion Differential signals resulting from common-mode signals reflected from an unbalanced impedance; typically, differential noise resulting from backward crosstalk arriving at a driver that has a different source impedance for the HIGH output than for the LOW output.

Complementary Metal-Oxide Semiconductor (CMOS) A popular form of integrated circuit technology with two polarities of isolated gate field-effect transistors. These can be configured to drive digital signals to either level without significant static power consumption.

Conductive Adhesive An adhesive to which metal particles have been added to increase electrical or thermal conduction.

Conductive Epoxy An epoxy to which conductive particles of copper, gold, silver, or thermally conductive particles of ceramic have been added to provide for electrical or thermal conduction.

Conductivity The ability of a material to carry electric current or heat.

Conductor Spacing The distance between adjacent edges of conductors on printed circuits or substrates made from ceramics or other insulators.

Conformal Coating A thin dielectric applied over the circuitry and components of printed wiring assemblies to afford protection from contaminants and moisture.

Coupler A component used to carry electrical energy from one circuit line to another.

Critical Line Length The length of line that is equal to one-half of the rise time of the signal divided by the propagation delay of the line.

Crossover A location where one conductor passes over another conductor. The two conductors are separated by an insulator which may be air or another dielectric material.

Crosstalk A type of interference caused by signals from one circuit being coupled into adjacent circuits.

Current Penetration The depth to which current will penetrate into the surface of a conductor at a specific frequency. Penetration is less at high frequencies. (*See also Skin Depth.*)

Decibel (dB) The standard unit for expressing transmission gain or loss and relative power levels.

Die A small uncased integrated circuit device, also called a chip.

Dielectric Constant The property of a dielectric that determines the amount of electrical charge (energy) that can be stored per unit volume of that dielectric. Also the ratio of the capacitance of that dielectric to the dielectric of air.

Dielectric Layer A layer of insulation placed between two conductor layers.

Dielectric Loss The time rate at which electrical energy is changed to heat in a dielectric when it is subjected to an alternating electric field.

Differential Impedance (Z_{diff}) The dynamic impedance between two conductors with opposite polarity (complementary) signals.

Differential Signals A pair of signals, one of which is always the complement or inverse of the other.

Dispersion The separation of waves into different frequencies and velocities.

Dissipation The loss of electrical energy, usually in the form of heat or radiation.

Dissipation Factor The ac loss in a circuit, numerically equivalent to the tangent of the loss angle. Also known as *loss tangent* or *tangent (tan) delta.*

Distortion A change in the waveshape of a signal; usually, the undesired degradation of a signal waveshape due to the nonuniform propagation delay, attenuation, or amplification of different frequency components of a complex waveform.

Distributed Element The distribution of an electrical component characteristic along a transmission line as a property of that line.

Double Termination The addition of a termination device at both the source and the far end of a line to establish a relationship to the line impedance at both ends of the line.

Drain Wire A conductor run in parallel contact with a cable shield to provide a lower impedance discharge path for currents in the shield. Typically used with metal foil and/or spiral-wrapped shields that do not provide a sufficiently low impedance along the length of the cable.

Dual Stripline A multiple-layer interconnection with two layers of signals between ground or dc reference planes. Conductors within each of the signal layers are run in the same direction, but at right angles to the other layer to reduce crosstalk between layers. Also called *X-Y stripline* or *offset stripline.*

Dynamic Impedance (Z_0) The ratio of the transient voltage change to the change in current when a transverse electromagnetic wave travels in a dielectric. Z_0 is lower than the 377 Ω of free space when the dielectric constant is higher, or the wave travels between two conductors. Also called *line impedance, characteristic impedance,* or *surge impedance.*

Electric Field A region in which there is a voltage potential. The level of the potential varies inversely with the distance from the source. The field strength is given in volts per unit distance.

Electroless Deposition The deposition without an electric current of an electrically conductive material from an autocatalytic plating solution onto either a dielectric or metal surface.

Electrolytic Plating A metal deposition process in which an electrolyte, that is, a solution of the metal to be plated, transfers cations from the anode to the cathode or the workpiece by means of an electric current.

Electromagnetic Compatibility The ability of electronic equipment to operate properly without causing unacceptable electromagnetic interference.

Electromagnetic Field A rapidly moving electric field and its associated moving magnetic field, placed at right angles to the electric lines of force and to their direction of motion.

Electromagnetic Interference (EMI) A disturbance in the performance of an electronic system caused by waves that can impair both reception and transmission of electrical signals.

Electromagnetic Shield A metal screen or metal foil that surrounds electronic devices or circuits to reduce electric and magnetic fields. The shield either absorbs or reflects electromagnetic energy depending on its electrical conductivity.

Electrostatic Discharge (ESD) The sudden transfer of electrical charge between two bodies that were at a different electrostatic potential. The discharge may result from direct contact, induced by an electrostatic field, or as an arc when the distance between the bodies is below the capability of the dielectric to withstand the electrostatic potential. Typical ESD damage may include dielectric breakdown or conductor burnout in the fine geometry of high speed circuits.

Emitter Coupled Logic (ECL) A high speed, low-signal-level digital technology that uses nonsaturated bipolar *npn* transistors with parallel common emitters inputs and emitter follower outputs.

Environment The combination of temperature, humidity, pressure, radiation, magnetic and electric fields, shock, and vibration that influences the performance of a circuit or system.

Environmental Seal A type of seal to keep out moisture, air, or dust that might reduce the performance of a circuit.

Epoxy A thermosetting polymer that is frequently used in electronic packaging to adhere devices to supporting structures and, if the epoxy is electrically conductive, to provide a conductive path.

Even-Mode Impedance (Z_{even}) The dynamic impedance of one conductor in the presence of another signal conductor with a signal transition of the same polarity. (*See also Odd-Mode Impedance.*)

External Leads The flat ribbons or round wires that extend from an electronic package for input or output signals, power, and ground.

Eye Patterns A method for observing the quality of a digital signal pattern by overlay (superimposing) of multiple-signal transitions, with respect to the clock reference for capturing the signal. The open area between transitions indicating the safer region for signal capture resembles an eye.

Face Bonding The process of bonding a semiconductor chip with its circuitry side facing the substrate. Common methods are flip-chip and beam lead.

Farad A unit of capacitance. When a capacitor is charged with one coulomb of electricity, it produces a difference of potential of one volt between its terminals.

Fatigue The weakening of a material caused by the application of stress over a long time.

Fatigue Factor The force that causes the failure of a material or device when placed under repeated stresses.

Fatigue Life The number of cycles of stress that can be sustained prior to failure for a stated test condition.

Fatigue Limit The maximum stress below which a material can presumably endure an infinite number of stress cycles.

Ferrite Shield Beads A ferrite material surrounding a signal conductor that is used to reduce overshoot and ringing by absorbing some of the high-frequency energy to increase rise time without any significant reduction of the dc signal level.

Film Adhesive A class of adhesives provided in dry-film form with or without reinforcing fibers contained in the film and cured by heat and pressure.

Fine-Pitch Technology Any interconnection process in which the pitch or distance between centers of the connections is between 0.01 and 0.04 in.

Flat Cable A multiconductor cable assembly whose thin flat conductors are laid out in the same plane.

Flip-Chip A semiconductor chip having all terminations on one side in the form of solder pads or "bump" contacts. Electrical and physical attachment is made by flipping the chip over onto pads on a substrate.

Flip-Flop A bistable logic memory device employing internal positive feedback to retain one of two states indefinitely (while powered), and capable of being set to either state by external signals. One or more flip-flops may be called a *register.*

Glass-to-Metal Seal A hermetic seal between glass and metal parts. The seal is made by fusing glass with a metal alloy that has nearly the same coefficient of expansion as the glass.

Glob Top The application of encapsulation material in the form of a glob over a chip after assembly and test to afford protection.

Green Tape An unfired flexible ceramic material made from a slurry and having a thin uniform thickness. The flexibility is due to included organic binders which are evolved during the subsequent firing process of the material after it has been cut to prescribed forms, at which point the ceramic becomes firm and hard.

Ground Bounce Transient ground shifts caused by transient current flow to ground.

Ground Strap A conductive strap for dissipating electrostatic charges from sensitive electronic parts. Also a strap attached to a ground line on a printed circuit board or substrate at one end and to a chassis or structure that is at ground potential.

Guard Lines Additional conductors or connector pins that are run beside or around critical signals to shield them from interference. The guard lines are usually grounded, or may contain static signals that will not interfere with the critical signals.

Hermetic Permanently sealed by fusion or soldering, such as metal-to-metal or glass-to-metal to prevent passage of moisture or other gases.

Hertz The international system (SI) unit for frequency, equivalent to one cycle per second of some recurring event.

High-K Ceramic A ceramic dielectric composition such as a titanate that has a high dielectric constant.

Hybrid Electronics A technology using thin and thick film circuitry, discrete integrated circuit devices, and other chip-type components with wire bonding to other device interconnect techniques to form functioning electronic circuits.

Hygroscopic Having a tendency to absorb and retain moisture from the atmosphere.

Hysteresis An effect in which the magnitude of a resulting quantity is different during increases in the magnitude of the cause than during decreases arising from internal friction in a material and accompanied by the production of heat.

Hysteresis (Electrical) A memory effect, using positive feedback to a digital input so that the device tends to stay in the prior state when input signals are in the threshold region. When the input signal level overcomes the feedback bias, the output "snaps" to the new state. Hysteresis tends to prevent unintended oscillations or amplification of noise when input signals are near the threshold region.

Impedance The total opposition that a circuit offers to the flow of alternating current or any other varying current at a particular frequency. It is a combination of resistance R and reactance X measured in ohms and designated by Z: $Z = (R^2 + X^2)^{1/2}$.

Impedance Match The condition that exists when the impedance of a component, circuit, or load is equivalent to the internal impedance of the source.

Inductance The property of a circuit or circuit element to oppose a change in current flow, causing current changes to lag behind voltage changes.

Inductive Coupling The interaction of two or more circuits by means of either mutual inductance or self-inductance common to the circuits.

Inner Lead Bond In tape-automated bonding, the connections made between the chip and the etched conductors on the tape.

Insertion Loss The difference between the power received at the load before and after the insertion of a component, connector, or device at some point in the line.

Integrated Circuit A small chip of semiconductor material containing an array of active and/or passive components that are interconnected to form a functioning circuit.

Interconnection The conductive path required to achieve connection from one circuit element to another or to the rest of the circuit system. Any mating methods such as terminals, pins, soldered joints, welded joints, wires, and so forth, constitute interconnections.

Interference Protection The measures taken to shield sensitive areas of electrical equipment from electromagnetic and radio-frequency interference.

Inverter A logic device that changes the input logic level from a high to a low or from a low to a high.

Ionic Contaminant Any polar contaminant that, when dissolved in water, increases the electrical conductivity of the water by virtue of putting ions into solution.

Isotropic Pertaining to a material whose electrical and optical properties are the same in every direction in the material.

Jacket A plastic, rubber, or synthetic covering over the insulation, core, or sheath of a cable.

Junction (1) A contact between two dissimilar metals. (2) A region of transition between p-type and n-type semiconductive materials in transistors and diodes.

Junction Temperature The temperature of a semiconductor junction.

K The symbol for dielectric constant, also the symbol for thermal conductivity.

Kiln A high-temperature furnace used for firing ceramics.

Kilo A prefix that indicates a multiple of one thousand.

Kirkendahl Voids Voids that are induced at the interface between two different metals with different interdiffusion constants.

Kovar Originally a Westinghouse trade name for an iron-nickel-cobalt alloy used in hybrid and microelectronic packages because its low coefficient of expansion is closely matched to glass used to form hermetic seals at the metal lead-to-glass interface.

Laminate Sheets of materials that are bonded together. A common laminate is that of printed wiring sheets with conductors formed on them which when laminated together form a multilayer board.

Laminated Ceramic Package An electronic package consisting of a multilayer cofired ceramic body, brazed leads, or leadless solder pads and a hermetically sealed cover.

Large-Signal Capture The nonlinear characteristic of digital signals that amplifies input signals that are less than the full-scale levels. Inputs above the threshold are amplified to a HIGH output level and inputs below the threshold are amplified to a LOW output level. (*See also Small-Signal Suppression.*)

Leadless Pertaining to electronic devices that do not have electrical leads extending from their enclosures but rather solder lands or bumps located on the top, bottom, or sides of the package.

Line Capacitance The capacitance associated with a given signal interconnection.

Line Discontinuity A point on a transmission line equivalent to a separate circuit having resistance, capacitance, and inductance.

Line Loading The connecting of external resistance, inductance, and capacitance in a transmission line.

Line Resistance The resistance presented by conductor lines in a package, measured in ohms per unit length.

Load Termination A component (typically a parallel resistor or diode) added at the far end of a line to establish a relationship of the load impedance to the line impedance.

Logic Design The selection and interconnection of logic units to achieve logical function in a digital system.

Loop The curvature of a wire between each end of the two attached wire bonds.

Loop Height The maximum perpendicular distance from the top of the wire loop to a point between the two attached wire bonds.

Loop Routing The connection of all sources and loads on a net into a loop with no ends.

Loss Factor The rate at which heat is generated in an insulating material, equal to its dissipation times the dielectric constant.

Loss Tangent *See Loss Factor.*

Low-Loss Dielectric An insulating material that has a low power loss over long lengths, making it useful for transmission lines. It usually has a low dielectric constant as well as low dissipation factor.

Low-Loss Substrate A substrate with high radio-frequency resistance and low energy absorption. It usually has low dielectric constant and low dissipation factor.

Low-Temperature Firing Ceramic A ceramic composition that can be fired at temperatures considerably below those used in normal kiln firing as for thick film processes. These compositions are usually glass-ceramic.

Mass Ground The use of common ground planes, structures, cable shields, etc., for a ground reference without regard to the paths in which the return currents flow.

Mega A prefix indicating a multiple of one million.

Megahertz (MHz) Millions of cycles per second.

Metallization An electrically conductive film deposited on the surface of another material to perform electrical and mechanical functions.

Metallized Ceramic A ceramic substrate that has been metallized with fired thick films or evaporated thin films or by some other deposition technique.

Microchip A general term for a semiconductor chip with a large number of components at fine-resolution feature sizes. Mid-1990s technology is typically over a million transistors at submicron resolution.

Microstrip A type of microwave transmission-line configuration that consists of a conductor over a parallel ground plane separated by a dielectric.

Microwave Integrated Circuit (MIC) A miniature microwave circuit using hybrid circuit technology to form the conductors and attachment/interconnect techniques for lumped components.

Migration An undesirable movement of metal ions such as silver and tin, from one location to another in the presence of moisture and under the influence of a dc potential. The result is a serious lowering of the insulation resistance of a surface or coating separating two conductors.

Miller Effect The ac coupled negative feedback from the output of an inverter to the input that limits the output slew rate, or rise time. Also called a *Miller integrator.*

Model A simplified representation of a physical component, circuit, or material that reasonably describes the response to a limited set of conditions. Models for electronic devices have been called *equivalent circuits.* Simple models may not adequately describe actual high-frequency characteristics.

Monolithic Microwave Integrated Circuit (MMIC) An integrated circuit similar to a monolithic semiconductor integrated circuit but for use in microwave functions and applications.

Mother Board Generally, a relatively large printed circuit board that serves as the electrical and mechanical interface between components, including devices mounted directly on the mother board, connectors for other printed circuit boards (called daughter boards), and interface connectors. Also called a *backplane.*

Multilayer Board A circuit board or substrate consisting of layers of electrical conductors separated from one another by an insulating material and interconnected together at desired locations all laminated together to form a solid mass.

Multilayer Ceramic Capacitor A ceramic substrate containing multilayers of thick film circuitry separated by dielectric layers and interconnected by vias.

Nano A prefix meaning one-billionth (1×10^{-9}).

Nanosecond A measure of time equal to 10^{-9} second.

Net A group of circuit nodes that are interconnected.

Noise Immunity A measure of how good a circuit is at rejecting extraneous signals.

Noise Margin The amount of extraneous voltage that a signal can tolerate before the signal is no longer recognizable as the intended logic level.

Non-Return-to-Zero (NRZ) The common method of representing binary data as either a HIGH level or a LOW level during a unit interval (bit time or clock period), without regard to other data. Also called *uncoded data.*

Odd-Mode Impedance (Z_{odd}) The dynamic impedance of one conductor in the presence of another signal conductor with a signal transition of the opposite polarity. (*See also Even Mode Impedance.*)

Ohm The unit of electrical resistance. The resistance through which one ampere of current will flow when a voltage of one volt is applied.

Ohmic Contact An electrical connection between two materials across which the voltage drop is the same in both directions.

Ohms per Square The unit of measure for sheet resistivity. It is the resistance of a square area measured between parallel sides of both thin and thick film resistive materials.

Overshoot The condition when a signal goes beyond its normal range or steady-state level.

Oxygen-Free High-Conductivity (OFHC) Copper Copper having a minimum copper content of 99.95 percent, used in electronic applications, especially very high conductivity foils for printed circuit boards.

Pattern Sensitivity A measure of the vulnerability of correctly capturing data as a function of the contents of the data; typically, the extent to which signal history affects the ability to capture serial data, or the extent to which parallel data patterns interfere with each other. Also called *intersymbol interference.*

Pico A prefix meaning one-trillionth (1×10^{-12}).

Pitch The nominal distance between centers of adjacent conductors.

Polarizability A measure of the electrical energy that a dielectric can store while under the influence of an externally applied electric field. Also called *permittivity.*

Printed Wiring A conductive pattern that provides point-to-point connections in a predetermined arrangement on a common insulating base.

Propagation Speed The speed at which an electromagnetic wave travels through a medium.

Pull-Down Resistor A bias resistor or current source connected between device inputs and ground (or other reference) to avoid a static charge or invalid input levels. It may be called a *pull-up resistor* if the bias is to a positive power supply.

Pulse Duration The time interval between a reference point on the leading edge of a pulse waveform and a reference point on the trailing edge of the same waveform.

Pulse Width The time between the leading edge and the trailing edge of a pulse.

Q Factor The relationship between stored energy and its rate of dissipation.

Quiescent Voltage or Current The dc voltage or current at a terminal with reference to a common terminal, normally grounded, when no signal is applied.

Quiet-Mode Impedance (Z_{quiet}) The dynamic impedance of one conductor in the presence of another signal conductor with no signal transition.

Random Access Memory (RAM) A memory in which information can be stored or retrieved.

Radio Frequency Electrical frequencies in the range from about 30 kHz to 3×10^5 MHz.

RC Network A network consisting entirely of resistors and capacitors.

Read Only Memory A memory in which information is permanently stored at the time of fabrication and is not erasable.

Reflection Diagrams A graphical method for evaluating the voltage and current characteristics of the source and load on a transmission line to estimate the response to signal transitions. This method is appropriate when other models, such as linear analysis or finite-element approximations, are not appropriate or available. Also called *Bergeron plots*.

RF Connector A connector to which a coaxial cable may be attached.

Ribbon Interconnect An electronic connection between circuits or between a circuit and its package output pin, in the form of a rectangular ribbon.

Ribbon Wire A flat, flexible wire having a rectangular cross section.

Ringing The condition of a signal overshooting and undershooting the final steady-state level a number of times following a logic level transition.

Rise Time The time interval for a signal transition to pass between two reference levels. Generically applies to either positive or negative transitions, but some authors may distinguish the negative transition time as *fall time*. The reference levels for high speed transitions are generally defined at the 20 and 80 percent levels, but there is a tradition of using 10 to 90 percent levels, and computer simulations with linear transitions typically define rise time at the full 0 to 100 percent levels.

Root Sum of Squares (RSS) A less-conservative method for combining the effects from multiple variables, compared to *worst case analysis,* which simply sums the worst possible combination. RSS computes the square root of the sum of the squares of each effect. RSS is usually valid if the most significant variables have a strong central tendency ("normal" distribution), and the variables are independent so that there is very little chance that they would all be at the worst limit.

Routing The process of defining the configuration of the interconnection between the circuit nodes on a signal net, typically restricted by rules for the sequence and placement of the interconnections.

Safety Ground An auxiliary ground connection, independent of signal and power returns, that provides a discharge path to ground before power or signals are connected, in order to avoid hazards to personnel or components due to static charge, leakage currents, or power faults.

Sapphire The single crystal form of aluminum oxide (Al_2O_3), used as microwave substrates.

Shielding Effectiveness The effectiveness of a shielding barrier such as a magnetic covering, in a measure of the lowering or reduction of field strength between a source and a receiver of electromagnetic or rf energy.

Small-Signal Suppression The nonlinear transfer characteristic of digital signals that maintains full-scale output signal levels despite small variations of the input signals when the input is near the full-scale levels. Also the forward voltage response of a diode to small changes in forward current.

Soft Substrate A substrate that is fabricated from a soft plastic material. Common soft substrates are low electrical loss materials (PTFE or polytetrafluoroethylene) having a low dielectric loss.

Source Termination A component (typically a series resistor or shield bead) added at a source to establish a relationship of the source impedance to the line impedance. Also called *series termination.*

Split Termination A parallel load termination consisting of two resistors in series, with the ends connected to power supplies. The junction of the two resistors is connected to the line. The load termination impedance is the parallel equivalent resistance. Also called a *Thevenin termination.*

Stripline A microwave transmission-line configuration in which the central or signal conductor is between and parallel to two ground planes.

Substrate The base material that serves as a support surface for metallizations that form interconnect patterns.

Superposition Principle The characteristic of linear functions that the response to one input is independent of other inputs or conditions. This means that complex inputs can be divided into separate parts for analysis, with the total result equal to the sum of the results from the separate parts.

Switching Noise A noise produced by an induced voltage at the circuit terminals when signal levels change.

Synchronous The situation where all signals change state in step with the system clock.

Tape Automated Bonding (TAB) A process in which etched leads supported by a film tape are automatically positioned over a chip so that the leads can be bonded to the pads of the chip to form electrical connections.

Temperature Aging The stressing of a material, component, or system at an elevated temperature for a specified length of time.

Temperature Coefficient of Capacitance (or Resistance) The amount of change in capacitance of a capacitor or the amount of change in resistance of a resistor with respect to a change in temperature.

Threshold Voltage The input voltage level at which the output logic level is no longer defined.

Transfer Impedance The ratio of the voltage induced on the signal conductors inside of a cable shield to the current flowing on the shield. Also called *surface transfer impedance.* (*See also Shielding Effectiveness.*)

Ultra High Frequency The band in the radio-frequency spectrum from 300 to 3000 MHz.

Undershoot The condition where a digital signal rings toward or across the nearest logic level threshold.

Vacuum Deposition The deposition of thin films of metal or dielectric onto a substrate in a vacuum chamber. The metal films are later etched to form conductor patterns.

Virtual Ground Plane The principle that there exists a zero potential plane at all points equally distant from two conductors with opposite polarity signals.

Voltage Standing-Wave Ratio (VSWR) The ratio of the maximum to minimum signal amplitudes that result when a reflected wave combines at either in-phase or opposite phase with a new incident wave. It is also the ratio to the line impedance at a line discontinuity, source, or load.

Welded Wire Board An interconnection technique for boards or backplanes that typically uses AWG 30 round insulated wires to interconnect circuit nodes by a welding process. Typically used for small quantities or rapid prototyping where the possibility of rework is desired compared to printed circuits.

Wirewrap Board An interconnection technique that typically uses AWG 30 round insulated wires to interconnect circuit nodes by stripping insulation and wrapping the bare ends of the wire around square posts, which are located at each circuit node. This technique allows twisted pair interconnections.

Index

ABOUT THE AUTHORS

STEPHEN G. KONSOWSKI has 35 years of electronic packaging engineering experience. Retired from the Electronic Systems Group of the Westinghouse Electric Corporation as an advisory engineer, he is now associated with Technology Seminars, Inc. of Lutherville, Maryland. He received both a Bachelor of Science in Physics and a Master of Science in Physics from the University of Detroit. He is a founding member of the International Electronics Packaging Society (IEPS) and one of its former directors. Mr. Konsowski has authored many technical papers and articles on electronic packaging, including chapters for the Second Editions of McGraw-Hill's *Electronic Materials and Processes Handbook* and *Electronic Packaging and Interconnection Handbook*.

ARDEN R. HELLAND is an advisory engineer at the Northrop Grumman Corporation's Electronic Sensors and Systems Division with over 35 years of experience in most phases of the electrical design of advanced electronic systems. He received an Associate of Science in Engineering from the North Dakota State College of Science in Wahpeton; a Bachelor of Science in Electrical Engineering (BSEE) from North Dakota State University in Fargo; a Master of Science in Electrical Engineering (MSEE) from the University of Pittsburgh; and a Master of Science in Management Science from the Johns Hopkins University in Baltimore. He is a member of the Institute of Electrical and Electronic Engineers (IEEE) Computer Society. Mr. Helland has authored many technical papers on the design and analysis of electronic systems and contributed to *CMOS/TTL Digital System Design* and *BiCMOS/CMOS Systems Design*, both published by McGraw-Hill. He authored the chapter on reliability of digital systems for McGraw-Hill's *Computer Engineering Handbook*.